Historic Ships 2023

Historic Vessels - Sustainable Futures

London, United Kingdom
28-29 November 2023

ISBN: 978-1-7138-9562-6

Printed from e-media with permission by:

Curran Associates, Inc.
57 Morehouse Lane
Red Hook, NY 12571

Some format issues inherent in the e-media version may also appear in this print version.

Printed with permission by Curran Associates, Inc. (2024)

For permission requests, please contact The Royal Institution of Naval Architects
at the address below.

The Royal Institution of Naval Architects
8-9 Northumberland Street
London, WC2N 5DA
United Kingdom

Phone: 020 7235 4622
Fax: 020 7259 5912

publications@rina.org.uk

Additional copies of this publication are available from:

Curran Associates, Inc.
57 Morehouse Lane
Red Hook, NY 12571 USA
Phone: 845-758-0400
Fax: 845-758-2633
Email: curran@proceedings.com
Web: www.proceedings.com

160 YEARS OF HISTORY, KNOWLEDGE AND EXPERIENCE

THE ROYAL INSTITUTION OF NAVAL ARCHITECTS

Historic Ships 2023

Historic Vessels - Sustainable Futures

28-29 November 2023

London, United Kingdom

In Partnership With

ROYAL MUSEUMS GREENWICH

NATIONAL HISTORIC SHIPS UK

TOGETHER WE WILL SHAPE OUR FUTURE

THE ROYAL INSTITUTION OF NAVAL ARCHITECTS

Historic Ships 2023

28-29 November 2023

National Maritime Museum, London

THE ROYAL INSTITUTION OF NAVAL ARCHITECTS
8-9 Northumberland Street
London WC2N 5DA

Telephone: 020 7235 4622
Fax: 020 7259 5912

ISSN No: 2752-6569

CONTENTS

THE FATE OF TRADITIONAL MALTESE BOATS

A Grech La Rosa, UCL, UK and **P Grech,** Malta

SUMMARY

Traditional Maltese vessels have long been part of Malta's cultural identity. These vessels were used for a variety of roles including transportation, work, fishing, defence and pleasure. Changes in society and the local economy resulted in these vessels no longer forming part of the county's working fabric and were replaced with more modern vessels. In light of this, the population of traditional Maltese vessels has been dwindling with critical aspects such as how these vessels were sailed being lost for good. A recent incident with the planned disposal of a vessel called the Maryanne led to public uproar and raised further awareness for the current situation.

This paper investigates whether traditional Maltese vessels are still considered an integral part of the fabric of society. Different stakeholders were questioned to understand what challenges were experienced when operating such vessels. The results were then synthesised into key points to provide a unified message for what has led to the current state of affairs and what could be done to rectify the situation. A proposal for actions to be taken to safeguard these vessels for future generations was also suggested.

1. INTRODUCTION

Malta is a small island located in the centre of the Mediterranean sea, synonymous for its touristic appeal and dense historic fabric. Malta's sea and iconic harbours are a central pillar of the country's economy, noting the island's otherwise limited natural resources. In particular, the fishing industry has sustained locals over the centuries and the vessel's used to facilitate this became part of the nation's cultural identity. These wooden, multicoloured boats were customised to the fisher's needs and became a canvas to express themselves. Each vessel proudly showed their family's skilled workmanship through the joinery, intricate carvings and special adaptations to improve their fishing procedures.

As the economy grew and the demand for fish increased, these traditional boats didn't remain the nation's preferred choice of vessel due to their inability to explore more distant waters and adapt to larger scale fishing practices [1,2]. Locals preferred vessels requiring less maintenance, resulting in most of the traditional stock of vessels being decommissioned. The skilled workforce needed to repair these vessels has dwindled and the knowledge of how these vessels were sailed is already lost. The few larger vessels that were converted to commercial pleasure craft have now also started being decommissioned, making many wonder what the fate of the remaining traditional boats will be. This once strong strand in Maltese identity is fading away at an accelerated pace.

2. TRADITIONAL MALTESE BOATS

Being an island nation, Malta had strong maritime and boat building industry. Vessels were made to fulfil different roles including transportation, work, fishing, defence and pleasure. These were made by locals using traditional techniques that had been passed on over generations as well as being influenced by practices of neighbouring countries. Over the years many of these vessels have been replaced by more modern alternatives resulting in the older, traditional vessels without a role to fulfil.

2.1 VESSEL TYPES

The most common types of traditional boats include the Luzzu, Kajjik, Fregatina and Dghajsa tal-Pass with various examples still found in local waters today, despite their dwindling population. Figure 1 shows an example of each of these types of vessels. Unfortunately only few examples of vessels such as the Firilla and Dghajsa tal-Latini are still in circulation with no known examples of vessels such as the Xprunara about [3,4].

(a) Luzzu [5]

(b) Kajjik [1]

(c) Frejgatina [1]

(d) Dghajsa Tal-Pass [6]

Figure 1: Examples of Traditional Maltese boats still in service

2.2 ADAPTATION

Similar to other cultures, each of these vessel types had their own technical features that were adapted over time to reflect the needs of their most recent role as well as the innovation of the time. Vessels were originally propelled by various types of sailing rig, depending on the particular type and use of the vessel. In time, mechanical propulsion, mainly in the form of inboard diesel engines, was adopted alongside the sails as a more reliable propulsion system which was easier to handle by a smaller crew. Over the years, mechanical propulsion became more and more popular which meant that sailing rigs were phased out completely. Unfortunately, no examples of traditional boats with their rigging are known to still exist on the island.

Interestingly, although the majority of these vessels were small craft operating in local waters, they were equipped with two or more propulsors to improve reliability and increase manoeuvrability for fishing operations. Various examples of these vessels have had engines from war time land vehicles installed that needed further modification to operate with an open water-cooled system.

Each of the vessel types highlighted were designed, built, and maintained for an extended lifespan to serve generations of fishers. Most remaining wooden traditional vessels were built more than 70 years ago. Figure 2 (a) shows Neptune 89, a wooden traditional luzzu, returning to port in 1951 while Figure 1(b) shows the same vessel on the hard 68 years later. The robustness of the design as well as the care shown to it over the years has permitted this vessel to still be in regular use till this day.

(a) (b)

Figure 2: Neptune 89 in 1951 and in 2019

Figure 2 also shows some traditional features found on luzzus including the eyes of Osiris, the distinct colour scheme as well as carvings and mermen. Each of these features have distinctive names and cultural significance. In particular, the eyes of Osiris were originally installed to ward off evil spirits and ensure safe passage. The popular use of these vessels meant that technical vocabulary was created to facilitate more accurate communication when discussing distinct parts of the vessel.

2.3 IDENTITY

The distinctive colourful look of these vessels soon became part of the national identity. With the island's fishing villages being full of luzzus and kajjiks over the last 100 years, numerous paintings, songs, postcards, postage stamps, coins, photos and artwork have been made, immortalising their presence in local history.

The identity of these vessels goes beyond that of a national fleet but also to local level where vessels hailing from particular fishing villages would sport a different coloured "moustache". This can be identified as the ochre yellow triangular section in Neptune 89's bow as shown in Figure 1(b) implying that Neptune 89 berths in Marsaxlokk. Other localities would be represented with other colours such as white for Saint Paul's Bay and red for Gozo [7].

Further distinction in identifying luzzus would be made by assigning names, typically related to Christian saints that the family adored or by carving decorative elements. When questioning senior fishers who served on such vessels in their younger years, an emphasis was always made on the pride they had for their vessel as it was considered to be a representation of themselves and their family. In fact, when a fishers family was in mourning, the boat's moustache was painted black as shown in Figure 3.

Figure 3: A vessel whose family is in mourning [source unknown]

3. ROLE IN SOCIETY

The large number of these vessels clearly demonstrated that they were relevant to society.

3.1 INDUSTRY

As highlighted earlier, the original purpose behind designing and building these vessels was to provide a service to society.

The Dghajsa tal-Latini was used as part of a service to ferry people to Malta's sister island, Gozo while the Dghajsa tal- Pass was originally a smaller passenger ferry to operate within the Grand Harbour which is why they could have shallower freeboards. No traditional boats are used for local commuting any longer but some services by independent parties are done with the dghajsa tal-pass in the grand harbour for touristic purposes.

The current industrial use of traditional vessels has been dominated by fishing and day charters for the tourism sector. Figure 4 shows a small Kajjik being used by an independent fisher while Figure 5 shows Neptune 89 on a day charter with tourists. Other infrequent uses of these include props in films such as the luzzu in Figure 1 (a) claiming the title role or skuba diving platforms.

Figure 4: A fisher trawling on a traditional Kajjik [8]

Figure 5: Neptune 89 on a day charter

3.2 PRIVATE USE

Since neither of these vessel types were designed for leisure, the uptake of recommissioning these vessels to leisure boats has been poor. The vessels are typically up to 25ft overall length with little space for lounging. Larger vessels have been converted into day boats but these often lack modern commodities that would render them attractive for a wider audience. With the advent of fibreglass, pastiche kajjiks and luzzus have been made that require less maintenance and offered more space for daytime cruising. Despite this new option, these vessels were still less popular than other day cruisers on the market and hence had limited uptake.

3.3 THE CURRENT STATE OF AFFAIRS

In order to investigate the fate of traditional Maltese boats, different stakeholders were interviewed to determine what their take on the current state of affairs is. People having a history with these vessels as well as fishers, day charters, enthusiasts and even some people from the general public were asked what they believe are the biggest challenges, threats and frustrations are. The same people were also asked whether they see a future for these vessels or whether their time has passed. The following information is a compilation of the various thoughts that were expressed, synthesied into discrete elements to provide a unified voice.

4. CHALLENGES

The continuous decline in population of traditional Maltese boats is a symptom of the changes experienced by the country. Maltese people still consider these traditional boats to be part of their national identity but the current social fabric has left little room for these vessels to thrive. The fishing and day charters sectors have been put under great strain by various aspects that may initially seem unrelated.

4.1 SPACE

The vast majority of traditional boats used to be berthed or stored on the hard in the fishing village harbour. Over the past 30 years, these locations have increased in popularity, primarily because of the picturesque look that these boats gave the village. This popularity has led to increased tourism which in turn commanded that the space originally attributed to these vessels for maintenance and safe keeping over the winter months be instead allocated to eateries, promenades and other business ventures [9]. Figure 5 and Figure 6 show the same locations in two different fishing villages that originally used to be a hub for traditional Maltese boats that now have minimal space allocated to them.

(a) Xatt is-Sajjieda Postcard [Source Unknown] (b) Xatt is-Sajjieda [10]

Figure 6: Several traditional vessels on the hard in Marsaxlokk

(a) Church Street Postcard [Source Unknown] (b) Church Street [11]

Figure 7: Traditional vessels in St Pauls Bay harbour in approx. 1965 vs the same location in 2016

The reduction in publicly allocated space to these vessels without providing a substitute made it difficult for many owners to store and maintain their boats. This introduced new hardships and served as a means to break an established community that shared resources and knowhow.

In the case of those fishers operating in Marsaxlokk, a hard standing facility was opened in 2007 to aid those operating under a fishing licence [12]. While this was a welcome step forward, the hard standing was on the periphery of the community and was therefore no longer part of the social mechanism of the village.

To this day, the country still does not have a dedicated facility to restore and maintain traditional boats. This is a space that has been desired by enthusiasts since at least 2003. The Malta Traditional Boat Association has long been striving to secure a sheltered space where an appropriate workshop could be set up to restore the vessels they have saved from destruction [13]. The National agency for museums, conservation practice and cultural heritage, also known as Heritage Malta, recently added one of the last remaining examples of a Dghajsa tal-Latini in its portfolio which it aims to restore. This vessel was craned down into a ditch instead of secured in an appropriate climate controlled facility [14,15].

Figure 8: A Dghajsa tal-Latini rescued by Heritage Malta being stored and restored in a ditch [14,15]

4.2 COMPETITION

The introduction of larger, more efficient vessels in the local market meant that those people operating traditional vessels were at a disadvantage to their competitors. With larger fishing companies dominating the market by securing a larger catch and day charter vessels experiencing great financial difficulties to maintain a competitive price per person; it's no surprise that various people upgraded their vessels to keep up with the times, resulting in more traditional vessels lacking a role to fulfil.

4.3 FUNDING

While competition is in itself good for a healthy economy, those people operating part of the country's national heritage and identity should receive aid. This would ensure that the fleet of privately owned vessels would be preserved and be compensated for the burden of not upgrading like their competitors.

The Government of Malta has regularly taken such a stand in fixed assets where privately owned buildings would be granted funds to restore or replace their traditional Maltese balcony [16]. This was not done with the traditional Maltese buses which ultimately led to them all being decommissioned from public service and the vast majority scrapped. As of September 2023, a new scheme has been launched to aid those who privately own a traditional Maltese boat after years of lobbying. A maximum of 5000Euro can be granted towards repairs for each project with a total budget of 80,000Euro for the nation [17]. While this is a welcome start, it is merely a token aid to try and safeguard this heritage.

In the case of vessels operating under a fishing registration, some funding applications were made available over time but the application process did not work to the strengths of the fishers. This meant that many often didn't make use of the scheme. An alternative, controversial funding option became available prior to 2007 through the EU which offered fishers a handsome sum in exchange for their fishing licence [18]. The policy resulted in the destruction of fishing vessels, traditional or not, as part of the settlement to ensure that the vessel would no longer be used for fishing [19].

4.4 RESOURCES

In light of the points described above, the number of traditional boats in circulation reduced. The reduced community element, less cashflow and tight competition meant that skilled labourers who were trained by their elders on how to go about making and repairing these traditional vessels focused their efforts elsewhere. This led to a shortage of skilled labour, making it even harder for those who still had a traditional fleet to keep up with their maintenance requirements [20,21]. A traditional boat building course was set up at MCAST but the uptake has been very poor. This links to the lack of incentives associated with the Maltese traditional boat industry. Maintaining a wooden traditional vessel is very labour intensive which makes it a less attractive option. Unfortunately, no boat yard in Malta is constructing new wooden traditional Maltese boats.

5. AN ONSLAUGHT ON CULTURE AND TRADITION

The challenges experienced by many people owning traditional boats led to them decaying or accumulating fines in different areas around the country. Instead of providing support to maintain these vessels or confiscating them to auction them off to the public which is a common practice done with impounded vehicles [22], various local authorities went as far as doing a public demonstration of breaking these traditional boats. Figure 9 shows an example of a traditional boat being broken in public, despite members of the Malta Traditional Boat Association willing to save the vessel at their own expense. Footage of this vessel was recorded as part of a promotional video celebrating its destruction by the Cleansing and Maintenance Division. At least 70 traditional vessels were recorded to have been broken in this manner [23].

Figure 9: Traditional vessel being broken in public [23]

Similar scenes were also seen by the authors as part of the scheme described in Section 4.2 which saw several vessels broken as part of a policy.

6. THE CASE OF THE MARYANNE

The Maryanne is a vessel that had a longstanding history in Malta's recent history by being a charter boat for one of the larger tourist excursion providers. Although the vessel was built in Sicily, it's characteristics are a close match to a traditional Dghajsa tal-Latini, of which there are only 5 vessels known to be available, most of which are unusable [14,15]. Traditional Maltese boats have long been influenced by the surrounding cultures.

For reasons not published, the Maryanne was dismantled of her outfit and was in line for breaking until this news was leaked and local enthusiasts including the Malta Traditional Boat Association tried to rescue it. Figure 10 shows the Maryanne on the hard awaiting to be broken.

Figure 10: The Maryanne on the hard awaiting breaking [Source Unknown]

Following public uproar, government entities stepped in to salvage the vessel. Once again, no appropriate facility exists to restore her and she was craned onto an exposed hard stand. Despite being a well utilised day charter boat for a substantial number of years, the condition of the vessel suggests that a significant amount of funds would be needed to render her sea worthy again.

This recent event demonstrates that although issues have been flagged over several years, the situation for traditional Maltese boats has not gotten any better and should things remain the same, it is likely that the last examples of these vessels will be purely limited to a museum and not out on the water where they belong. As highlighted earlier, this is already the case with traditional rigging as very few records of how these vessels were rigged and sailed are available.

7. TRADITIONAL MALTESE BOATS IN A MODERN SOCIETY

Traditional Maltese boats, especially luzzus and kajjiks are still widely recognised as part of the national identity. The tourism industry regularly uses images of them as part of their promotional material were some companies go as far as using luzzu in their name. In additional to this, their colour scheme and design motifs are often interpreted as other forms of artwork including jewellery, pottery, paintings, etc.

Acknowledging this, action must be taken to safeguard the remainder of these vessels by:

- Reintegrating these vessels into society, especially in the fishing villages where they original came from [24]
- Celebrating these traditional vessels by means of public events to showcase the skill, craftsmanship and dedication
- Introducing at least 2 facilities that are appropriately equipped to manage, maintain, train and build traditional Maltese boats
- Increasing the funding for these vessels
- Introduce incentives to make use of traditional vessels as part of commercial ventures

8. A FUTURE FOR MALTESE BOATS

While conserving the remaining stock of traditional vessels is essential to safeguard national identity, innovation is an alternative approach to instil new life in these vessels. These may involve retrofitting already existing vessels with new technology or developing a new iteration to the traditional vessels that would maintain the fundamental elements of the design but apply new purpose.

A local architectural studio, Mizzi studio, has applied this approach to the traditional Maltese bus where the essence of the original design was extrapolated to a modern context [25]. These vehicles would be able to cater for the needs of a modern society while staying true to their heritage.

Figure 11: A contemporary take on the traditional Maltese bus [25]

In light of this, a think tank to design and develop the new luzzu should commence. Further work in this research will explore the primary tasks that would be appropriate for a new design to undertake by first understanding the maritime needs of the modern, local society.

8. ACKNOWLEDGEMENTS

The authors would like to acknowledge the work carried out by all those who strive to preserve Malta's maritime heritage.

9. REFERENCES

1. Vassallo History, 'Fishing in Malta', *Vassallo History*, [cited 2023 Oct 24]. Available from: https://vassallohistory.wordpress.com/fishing-in-malta/

2. GRILLO, F., 'Killing tradition and innovation – Francesca Grillo', *Times of Malta* [Internet]. 2023 May 9 [cited 2023 Oct 24]; Available from: https://timesofmalta.com/articles/view/killing-tradition-innovation-francesca-grillo.1030307

3. MUSCAT, J., 'Traditional Boats of Malta', *Malta Archaeological Review*. 2002;

4. MUSCAT, J., 'The Gozo Boat'., *Vassallo History*, 2013 [cited 2023 Oct 24]. Available from: https://vassallohistory.wordpress.com/gozo-ferry-since-1885/the-gozo-boat/#:~:text=In%20the%2013th%20century%20the,a%20brigantine%20or%20a%20fregata.

5. BARANI, R., CAMILLERI, A., Luzzu [Internet]. 2021. Available from: https://variety.com/2021/film/reviews/luzzu-review-1234896755/

6. Foldiphoto. Dgħajsa [Internet]. 2022. Available from: https://commons.wikimedia.org/wiki/File:Dg

7. AKRIVOPOULOU, E., MUÑOZ, I., 'Artistic and symbolic elements of the Maltese luzzu', *Archaeonautica*. 2021 Jul 1;(21):337–9.

8. MALLIA, J., 'Throwing it back! Check out these glorious Malta holiday photos from 1988'., *GuideMeMalta*. 2021 [cited 2023 Oct 24]. Available from: https://www.guidememalta.com/en/throwing-it-back-check-out-these-glorious-malta-holiday-photos-from-1988

9. TESTA, M., 'Pietagrave landmark to be restored and converted'., *Times of Malta*. 2007 [cited 2023 Oct 24]. Available from: https://timesofmalta.com/articles/view/piet-agrave-landmark-to-be-restored-and-converted.7245

10. Google Maps. 45 Xatt is-Sajjieda. 2016.

11. Google Maps. 80 Church Street. 2016.

12. VASSALLO, R., 'Marsaxlokk residents find 'Hard Standing Facility' hard to stand' - MaltaToday.com.mt., *maltatoday*. 2011

13. VASSALLO, A., 'Association set up to safeguard traditional Maltese boats', *TVM News* [Internet]. 2020 Dec 23 [cited 2023 Oct 24]; Available from: https://tvmnews.mt/en/news/titwaqqaf-assocjazzjoni-biex-issalva-opri-tal-bahar-tradizzjonali-maltin/

14. Times of Malta. 'One of the last surviving "dgħajjes tal-latini" will be restored', *Times of Malta*. 2020 [cited 2023 Oct 24]. Available from: https://timesofmalta.com/articles/view/one-of-the-last-surviving-dghajjes-tal-latini-will-be-restored.835058

15. Gozo News. One of the last surviving examples of a 'dghajsa tal-latini' to be restored, *Gozo News* [Internet]. 2020 [cited 2023 Oct 24]; Available from: https://gozo.news/87130/one-of-the-last-surviving-examples-of-a-dghajsa-tal-latinito-be-restored/

16. Planning Authority., 'Restoration grant scheme for facades of grade 1 or grade 2 scheduled residences and heritage residences within ucas', 2023, *Irrestawra Darek* 2023 (idgs23)

17. Government of Malta. Skema ta' preservazzjoni tad-dgħajjes tradizzjonali. 2023.

18. Government of Malta. Malta's National Strategic Plan For Fisheries 2007-2013, *Malta's National Strategic Plan for Fisheries.*

19. JENTOFT, S., FRANZ, N., BARRAGAN PALLIDINES, J M., CHUENPAGDEE, R., 'Are the Small-Scale Fisheries Guidelines Sufficient to Halt the Fisheries Decline in Malta? In The Small-scale Fisheries Guidelines Global Implementation Eds', *Springer*; 2017.

20. CARABOTT, S., 'The Maritime Museum shares a rare find', *Times of Malta*. 2018 [cited 2023 Oct 24]. Available from: https://timesofmalta.com/articles/view/the-maritime-museum-shares-a-rare-find.695052

21. CINI, G., 'Master craftsman calls for help for revival of boat building tradition', *Times of Malta*. 2003 [cited 2023 Oct 24]. Available from: https://timesofmalta.com/articles/view/master-craftsman-calls-for-help-for-revival-of-boat-building-tradition.137972

22. AGIUS, R., 'Cars seized due to foreign number plates to be auctioned off', *TVM News*. 2016 [cited 2023 Oct 24]. Available from: https://tvmnews.mt/en/news/cars-seized-due-to-foreign-number-plates-to-be-auctioned-off/

23. GALEA DEBONO, F., 'Imagine Marsaxlokk without a luzzu': Maltese fishing boats are slowly vanishing', *Times of Malta*. 2022;

24. VEGAS MACIAS, J., LAMERS, M., TOONEN, H., 'Connecting fishing and tourism practices using digital tools: a case study of Marsaxlokk, Malta', *Maritime Studies*. 2023 Jun 1;22(2).

25. MIZZI, J., Malta Bus Reborn [Internet]. Available from: https://www.mizzi.co/malta-bus-reborn

9. AUTHORS BIOGRAPHY

Dr Andrea Grech La Rosa CEng MRINA is a Lecturer (Teaching) of Engineering Design and Industry Engagement at University College London in the department of Mechanical Engineering. He is a Naval Architect with interests in many areas including sustainability, technological advancements and practical application. Andrea is a keen enthusiast for traditional Maltese boats and has dedicated a substantial amount of time aiding in the upkeep and restoration of Neptune 89, a vessel he has sailed on since birth. He has also participated wherever possible in providing support to other owners of traditional vessels to safeguard their future for generations to come. Andrea's frustration of the current state of affairs has led him to cowrite this paper in the hope that it will get to the right people to instigate change.

Mr Philip Grech is a business owner currently involved in household heating. His interests include our natural environment and sustainability. Philip's previous business involved chartering the traditional Maltese 'Luzzu', Neptune 89, for day trips. This vessel was built for his family in 1951 for fishing purposes and refitted as passenger vessel by Philip with assistance from Dr Andrea Grech La Rosa. Although Neptune 89 is still used as a commercial passenger vessel, the ultimate sale of the vessel has highlighted the challenges of owning and maintaining such vessels first hand to Philip and prompted him to cowrite this paper in the hope of raising awareness.

AN INTRODUCTION TO THE SHIPS PLANS COLLECTION AT THE NATIONAL MARITIME MUSEUM AND ITS USES FOR PRESERVING HISTORIC SHIPS

A Grover, Royal Museums Greenwich, UK

SUMMARY

Royal Museums Greenwich is the National Repository for Admiralty ships plans and technical records. The institution also holds significant merchant collections, comprising of ships plans, photographs, business, and technical records.

Despite its size, it is arguably an underutilised and relatively unknown archive outside model makers and a small group of maritime researchers. This paper gives an overview of the collections showing how are used as a key reference for the preservation and understanding of Historic Ships. It highlights specific examples where the collections have been employed, including the building of replica vessels, and recently built warships.

It touches on the applications of the collection as an educational resource to facilitate further shipbuilding knowledge. This includes rediscovering lost shipwright skills and past techniques to inform future preservation. It can also potentially inform more greener ship designs through the extensive sailing ship archive from more hydrodynamic hull designs to cleaner propulsion systems.

NOMENCLATURE

CAD – Computer Aided Design
HMS – His/Her Majesty's Ship
NMM - National Maritime Museum, London
IWM - Imperial War Museum London
MoD - Ministry of Defence
MSA - Merchant Ship Plan Collection
RMG - Royal Museums Greenwich
SNR – Society for Nautical Research
VATC – Victory Advisory Technical Committee
Victory (1765)- The name of a ship is italicised with the vessel's launching following it in brackets.

1. INTRODUCTION

The National Maritime Museum holds one of the largest technical collections in the UK, comprising 1 million plans, 286,000 photographic negatives, and 1 million photographic prints. It covers a date range from the late 1600's to the 1980's. The museum is the National Repository for the Admiralty technical records and ships plans. It holds significant collections from private shipbuilders, designers as well as artificial collections from collectors and ones created by the museum.

The size of the collection means that it has greater potential for researchers than is currently being utilised. The purpose of my paper is two-fold to provide a historical narrative of the museum's collection and the institution's involvement with Historic Ships. The second aim is to highlight the collections potential for academics and maritime professionals who may be unaware of the extent of the collections and their variety of uses.

This paper will be split into 4 parts. The first part will look at the founding of the National Maritime Museum. It will include details of the first acquisition of ships drawings and a broad overview of the ships plans collections held by the museum with specific examples highlighted.

The second section of this paper will highlight the different levels of assistance that the museum has given for the preservation of historic ships. The museum's involvement with ship preservation spans from as little as providing copies of plans for a project up to sitting on advisory committees overseeing the restoration of an historic vessel.

The third part will focus on the use of the ship plans as an educational and museological resource. This section covers the rediscovery of dormant shipbuilding knowledge with the case studies of three different wooden replica ships. There is the digital element where the plans have been used by a variety of organisations to better portray certain aspects of a ship that are difficult with the plan on its own. The drawings have also been useful for underwater archaeologists identifying environmental issues on sunken ships.

The final section will briefly examine how the collection can potentially inform or inspire greener ship design particularly in the areas of propulsion and hydrodynamics. This will provide a historical context to some of the design modifications that have previously been attempted.

2. OVERVIEW OF SHIPS PLANS COLLECTION

Before the museum's founding in 1934, there were small maritime collections in other institutions, but none were deemed suitable to hold a national collection. The main person pushing for an independent national naval museum in Britain was Sir Geoffrey Callendar. He was Treasurer of the Society for Nautical Research, lecturer in naval history at the Royal Naval College, Greenwich, and later became the first Director of the museum. The establishment of a national naval museum had been one of the society's chief aims since its founding [1].

As far back as 1927 when the idea of establishing the museum became more solid, ship plans were desired for the museum to "illustrate the development of naval construction from the seventeenth century to the present" [2].

The first transfer of Admiralty material to the museum started in 1938. Callendar reported at a Museums Trustees meeting that about 1,200 'Sailing Navy' plans dealing with 320 ships were delivered by the Director of Naval Construction. According to the Minutes of the Trustees 9th May 1938 it was proposed that a selection of the drawings be exhibited as soon as possible as 'they demonstrated very clearly the method of construction of ships'[3]. Despite objections from the Department of Naval Construction, the first transfer of Admiralty ship plans to the museum commenced that year. A total of over 4000 drawings (including the 1200 already delivered) dating up to 1837 would be transferred to the museum [4].

There was an expectation of 4,400 naval plans to eventually be delivered but this increased to nearly 8,000 drawings [5]. In 1961, the museum was appointed a place of deposit for Admiralty records, under the 1958 Public Records Act. This meant the further transfer of ships drawings, contracts and building specifications, and Ships' Covers to the museum. As a result of these transfers, the museum has a virtually unbroken record of technical and historical development of warships from the late-seventeenth century to the late 1960s. After this date, plans become rare due to ships being sold to foreign nations, changes in methods of design with the introduction of CAD and external designs based on MoD specifications, and the 30-year Rule for confidential documents from the Ministry of Defence.

There was an agreement in place that additional transfers would be made to the museum every twenty years. A large collection of drawings from the Naval Historical Branch and the Director General Ships Department in Bath were transferred to the museum in the 1970's. The latter included an array of Department of Naval Intelligence drawings of foreign ships along with plans surrendered to the Admiralty at the end of both world wars. Many countries from around the world are represented in this collection. The most numerous, however come from Germany.

Out of all the ships plans collections held by the museum, the Admiralty collection generates the most interest. If a specific ship is not represented, then it is likely we will have drawings of its sister ship or one similar enough to answer the enquiry. Most of these drawings are rolled but we also have a significant folded collection, the latter includes docking arrangements and watertight compartments on board a vessel. You can see a progression in the materials used by the Admiralty in the composition of the drawings. This ranges from cartridge paper through to the introduction of tracing paper to linen backed paper and acetate and the use of paper dyeline for their drawings in our collection. This reflects the change in format and processes used for replicating plans used by the Admiralty and naval dockyards.

The museum holds a substantial merchant ship plan archive, but this was established later in the 20th Century. Frank Carr, Director of the museum in the 1950's was aware of the accusation that the museum paid too much attention to Royal Navy affairs and hardly any to merchant concerns [6]. The Museum redressed this imbalance, by sending correspondence to various shipyards asking for the donation of original drawings. Some shipbuilding firms complied and sent the museum original plans and copies. Other museums and archives around the country were also doing the same thing, but several merchant ship plans collections were acquired by the museum through this process. The first set of drawings received were compiled into an artificial archive by the museum curators known as the MSA collection. Despite receiving plans from a variety of sources there were not enough drawings individually to warrant their own company collections. However, a large number of our merchant collections, were acquired later due to the rapid contraction of the British shipbuilding industry. Many of these new collections, unlike the MSA archive were large enough to have their own identity. These were rescued by the Museum and other institutions for posterity when the companies that generated them ceased to operate. In one case a collection was rescued before the buildings in a shipyard were being demolished, showing how close some archive came to being lost.

The museum features collections from various merchant shipbuilders throughout the UK such as Camper & Nicholson, Alexander Stephens, and John I. Thornycroft. Collections from yacht designers such as Robert Clark and Laurent Giles are also housed within the museum. These latter archives were presented to the museum due to preservation concerns or when the original company was downsizing. These merchant collections represent everything from rowing boats up to passenger liners and tankers. One collection that bears mentioning is the William Denny & Brothers of Dumbarton collection. Virtually the entire surviving ships plans archive dating from 1844 to 1963 were presented to the museum after the company went into voluntary liquidation in 1963. It consists of 58,500 ship and engine drawings and just over 4000 technical records. A previous curator, David Lyon, created the *Denny List* which gives details of the ships and plans represented in the museum facilitating public access to the collection. The Denny collection is the only ship plan archive to have a published catalogue, although the collection itself is so large that it waits cataloguing at item level. This illustrates the challenges of sorting numbering and cataloguing large ship plan collections held in the museum's care. The issues with staffing to catalogue large collections means curators are put under more pressure to provide access to the archives for researchers.

Both the Admiralty and mercantile collections have a huge variety of ship plan types. This ranges from the general arrangement plans of decks, profiles, sheer lines, sail, and rigging plans as well as detailed drawings that cover boilers, magazines, and engines. In both the Admiralty and mercantile archives, there is a lot of crossover with the drawings. Private yards are building ships for the Admiralty and there are collections that illuminate certain aspects not found on the Admiralty drawings. One example of this is the Vickers Ordnance collection, which provides detailed plans of the weapons that were being fitted to Admiralty warships.

Despite the size of our ships plans collections, it is incomplete. In both the Admiralty and merchant collections there are gaps. However, there is usually an option for finding alternative ships from the same class or design. The Admiralty technical records collection provides additional data and can inform the motivations behind specific choices in a warship's design. The most frequently consulted technical records are the Ship's Covers. This is a bound volume containing the surviving correspondence between the Admiralty and the shipbuilder of a specific vessel's construction. The ships plans are supported by various photographic collections. Part of this collection was created in reference to or to work with the ships plans and provides another dimension to the vessels represented in our archives. This can focus on the ship's construction on the slipways but also interior views as well as depictions of the vessels in port and at sea.

3. THE NATIONAL MARITIME MUSEUM AND ITS INVOLVMENT IN THE PRESERVATION OF HISTORIC SHIPS

The museum's involvement in preserving historical ships has varied depending on the ship in question and the organisation involved. This section uses individual case studies to illustrate the different levels of participation by the Museum.

In some cases, the museum's involvement has simply been providing copies of a ship's drawings to be used as a working archive for a vessel's restoration. These requests can come from individuals who have recently purchased a small boat to restore, the company restoring the vessel, as well as private trusts and museums that look after larger ships. Examples of this range include the sailing frigate HMS *Trincomalee* (1817), HM submarine *Alliance* (1945), the motor cruiser *Malahne* (1937), and various lifeboats.

Some researchers will prefer to visit the museum where we can show them related drawings and technical material. The museum can give consultative advice on the best drawings to use. This consultation service can occur during the planning process for a ship's preservation or during ongoing work on a vessel.

One recent example of our assistance during the preservation planning process is for the minesweeper HMS *Bronington* (1953), which sunk at its moorings in Birkenhead in 2016. The *Bronington* Trust visited the museums archives, consulted, and bought copies of *Bronington*'s drawings. These plans and the technical records will provide the necessary information needed for its potential refloating and transfer for restoration in dry dock in Birkenhead.

One example of our assistance for ongoing preservation work is for the sloop HMS *Gannet* (1878). The museum hosted the rigging company T.S. Rigging, who were commissioned to undertake rigging maintenance work on *Gannet* in Chatham Dockyard. The team consulted the drawings, photographs, and technical records to get a detailed understanding of the rigging and related fitting arrangements on board the ship.

The museum drawings have been consulted regarding the prevalence of hazards on board shipwrecks. This can vary from fuel and ordnance to dangerous construction materials used on board various ships, the most common of these being asbestos. Many of these investigations come from law firms representing former crewman or ship constructors who are suffering health problems related to asbestos exposure. The drawings can sometimes suggest where asbestos may have been used, but the detailed identification relies on onsite surveys and sampling [7].

As I stated previously there are gaps in the collection where ships are not represented. However, the technical collection can supplement the plans by providing additional construction correspondence, design plans and sketches. This was the case with the monitor HMS *M.33* (1915), where no plans survive in our collection, but the ship's cover did. The cover was also consulted by Hampshire County Council when they undertook the external restoration of M33 post-1994 [8].

Replica ships have relied on surviving primary sources to try and ensure a degree of historical accuracy. While these are compromised by the needs to fulfil modern safety standards, in-built longevity and alternative propulsion methods, the desire to remain as authentic as possible is still evident. Drawings and technical records have been supplied by the museum for the building of fully functional replica vessels. This has included HMS *Bounty* in the 1960's and HM Bark *Endeavour* built in the 1980's.

The museum has also assisted in non-British replica vessels. One recent example was the French frigate *Hermione* (1779), where the museum provided copies of its captured sister ship *Concorde* to assist with the construction. The museum has assisted with smaller replicas. One example is a fully functional 40ft coastal motorboat (CMB). The Portsmouth Naval Trust purchased copies of drawings from the John I. Thornycroft collection and consulted the ship's Cover to build this small craft.

The museum had a brief but more hands-on involvement with the preservation of the broadside ironclad frigate HMS *Warrior* (1860). The reason for this is likely that, most of *Warrior*'s interior, fittings and equipment had been removed after its conversion to an oil jetty pontoon in Pembroke [9]. This meant *Warrior* needed a complete internal rebuild rather than just preservation. As early as 1967, the museum was involved in preservation proposals and along with Lloyd's Register, participated in initial hull surveys of the vessel in the 1960's [10]. The museum provided copies of *Warrior*'s drawings in addition to other ironclads from the same era to supplement the fittings used on ships of the period. There were some aspects of the vessel, such as the engines and boilers where the information on *Warrior*'s drawings were limited. Luckily this dilemma was solved by looking at the museum's full and detailed drawings of *Warrior*'s sister ship HMS *Black Prince* [11].

A further testament to the museum's involvement in preserving ships is that it currently looks after the *Cutty Sark* (1869), one of the last clipper ships in existence. The museum initially provided informal advice on the funding and preservation when *Cutty Sark* arrived in Greenwich in the 1950's [12]. The drawings and other materials relating to *Cutty Sark* held by the museum in the 1950's, however, was very limited and the museum was not involved with the ship's restoration and preservation. The museum took ownership of the *Cutty Sark* in 2015 and employs ship keepers and curators for the ongoing maintenance, research, and interpretation of the ship. The previous trust had accrued a large amount of material relating to *Cutty Sark*, which is now held by the museum, but a lot of material is held by various archives around the country. Whilst the museum is an obvious place to search for information and archives relating to historic vessels, there is material stored in other institutions like museums, galleries and universities that should be consulted.

One ship where the museum has been heavily involved in its preservation is HMS *Victory* (1765). *Victory* was initially preserved through the campaigning and fundraising efforts of the Society for Nautical Research in 1922. The main extent of the museum's involvement has been as a member of the *Victory* Advisory Technical Committee (VATC). This was an advisory committee originally set up in 1929 by the Admiralty Board with the Society of Nautical Research to advise on all future repairs and continued maintenance of the flagship. This first committee disbanded in 1939 and remained dormant until 1955. The museum's director Frank Carr became aware of a proposed plan by the Admiralty to make Victory's deck watertight and improve air circulation. This plan involved decking over the skid beams in the waist of Victory, which would alter the appearance of the weather deck from its Trafalgar form [13]. The museum protested to the Admiralty arguing that the VATC needed to be reformed as there had been no consultation with them or the SNR on this plan. The Admiralty agreed and the VATC was formally reconstituted as a single body in May 1955 with representatives from the National Maritime Museum, the Admiralty, and the Society for Nautical Research [14]. This would continue until the committee was disbanded in 2012 after the transfer of ownership of *Victory* from the Ministry of Defence to the National Museum of the Royal Navy. Its purpose was to advise on all aspects of Victory's maintenance, repair, and restoration, with the object being to retain the appearance and state of the ship as it was at the time of the Battle of Trafalgar [15]. The museum would be part of an Interpretation sub-committee, assisting with research and analysis on parts of the restoration and reinterpretation of *Victory* to make sure that this work is carried out accurately [16]. Examples of this work include consulting the Sheerness Dockyard fitting books to ascertain if there was a flooding system for the powder magazine and the reconstruction of the after hold storerooms. The museum's ship model collection would also be consulted for the internal colour scheme on Victory [17]. The VATC helped to draft a conservation management plan of the ship and was regularly kept updated with ongoing maintenance work on board the ship.

4. THE USE OF THE SHIPS PLAN COLLECTION AS AN EDUCATIONAL RESOURCE

The historic collection of ships plans is a repository of knowledge by past architects and designers. As technology advanced, some of their skills, understanding of materials, propulsions, and the challenges of working with composite materials has been lost. However, viewing the collections with an appreciation of past practice is an opportunity for us to refresh our knowledge and skills. This can range from the use of wooden tree nails to fasten wooden planks to the frame, how to attach frames to the beam of a ship and the correct application of iron plates to a malleable wooden frame. With this background information, the drawings can allow researchers to expand on these skills and techniques in the case of specific ships.

One key example of how this past shipbuilding knowledge is being refreshed is through the building of replica wooden ships. Sean McGrail states that building replicas allow the builders and public to appreciate the knowledge and skills of shipbuilders. It can also lead to an appreciation and better understanding of the skills needed to design, build, and operate such ships [18].

As mentioned previously, three of the main replicas that the museum has been involved with are the *Bounty*, the French frigate *Hermione* and the *Endeavour*. The drawings can provide a foundation of understanding into how these ships were built. In the case of *Endeavour*, several of the original drawings still exist in the museum's collection, and they provided the basis for the full-size patterns needed to make the replica [19]. All three replicas had to incorporate modern practices, such as the installation of an auxiliary engine, into their construction. Part of this was to meet modern health and safety requirements, but in the case of *Endeavour*, longevity and supply aspects were also considered. Alternative woods were chosen for the hull over oak, the wood used on the original *Endeavour*. Oak is susceptible to rot, and the vessel needed to last at least 40 years, but it was not available for an affordable price, or the quantity needed to construct the hull [20]. Despite these understandable modern alterations, efforts were taken to not compromise too much on the original method of construction.

The ships plan collection has become a resource for future ship design, where it is important to learn lessons from past proposals. One example of this relates to the building of the Queen Elizabeth class fleet aircraft carriers. The museum was visited by representatives of Thalis and BAE Systems to look at aircraft carrier drawings. This included the design drawings for the cancelled fleet aircraft carrier CVA-01 from 1966, the last fleet carrier to be designed for the Royal Navy. The representatives looked at the Colossus class aircraft carrier's plans as part of their visit. The drawings gave key information into what the size of a fleet carrier should be along with the absolute minimum requirements for the wartime functionality of an aircraft carrier.

From an educational perspective the plans have been used to give an alternative approach on well-known stories. For instance, there have been cases where the plans have been repurposed for use on a digital platform. A *Mariner's Mirror* podcast used drawings of the K class submarine and superimposed a working 3D digital reconstruction of the vessel's armaments, accommodation, ventilation, and propulsion systems on to the digital drawings. This gave a good interpretation of how the major systems operated and the crew's accommodation.

The digitised plans from the collection are also used by model makers to create 3D renderings not only as 3D models, but also to 3D print fittings for their models. A good example of the use of 3D in the commercial context is the computer game World of Warships. The company behind this game has bought copies of our plans to make 3D models of warships with a relatively realistic interpretation of how the vessels engaged in combat. This includes vessels which were never built, like the G3 class battlecruiser. If one plays the game and sees a British warship and certain German warships, there is a high likelihood the drawings used to create that model came from the museum's ship plans collection.

The museum's collection is well used by marine archaeologists for the excavation, surveying, and identification of ships. These include commercial archaeology companies, the MoD, and the US Coastguard. A recent example of this is HMS *Erebus* (1826), one of two bomb vessels used to search for the Northwest Passage in 1845, led by Sir John Franklin. The position of the hatches and masts, and the measurements of the fittings on the drawings were key details for Parks Canada in identifying *Erebus* when it was found in 2014. Another example is the wide range of plans from different classes of warships used by the maritime archaeologist Innes McCartney. He argues in his book on the wrecks from the battle of Jutland in 1916 that the plans of many of the vessels suck have 'proved invaluable in deciphering much of the archaeology recorded'[21].

A second consideration is that the plans have proved key to conducting underwater surveys. One example is a survey conducted on the battleship HMS *Vanguard* (1909) in Scapa Flow, Orkney, which was destroyed by a magazine explosion in 1917. The plans of the ship and armaments provided the foundation for assisting in identifying parts of the debris field. The museum's plans collection were consulted for surveys undertaken on the battleship HMS *Royal Oak* (1914) also sunk in Scapa Flow.

I previously mentioned that plans provided a working archive for the maintenance or restoration of historic ships. There is, however, an interpretation aspect as well. The plans provide a visual record of a ship's layout and subsequent alterations during its lifetime. For instance, the light cruiser HMS *Belfast* (1938) underwent a number of refits and structural changes before becoming a museum ship at the Imperial War Museum. Nigel Steel, speaking for the IWM states that the drawings of HMS *Belfast* are important for 'understanding and charting the archaeology of the ship'[22]. The plans are essential in interpreting and understanding all of these changes over a vessel's life and can also give an insight into decisions made by previous staff about how a ship is to be restored and presented to the public. Nigel Steel states that the sickbay on HMS *Belfast* was vastly different in the 1970's from what it was in the ship's wartime configuration. He found this interesting 'both through what it tells us about *Belfast* but also how our predecessors as professional curators have gone about balancing visitor insight against the practicality of a visit' [23].

The plans proved useful for the reinterpretation of the light cruiser HMS *Caroline* (1914) as a museum ship. *Caroline* has had a longer life than *Belfast* but had a shorter active service than the latter and experienced a lot of changes and structural alterations on board after being converted into a drill ship. This included significant alterations to its superstructure including the construction of a large deckhouse amidships. Andrew Baines, speaking for the National Museum of the Royal Navy stated that with the drawings they were able to identify the correct layout for *Caroline*, and establish the pattern numbers for each piece of furniture used on the ship. To quote Andrew Baines, 'That, combined with the contract and spec and ship's cover, it allowed us to understand the archaeology of what we were uncovering, and make informed decisions on what we should keep, and what was a recent addition to the ship'[24].

5. INSPIRATION FROM THE COLLECTION FOR POTENTIAL GREEN SHIP TECHNOLOGY

The current concerns about the climate, excessive carbon footprint and the need to create cleaner alternatives brings our predecessor's work back into the spotlight. The historic collections provide inspiration for future greener vessel design, propulsion and in some cases stop us re-inventing the wheel. There are various aspects of the ship plan collection that can potentially inform environmentally sustainable ships. The first is arguably through the most environmentally friendly method of propulsion, wind power. The museum has an extensive archive of merchant sailing ships of all different sizes with a variety of sail arrangements. Two of the key archives the museum hold are the Alexander Stephens Company and Barclay & Curle ship plans collections. These both include detailed rigging drawings for large 19th Century barque and clipper ships. The collections also include numerous hull drawings providing the foundation for the most efficient shape to use on deep sea ships.

If the shipbuilding industry is reutilising wind power for larger ships, then there will be the accompanying need to use the oceans currents more effectively to move the ships around the seas. The trade winds and currents were vital on the sailing routes in the 18th Century. Like the building of steam ships over wooden vessels however, the knowledge of these original routes will have been less prevalent in the mid to late 20th Century. Detailed historical data about trade routes can inspire new ones. They will not provide the most direct route, but it will potentially be more environmentally efficient. By combining this data with modern weather prediction and navigation technology, this could potentially make wind powered ships more viable.

The ships plan collection can provide a historical context to what may seem an innovative idea but has already been considered by previous naval architects and ship constructors. An example of this lies in propeller design. Shipbuilders today must consider variables such as material composition, number of blades and rate of revolutions. The Admiralty and private ship designers in the 19th Century were doing the same thing, experimenting with different pitches, number of propeller blades to try and find the most efficient form. One example of the propeller experiment can be seen on the drawings for the steam sloop HMS *Rattler* (1836). The plans show numerous propellers with non-standard shapes that were used in trials in the 1840's. These trials were of interest to Isambard Kingdom Brunel, who not only helped instigate Rattler's trials but would use much of the subsequent trial data in his propeller design for the ship SS *Great Britain* (1846). The innovation in propeller design does not abate after Rattler but is less publicised. An example of later propeller experimentation is with the Cressy class armoured cruisers in the late 19th Century. The Admiralty were trying out different types of three and four bladed propellers to find the most efficient form and made notes on the degree of blade deterioration on both types. Admittedly, ships designers from 19th Century were limited by the technology and materials of their time and there is a greater variety available today. However, these drawings, and other models within the collection can provide a historical basis and inspiration for a current problem that is rooted in the 19th Century.

The collection can also inform on the installation of modifications on board vessels to make them more environmentally friendly. One of these is Flettner rotors. These were designed initially in the 1920's by German engineer Anton Flettner, who installed the rotors on two separate ships *Buckau* (1920) and *Barbara* (1926). The rotors worked satisfactorily to convert wind energy into propulsion power, and can potentially supplement other propulsion systems, leading to savings in energy costs and lower fuel consumption. Despite this the rotors never caught on initially due to high capital costs and cheap fuel sources. Due to advances in technology and a greater focus on environmental sustainability, many shipbuilding companies are now exploring the installation these rotor sails on ships as small as catamarans to LR2 class tankers.

A second potential modification is the use of other types of sails on modern steel ships. One example from our collection is a model of the Japanese tanker *Shin Aitoki Maru* (1980). This Japanese tanker was fitted with two sails in the 1980's as a response to the 1970's oil crisis at the time. A second example from the museum's collection is the Japanese built cargo ship *Yamakuni Maru* (1985). The ship was fitted with a bow sail during construction in 1985 which according to Lloyds Register, remained on the ship until at least 2009. These sail modifications are still being fitted to ships today. One example is from an article in the BBC News from late August 2023 for the building of a British cargo ship named *Pyxis Ocean* fitted with rigid sails for its maiden voyage. The belief is that these sails can potentially help to reduce the vessel's lifetime carbon emissions by 30% [25].

6. CONCLUSION

The National Maritime Museum holds an extensive collection of ships plans and technical records. It consists of drawings for both naval and merchant ships that are supported in many instances by our photographic and ship models collections. Its size and date range show there is a greater potential yet to be fully utilised by researchers.

The archive has proved useful for preserving numerous historic vessels and the museum's involvement has varied depending on the ship. This has ranged from simply providing copies of a ship's drawings such as HMS *Trincomalee* through to consulting on a ship's ongoing maintenance such as HMS *Victory*. The drawings have provided the primary historic foundation and authenticity for the construction of replica ships including HM Bark *Endeavour.*

The collection has proved to be a useful educational resource. The drawings have assisted informing past shipbuilding knowledge, in the form of replica vessels and provided useful data for future ship construction such as the *Queen Elizabeth* class aircraft carriers. The drawings provide a different perspective on well-known stories. A good example of this from a digital aspect is the video game World of Warships. This provides a realistic visual and relatively accurate combat interpretation of 20th Century warships. The plans have proved useful for the interpretation by the custodians of museum ships. It provides a visual record of the ship's alteration but can help illustrate decisions taken by previous staff.

Finally, the collection can suggest creative ideas for further green ship design. The sailing ship archives is a treasure trove of environmentally friendly designs in terms of propulsion and hydrodynamics. The propeller designs experiments from *Rattler* and *Cressy* in terms of pitch and blade numbers are still relevant today. They can provide inspiration to an industry that does not have the same limitation in technology as the 19th Century. The Flettner rotors and sail modifications can provide incentives for shipbuilding companies to reduce their carbon emissions and fuel costs. The ideas mentioned above are not new, but they illustrate how past naval architects, designers and ship constructors had to manage their resources with the technology of the period. It can also show how some of their experiences can be used as inspiration where it supports a more carbon friendly solution to the shipping industry.

7. ACKNOWLEDGEMENTS

My thanks go to my museum colleagues Andrew Choong, Jeremy Michell, and Bob Todd for proofreading the drafts, providing suggestions, and patiently putting up with my endless questions. My thanks must also go to Robert Rumble and Nigel Steel from the Imperial War Museum, London, Nicholas Ball from Chatham Historic Dockyard Trust and Andrew Baines from the National Museum of the Royal navy for their useful insights on the ships they manage.

8. REFERENCES

1. LITTLEWOOD, K., and BUTLER, B., 'Of Ships & Stars: Maritime Heritage and the founding of the National Maritime Museum in Greenwich', *National Maritime Museum* 1998 p.45

2. LITTLEWOOD, K., and BUTLER, B., 'Of Ships & Stars: Maritime Heritage and the founding of the National Maritime Museum in Greenwich', *National Maritime Museum* 1998 p.49

3. National Maritime Museum Meeting of the Trustees, 9th May 1938.

4. LITTLEWOOD, K., and BUTLER, B., 'Of Ships & Stars: Maritime Heritage and the founding of the National Maritime Museum in Greenwich' *National Maritime Museum*, 1998 p.108

5. Figure retrieved from Mimsy collections database on 13 September 2023.

6. LITTLEWOOD, K., and BUTLER, B., 'Of Ships & Stars: Maritime Heritage and the founding of the National Maritime Museum in Greenwich' *National Maritime Museum*, 1998 p.148

7. Email correspondence with Andrew Baines, Deputy Executive Director of Museum Operations at the National Museum of the Royal Navy 30 June 2023

8. Email correspondence with Andrew Baines, Deputy Executive Director of Museum Operations at the National Museum of the Royal Navy 30 June 2023

9. MAY, R., 'HMS Warrior Owners Workshop Manual 1860-date', *Haynes Publishing*, 2017 p.151

10. MAY, R., 'HMS Warrior Owners Workshop Manual 1860-date', *Haynes Publishing*, 2017 p.153

11. MAY, R., 'HMS Warrior Owners Workshop Manual 1860-date', *Haynes Publishing*, 2017 p.162

12. LITTLEWOOD, K., and BUTLER, B., 'Of Ships & Stars: Maritime Heritage and the founding of the National Maritime Museum in Greenwich', *National Maritime Museum*, 1998 p.155-157

13. MCGOWAN, A., 'HMS Victory: Her Construction, Career and Restoration', *Chatham Publishing*, 1999. p.122

14. MCGOWAN, A., 'HMS Victory: Her Construction, Career and Restoration', *Chatham Publishing*, 1999. p.122

15. COAD, J., 'The origins of the Victory Advisory Technical Committee', 2011

16. LAVERY, B., and MICHELL, J., 'Discussion Notes on the Development of the Victory Advisory Technical Committee' dated 13 May 2008

17. GOODWIN, P., 'The Application and Scheme of Paintworks in British Men-of-War in the Late Eighteenth and Early Nineteenth Centuries', *The Mariner's Mirror*, 99:3, 287-300, 2013 DOI: 10.1080/00253359.2013.815993

18. MCGRAIL, S., 'Experimental Archaeology: Replicas and Reconstruction' from Bennet, J, ed. Sailing into the past: learning from replica ships', *Annapolis Naval Institute Press* 2009 p.22

19. MACARTHUR, A., 'HM Bark Endeavour' from Bennet, J, ed. Sailing into the past: learning from replica ships', *Annapolis Naval Institute Press* 2009 p.168

20. Australian National Maritime Museum, Endeavour Official Souvenir Brochure, HM Bark Endeavour Foundation, *Ticktock Entertainment*, 2002 p.2

21. MACARTNEY, I., 'Jutland 1916: The Archaeology of a Naval Battlefield', *Bloomsbury* 2016 p.19

22. Email correspondence with Nigel Steel, Head of Content, HMS *Belfast* 20 June 2023

23. Email correspondence with Nigel Steel, Head of Content, HMS *Belfast* 20 June 2023

24. Email correspondence with Andrew Baines, Deputy Executive Director of Museum Operations at the National Museum of the Royal Navy 30 June 2023

25. SINGLETON, T., 'Pioneering wind-powered cargo ship sets sail' *BBC News*, (https://www.bbc.co.uk/news/technology-66543643) Accessed 24 August 2023

9. AUTHOR'S BIOGRAPHY

Alex Grover currently holds the current position of Assistant Curator of Historic Photographs & Ships Plans at Royal Museums Greenwich. He has joint responsibility for looking after the Historic Photographs, Ships Plans and the Orders & Medals collections, along with full responsibility for the Ship Badges collections held at the museum. He was one of several contributors to the publication 'Fish & Ships: A Nautical Miscellany (National Maritime Museum, Greenwich, 2023). He has given talks on a variety of maritime subjects including the mutiny on the HMS Hermione, The 1897 Arctic Balloon Expedition and the life of Emma, Lady Hamilton.

DOVER BRONZE AGE BOAT: THE BOAT, THE REPLICA, THE LEGACY

C Holt, Dover Museum and Bronze Age Boat Gallery

SUMMARY

Hailed as one of the greatest archaeological discoveries of the 20th century, the Dover Bronze Age boat is one of the oldest sea-going boats in the world.

Thirty years after its discovery, it is important to remember its incredible but little-known story. This paper will summarise over two decades of fascinating international research and experimentation. It will also document first- hand discussions with key members of the team that made the replica. I will also reflect on first-hand experience of maintaining and operating the replica boat, as well as being responsible for the conservation of both the original boat and decision making for the long-term future of the replica.

This discussion will be broken down into three main sections: "**the Boat**", "**the Replica**" and "**the Legacy.**"

1. THE BOAT

In September 1992, archaeologist from Canterbury Archaeological Trust were monitoring roadworks linking the port of Dover with the Channel Tunnel. They discovered a number of oak planks, carefully fastened together, buried 6m below the modern streets of Dover and about 200m inland from the present shore. They had discovered the perfectly preserved remains of a large prehistoric sewn-plank boat of complex design, of which a nine-meter length was exposed. [1]

The international importance of the discovery was immediately recognized, and a rapid rescue excavation was undertaken to record and salvage the boat. This unanticipated discovery, at the heart of a major construction programme, the unique unprecedented nature of the find and its position at the base of a narrow shaft (Figure 1), presented a daunting challenge to the project team.

Figure 1: Location of boat at the bottom of narrow shaft.

It was apparent at an early stage that it would not be possible to lift the complete remains of the Dover boat within the short window of time available.[2] The archaeologists were also aware that past attempts at excavating similar boats (the Ferriby boats) in one piece had been unsuccessful. Nor would dismantling the vessel into its component parts provide a solution, for the methods of joining and caulking the individual timbers appeared to represent arguably the most significant aspect in the vessel's technology. The decision was therefore taken to cut the vessel into thirty two pieces and remove them by crane. This approach also made it easier to carry out recording and conservation later. [3]

About 9.5m of the vessel was recovered, including one end of the boat, probably the bow. The presumed stern-end of the boat (the northern end) had to be left in-situ as it lay of outside the area of excavation.[4]

Following excavation, the work of recording, assessment, analysis, conservation, reassembly, and display would begin. A large multidisciplinary team of experts, consisting over 125 individuals, was brought together to study this remarkable discovery.

1.1 CONTEXT OF THE BOAT

It was clear that the boat was not a wreck, but that it had been partially dismantled, and its timbers deliberately cut through, at the time of its abandonment. Paleoenvironmental evidence indicates the boat was abandoned in a shallow, braided freshwater river, completely unsuitable for a boat of its size and complexity.[5] This suggests the boat had been moved some distance from the sea, perhaps being carried, or dragged.[6]

The boat had suffered damage and it is clear from the wear marks and repairs that the vessel was fairly old and leaking when abandoned. [7]

Despite this the structure of the vessel appeared to be quite sound there were no traces of rot nor of marine organisms infesting the timbers and many of the tool marks were still very fresh. This evidence appears to suggest that the boat could have been repaired. The condition of the missing upper parts of the vessel remains unknown, however, and it is possible that these parts may have been in a poor enough condition to have caused the boat to have been abandoned.[8]

It was found in association with fragments of pottery, shale, animal bones, chalk blocks and flints, which perhaps suggest a symbolic element to the act of disposal of the boat.[9]

1.2 ANALYSIS OF THE BOAT

Radiocarbon dating of the timbers suggested a date of 1550BC (3,500 years ago). It was made from oak trees at least 1.2m in diameter and about 350 years old when they'd been cut down to be fashioned into planks.[10]

From the outset it was clear that the structure of the Dover boat was incomplete and distorted. The boat planks had been compressed by the weight of the overlying sediment during their long burial. Analysis of tree rays and rings in cross-section showed that they had originally been up to 50 per cent thicker. [11] By compensating for this compression, it was possible to postulate the original thickness of the oak planks, a crucial factor in understanding the boats original form. [12]

The pieces of misshapen wood were problematic to record, and it was difficult to ensure that there was an accurate record not only of what had been found, but also the original shape of the boat. [13]

When dealing with an incomplete and distorted vessel from the ancient past, it is necessary to recognise that there are limits to the evidence. Therefore, two levels of study would be needed firstly to record, analyse and describe the vessel as found, and then to reconstruct a hypothetical model of what it was probably like when originally built.

The post-excavation analysis of the boat included detailed data on the boat timbers by Peter Marsden, Richard Darrah, and Damian Goodburn and papers written by Owain Roberts on reconstructing and testing the vessel. The surviving boat timbers were recorded in meticulous detail, accompanied by many hundreds of 1:1 drawings (Figure 2) and thousands of photographs.

Figure 2: Caroline Caldwell recording the timbers.

The surviving remains of the boat consist of four planks hewn from logs of huge, straight-grained oak trees without side branches. Two flat planks form the bottom, each carved out of a half log, leaving upstanding cleats and rails allowing its jointing with other boat timbers. These bottom planks were joined together along a central butt joint, with transverse timbers and wedges hammered through the cleats and central rails. Curved side planks were stitched to the bottom of the boat with withies of twisted yew. These side planks also possess side cleats carved out of the solid wood. The timbers forming the end of the boat play into a Y-shape, intricately carved from the main planks. This originally would have held a carved wooden board. There was clear evidence for the presence and some dimensions of this missing end board; there was also evidence for at least two other main structural timbers. On the top of the carved side planks was another row of stitches suggesting that there were two further side planks. The boat had been made waterproof by pressing a mixture of beeswax and animal fat into the stitch-holes and along the seams, where the stopping was overlain by pads of moss wadding, compressed, and held in place by long thin laths of oak under the yew stitches [14] Together these elements formed a broad-beamed, flat bottomed boat of unique design, employing a woodworking tradition now long forgotten. (Figure 3).

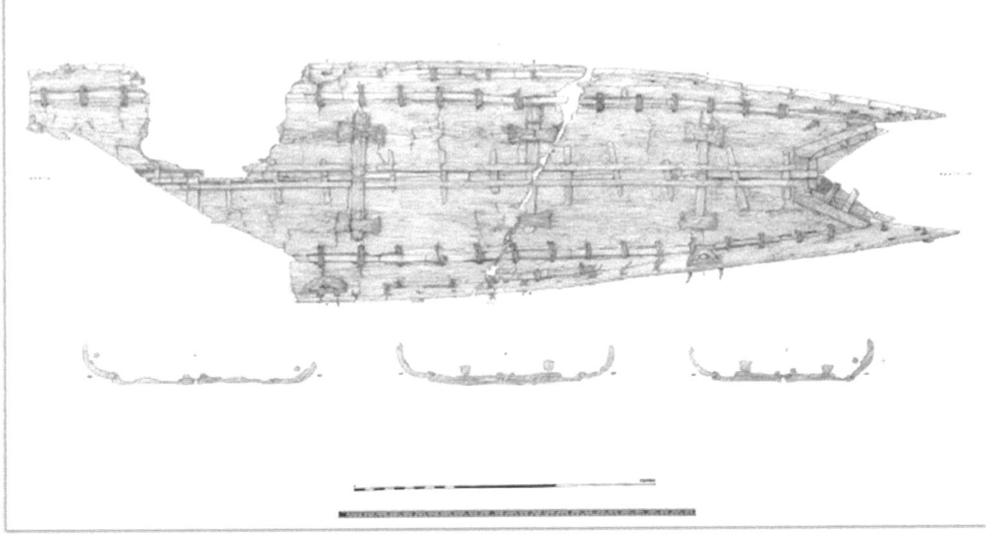

Figure 3: Drawing of the surviving timbers

The boat had clearly been used extensively. Tool marks on its bottom (outboard) surface were differentially worn away, suggesting it had been beached regularly on a sand or gravel shore. [15]

As noted earlier the boat has suffered damaged during its working life. Evidence of repairs show that the boat leaked. Extra stopping was added to the central seam between the bottom planks as these planks were not held as tightly together as was needed. The flexing of the bottom planks was evidently a problem, and the planks were tightened in two ways by extra moss caulking being stuffed in some wedge holes and by a flint flake being pushed under a transverse timber. [16]

It is believed that the original the boat could have been between 11m and 18m long and as wide as 3m.

The boat's structure suggests it is a product of a society confident in its technology and organised in its planning. The proportions of the boat show an experience and understanding of the demands to be made on the hull when going to sea, those of strength, stability, and speed.[18]

The technology used in the construction shows a conscious application of certain mechanical and structural engineering concepts, for example the use of tension fastenings, the use of rails and flanges to stiff inflexible areas of planking and the use of transverse linking members set at opposing angles through the rails to solve the problem of parallel movement and separation of the two bottom planks.[17]

The size and complexity of the boat, suggest that it would have travelled on the open sea, perhaps plying along the southern coast of Britain and across to continental Europe. Such voyages were well within the vessel's technological capabilities.

There was no evidence of a mast or of the use of oars, and it seems probable that it was paddled.[19] A crew of 16 paddlers might achieve speeds of around 4-5 knots with a range of perhaps 56km (30 nautical miles) a day.[20]

The overall shape and characteristics of the boat suggest it was intended to carry freight of some kind. The cargo capacity indicates that the boat could carry about 2 tonnes of cargo, which would have helped with stability. [21]

Although the team who excavated, conserved, and researched the boat were able to make huge increases in our knowledge of the boat there were still questions that could not be determined through the study of the ancient timbers alone. For example, there have been extensive debates about the original form of the boat and its capabilities.

2. THE REPLICA

During the process of interpretation and recording it became evident that it was necessary to create a physical reconstruction of the boat, to understand and be able to show the wider public the shape and style of the original vessel.

2.1 MID-SECTION RECONSTRUCTION

In the summer of 1996, an exact replica of part of the Dover Boat was produced.

The objective of this reconstruction was to rebuild this ancient vessel using the same raw materials and tools, available in the Bronze Age, as accurately as possible, both using and demonstrating the practices involved in the original construction and, where possible, replicating original techniques. This would produce a much more understandable model of the Dover boat and enable the team to learn much more about its original construction in the process.

The main aims of this reconstruction were to:

- Recreate a 3m length of the central section.
- Explore how it would be possible to remove cubic meters of oak with simple bronze and wooden tools.
- Show how the boat shape could have been marked up.
- Investigate stages of construction of the original boat.

- Explore how long the original boat may have taken to construct.
- Demonstrate how the sets of cleats, rails and wedges interact to hold the timbers rigid and control the shape of the boat.
- Investigate how the waterproof joints worked.
- Investigate the sourcing, preparation and fitting of the yew withy stitches. [22]

The process of how the original boat components were put together was worked out by a process of trial and experiment during the reconstruction. (Figure 4)

Figure 4: Midsection reconstruction experiment

The project helped us to understand how the complex system of cleats, rails, stitches, and wedges held the boat together and made it strong enough for sea travel.

Many valuable lessons were learnt during this reconstruction experiment which has provided a wealth of information about the techniques and sequence of construction, the tools used and the sophistication of woodworking techniques in the Middle Bronze Age.

To Summarise:

- The bronze tools worked effectively in shaping the green oak, their main drawback being their lack of weight.
- With wooden wedges and bronze tools, 2.5 tonnes of wood were removed, and the timbers shaped in eighty- five-person days.
- A sequence of marking out and shaping processes were demonstrated, which would have allowed the original boatbuilders to control the shape of their boat.
- On the cleats and stitch holes, a better finish was produced – and with ease – than that found on the original boat, suggesting that the original boat builders were not trying to produce a high-quality finish but that this was a functional boat.
- The time taken to axe notches in the timber suggested that the time taken to fell a 1.1m diameter tree would be two person days.
- To shape and assemble the complete boat, it is estimated that a team of ten skilled workers would have been needed, each working eight hours per day for twenty-five days.
- The finding and collection of the raw materials would have taken more time than the building of the boat.
- The reconstruction confirmed that the transverse timbers stop the longitudinal movement of the two boat halves.
- The flexibility of the transverse timbers meant that they did not resist the upward movement of the central joint completely; this bending was contained by the transverse timbers bearing against the top of the cleat holes and not the rails.

- The friction joint created by the wedges was sufficient to hold the boat halves together, assisted by the transverse timbers.
- The competence of the original boat builders suggest that they were very skilled, and this is unlikely to be the first of this type of boat that they had built [23]

Despite the advances in knowledge and understanding that this reconstruction provided it was still felt that that midsection reconstruction did not answer satisfactorily the shape and capabilities of the original boat.

2.2 HALF-SCALE REPLICA

After 15 years of international research, archaeologists and researchers had a much better understanding of how the original shape of the boat was achieved and were able to propose a full reconstruction of the Dover boat. This second reconstruction experiment was undertaken early in 2012.

Unlike the first reconstruction experiment which focused on the methods by which the boat was built, this replica aimed to build a complete boat to test its capabilities. This ambitious project provided excellent opportunities to demonstrate the challenges and potentials of experiments in reconstructing the partially preserved remains of archaeological structures.

It was imperative that the replica of such a significant boat be built to the highest academic standards. As only parts of the boat survive, the team used all the available evidence from the Dover boat as well as the expertise developed with other prehistoric boats to reconstruct the missing parts. Using the original plans, the corrections from compression, and the ancient timbers themselves to produce the best possible hypothesis for the original form of the vessel. This would produce an accurate replica according to the evidence.

The replica was designed by renowned maritime archaeologist Ole Crumlin-Pederson and built by a team of experts, led by ancient timber specialist Richard Darrah. The half-size replica of the Dover Bronze Age boat was built in the grounds just outside Dover Museum.

A severe challenge was the fact that several of the key dimensions of the original vessel and the shape of the missing northern end could not be determined from the evidence excavated in 1992. Consequently, it had not been possible to point to one single version of the Dover boat as the most likely original one, and instead three hypothetical alternatives were suggested by Owain Roberts (Figure 5).[24]

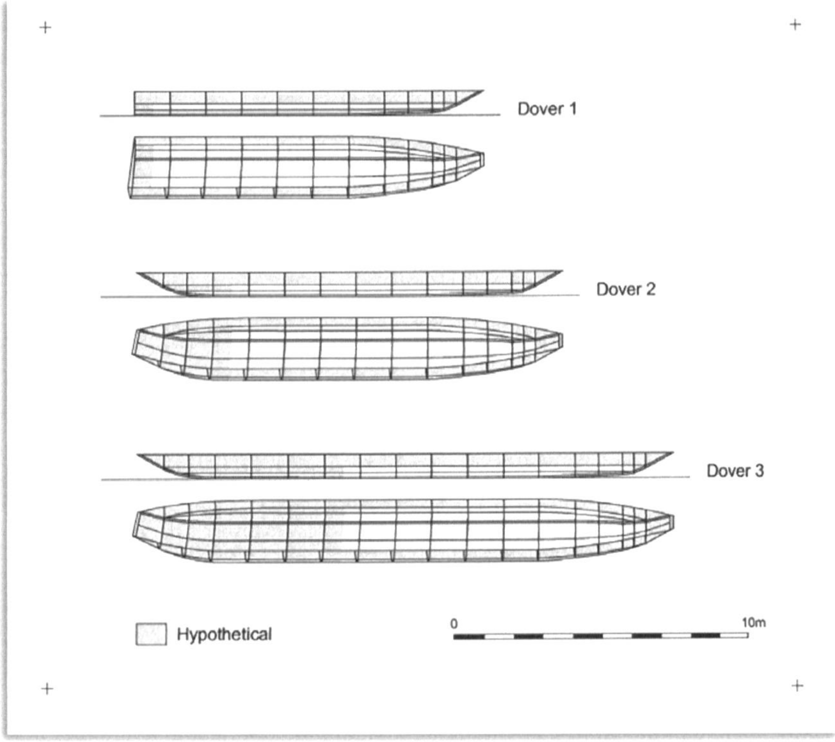

Figure 5: Hypothetical reconstructions of the Dover boat

Since it would be unrealistic to achieve experimental investigation of all three hypothetical models, a choice needed to be made between these.

As only one end of the Dover Boat was excavated, we may hypothesise that the missing north end might have used another complex and unique solution to be strengthened and made watertight. It might even have been as complicated as that used in the central seam between the bottom planks.[25] However, no physical evidence of this survives. For this reconstruction it has been assumed that the boat was equal ended, similar to comparative Bronze Age boats such as those found at North Ferriby. Ole Crumlin-Pedersen believed that steaming the timbers and creating a sheer would make the boat more stable and more capable of sea travel. (Figure 6).

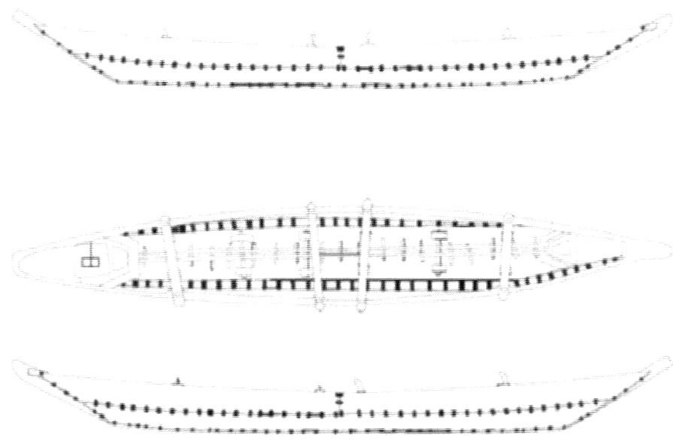

Figure 6: Plan of the Replica as constructed.

From the physical evidence of the original timbers, we know that the upper planks are missing as there is a stitched seam joining the planks on either side of the boat to the bottom planks, and there is a similar set of cut stitches on the curved plank's upper edge. The type of stitches and the shape of the joint suggest that this upper timber was the same thickness. There are cleats surviving on the curved planks, and these work if we consider that there were matching cleats on the other side planks as well. These and many other details on the surviving boat hint at the shapes of the missing pieces.[26]

It was initially intended that this reconstruction would be a full sized 1:1 replica of the original Dover boat, however this was dismissed for a several reasons. Firstly, the team were unable to source large enough timbers, and this would mean that the replica would need to be of composite construction which would not therefore be a faithful replica of the original vessel. There were also concerns about how a replica of this size would be stored and maintained into the future. It was ultimately decided that a 1:2 scale replica would be the most suitable alternative. In three dimensions a half scale replica is just one-eight the volume of the full scale. Building the replica at 1:2 scale would allow the team to experiment with ideas and possibilities at a relatively modest cost compared to a full-sized vessel.[27]

The half scale reconstruction began with splitting five large oak trunks in half down their lengths. These were used to hew out two bottom planks (each 6m long), the two ile planks (7 m), four side planks (4m) and two end pieces (2m long) from the solid wood. Each of these timbers was hewn flat, hollowed in parts and have large cleats left standing 15cm above the timber surface.

The timbers were cut to the correct shape and then fit together by heating and bending the timbers. There was an element of the unknown here; until they had fitted and joined the bottom planks and curved side planks, they could not know the exact shape of the end pieces. Like the original boat builders, the team placed the timbers together, to see if they fit and reshaped them if they did not. The pieces of the boat were then stitched together with twisted yew branches, it was then packed with moss waterproofing on the planks and under the stitches. [28] Due to the size difference between the original and the half-scale replica the yew withies proved to be too big to get the required number of stitches and so willow was used as a viable alternative.

Accurate recording of the tool marks enabled the team to recreate the sequence of carving the individual pieces. This stage allowed the team to ascertain whether axes or adzes were used, and in which direction. In some cases, the tool marks on the reconstruction are so similar to the original tool marks that we can be certain that the same type of tools were used in a similar way. It also revealed more about the process of hewing in that all those doing the hewing worked in a similar way. The careful recording of the process of reconstruction gives us great insight into the details of the boat's construction, but also hints at how the boat builders worked.[29]

Once constructed the team undertook preliminary testing of the vessel in the water. By testing such a replica on the water, the range of probable interpretations could be limited, and a host of new data acquired. The final phase of the experiment was to launch the boat into the sea and measure how it performed. The replica boat made its maiden voyage in Dover Harbour on 7th September 2013, crewed by a team of eight. It sat high in the water and rode the swell and waves of the Channel with ease, proving to be very stable, highly manoeuvrable, and seaworthy. (Figure 6).

Figure 7: Maiden voyage in Dover Harbour

2.3 LESSONS LEARNT

It has now been over 10 years since its launch which provides us with an ideal opportunity to reflect on the lessons that we have learnt from it, including its sea capabilities, propulsion method, and maintenance.

As the replica is at half the scale, many of the boat's elements are necessarily more robust than at full scale, as the weight of the structure increases faster than its strength when scaled up, this must be taken this into account when reviewing the capabilities of the boat in open water. [30]

As already mentioned, there was no evidence of a mast or oars on the original boat, and it seems likely that it was paddled. No paddle was found in association with the boat and therefore the design for the paddles for the replica were based off other comparable archaeological examples. Trials on the replica paddles have found them to be ineffective and too narrow. A lot of experimentation on the use of paddles has taken place, including the use of the stern oar, how best to organise paddlers, and how long the replica can be paddled for. It was estimated that on a good day, and with 6 hours of favourable tide, the original boat could have achieved a distance of more than 30 nautical miles under continuous paddling.[31] The replica has travelled similar distances to those estimated, travelling 29 miles in approximately 5 hours. It is important to note when looking at the use of paddles that although the replica is built at a half scale the paddles and indeed the paddlers are at full scale.

There were concerns for the boat's capabilities in open seas especially the flexing of the centreline. Use of the replica has shown that the centre seam does leak, despite extra caulking being added. Although the boat does take on water this can be easily managed by bailing.

There is little doubt about the original boats ability to maintain good passage times at sea in the right conditions. The calculated strength of the minimal reconstruction demonstrated that the speed, power and stability would meet the demands of seagoing.[32] The replica has completed many journeys along the Southeast coast of England. It works well in good weather and sea conditions, achieving speeds of between 2-4 knots, but has proven difficult to steer in waves and with the wind on its nose. The replica is slow in the water and requires a lot of energy to get moving, and to stay on course.

As discussed earlier when constructing the replica, the decision was made to use willow for the stitches instead of yew. This has caused some challenges to the on-going maintenance of the replica but also given us the opportunity to learn a lot about the process of stitching the boat.

Willow is not as long lasting as yew, and therefore the stitches have needed to be replaced regularly. The yew stitches on the original boat all seem to have a similar amount of wear suggesting that they were all fitted at the same time. There is also no physical evidence to show that they have been replaced. However, the on-going maintenance of the replica vessel has taught us that the stitches can be easily removed and replaced leaving no tangible marks on the boat. This may mean that the stitches in the original boat could have been replaced however the difficulty in accessing the stitches on the bottom outboard of the boat suggest that this is still unlikely to have been attempted.

It is interesting that the replica has been in service for 10 years as from our understanding of the original boat, it is estimated that it was also probably in use for approximately 10 years also.

2.4 FUTURE OF THE REPLICA

At the time of writing the replica is still in use. A small team of volunteers take the boat for regular trips around Dover Harbour and longer sea voyages. However, the replica is requiring more extensive maintenance every year to keep it seaworthy. The centre seam is leaking considerably forcing the team to consider the use modern materials such as Sikaflex to caulk this seam sufficiently. As we know from analysis of the original boat this was a significant issue and may have been a contributing factor in the decision to abandon the vessel. A similar dilemma now presents itself to Dover Museum and the Bronze Age Boat Trust and decisions need to be made on what the future of the replica should be once it is permanently removed from the water.

What is clear is that this is a unique resource which can still help us learn more about the original vessel and is a key tool to improving the interpretation of Bronze Age transport to the public.

Earlier this year (April 2023) a photogrammetric survey and laser scan of the replica was undertaken. This has produced accurate 3D models of the replica boat which can be studied for a variety of purposes. These will also be used to help improve interpretation of the original boat as part of a programme of gallery updates.

The hope is that the boat can continued to be used until late 2024. At this time, it will be removed from the water and once again be the focus of a project outside Dover Museum. The timbers will be carefully taken apart, cleaned, recorded, and then put back together. This process will give us a unique opportunity to study how the timbers have warped and changed in the 10 years since its construction they will also give us the ability to engage with specialist interest groups and the local community. This work is anticipated to take place between October 2024 and February 2025.

The replica boat will then be on static display at the Institut national de recherches archéologiques preventives (INRAP) in France to be part of an exciting temporary exhibition on the European Bronze Age.

As of the time of writing there is no confirmed plan for what will happen to the replica on its return to Dover in 2026. It is vital that this replica is conserved as an essential part of the Bronze Age Boat archive.

There are currently three proposals for how this might be achieved:
The first proposal would see the replica go on static display in the newly refurbished Bronze Age Boat Gallery at Dover Museum alongside the original boat. This would provide a greater layer of interpretation to the original boat but there are some concerns that it may detract from the original or cause confusion to visitors.

The second would see the replica boat as a stand- alone static exhibition at a venue near to the ferry terminal. This would allow the replica to be seen by a large audience and it could be utilised as a symbol of our transmarche maritime past, and an advertisement of the original vessel housed at Dover Museum.

The third proposal would see the replica as a focal point of a proposed a new Maritime Museum as part of a redevelopment of Dover harbour. This would have similar outcomes to proposal two but would also allow the boat to be viewed in context with the extensive maritime heritage of Dover district.

All these proposals are in the very early planning stages, and it is unclear which will be the most appropriate. What is clear is that making the archives of both the original excavation and of the two reconstruction experiments publicly accessible is essential. Sadly, several key members of the original team and important collaborators have already passed away in recent years. Others are now reaching retirement and therefore it is vital that this information is correctly catalogued, digitized and made publicly available, and that this expertise are passed on to the current and future generations.

3. THE LEGACY

The final section of this paper will examine the legacy of the Bronze Age Boat and its replica.

The Dover boat is one of a few rare survivals of prehistoric boats ever found in northern Europe, and certainly one of the most complex and well preserved. The discovery of the boat and reconstruction experiments have made a profound impact on our understanding of prehistoric maritime transport and the technological skills of our Bronze Age ancestors. [33]

The archaeological record is unequivocal about the significance of water transport to social and economic contact in the Bronze Age. The study of the Dover Boat has given us a unique opportunity to further our understanding of the technological basis of this contact, acting as both physical evidence and a symbol for our growing understanding of transmanche connections. [34]

The discovery of the boat has in part inspired a renaissance of study and research in the transmanche area. Terrestrial archaeologists have become more willing to incorporate the maritime dimension into their studies, whilst maritime archaeologists have become more sensitive to the cultural context of the vessels they study. This has been reflected in two major international conferences held in Dover, but also many publications in France, Belgium, the Netherlands, and the UK. [35]

The replica boat also acts as a focus for reflection on our mutual heritage on both sides of the Channel. Built as part of a large international research project (BOAT 1550BC) which brought together seven institutions in three countries to share with the widest possible audience the remarkable story of the transmache community it continues to promote cross Channel collaboration.

4. CONCLUSION

The discover of the Bronze age boat has undoubtedly provided major contributions to our understanding of prehistoric maritime transport and history. New insights and better understanding of our mutual maritime heritage continue to emerge. However, there is still much to learn about the Dover Bronze Age Boat, research continues into this remarkable vessel and the lost world it was once part of. The making of a new film documenting both the original boat and the replica along with further analysis and publication on the replica will be a priority for the team over the next 3-5 years to complete as far as possible the story of this unique and extraordinary vessel from 3,500 years ago.

5. REFERENCES

1. CLARKE, P., 'Introduction', *The Dover Bronze Age boat*, 2004

2. PARFITT, K., 'Discovery and excavation', *The Dover Bronze Age boat*, 2004

3. CLARK, P., 'Introduction', *The Dover Bronze Age boat*, 2004

4. CLARK, P., 'Introduction', *The Dover Bronze Age boat*, 2004

5. CLARK, P., 'Introduction', *The Dover Bronze Age boat*, 2004

6. CLARK, P., 'The Bronze Age Boat: Discovery, Beyond the Horizon Societies of the Channel and North Sea 3,500 Years Ago', 2012

7. MARSDEN, P., 'Description of the boat', *The Dover Bronze Age boat*, 2004

8. MARSDEN, P., 'Description of the boat', *The Dover Bronze Age boat*, 2004

9. CLARK, P., 'Introduction', *The Dover Bronze Age boat*, 2004

10. CLARK, P., 'The Bronze Age Boat: Discovery', *Beyond the Horizon Societies of the Channel and North Sea 3,500 Years Ago*, 2012

11. CLARK, P., 'The Bronze Age Boat: Discovery', *Beyond the Horizon Societies of the Channel and North Sea 3,500 Years Ago*', 2012

12. CLARK, P., 'Introduction', *The Dover Bronze Age boat*, 2004

13. DARRAH, R., 'The Reconstruction of the Dover Boat', *Beyond the Horizon Societies of the Channel and North Sea 3,500 Years Ago*, 2012

14. CLARK, P., 'One step at a time: The Dover Bronze Age boat experimental research programme', 2008

15. MARSDEN, P., 'Description of the boat', *The Dover Bronze Age boat*, 2004

16. MARSDEN, P., 'Description of the boat', *The Dover Bronze Age boat*, 2004

17. ROBERTS, O., 'Reconstruction and Performance', *The Dover Bronze Age boat*, 2004

18. ROBERTS, O., 'Reconstruction and Performance', *The Dover Bronze Age boat*, 2004

19. CLARK, P., 'Introduction', *The Dover Bronze Age boat*, 2004

20. CLARK, P., 'The Bronze Age Boat: Discovery', *Beyond the Horizon Societies of the Channel and North Sea 3,500 Years Ago*, 2012

21 ROBERTS, O., 'Reconstruction and Performance', *The Dover Bronze Age boat*, 2004

22. GOODBURN, D., 'Assembly and construction techniques', *The Dover Bronze Age boat*, 2004

23. DARRAH, R., 'The reconstruction experiment', *The Dover Bronze Age boat*, 2004

24. CRUMLIN-PEDERSEN, O., 'Dover', *Boat Reassessment 2007-2011*, 2011

25. DARRAH, R., 'The Reconstruction of the Dover Boat', *Beyond the Horizon Societies of the Channel and North Sea 3,500 Years Ago*, 2012

26. DARRAH, R., 'The Reconstruction of the Dover Boat', *Beyond the Horizon Societies of the Channel and North Sea 3,500 Years Ago*, 2012

27. CLARK, P., 'One step at a time: The Dover Bronze Age boat experimental research programme', 2008

28. DARRAH, R., 'The Reconstruction of the Dover Boat', *Beyond the Horizon Societies of the Channel and North Sea 3,500 Years Ago*, 2012

29. DARRAH, R., 'The Reconstruction of the Dover Boat', *Beyond the Horizon Societies of the Channel and North Sea 3,500 Years Ago*, 2012

30. DARRAH, R., 'The Reconstruction of the Dover Boat', *Beyond the Horizon Societies of the Channel and North Sea 3,500 Years Ago*, 2012

31. ROBERTS, O., 'Reconstruction and Performance', *The Dover Bronze Age boat*, 2004

32. ROBERTS, O., 'Reconstruction and Performance', *The Dover Bronze Age boat*, 2004

33. CLARK, P., 'One step at a time: The Dover Bronze Age boat experimental research programme', 2008

34. CLARK, P., 'One step at a time: The Dover Bronze Age boat experimental research programme', 2008

35. CLARK, P., 'One step at a time: The Dover Bronze Age boat experimental research programme', 2008

6. ACKNOWLEDGEMENT

In the thirty years since the discovery of the boat, many individuals and organizations have contributed to making the Dover boat and the Boat 1550BC projects successful. I would like to thank Keith Parfitt for discovering the boat and the incredible knowledge he has shared. The late Richard Darrah for the tangible reconstructions of the boat and the extensive research which accompanies it. Peter Marsden for the kind donation of his research notes. The late (and much missed) Peter Clark for his incredible passion and wealth of knowledge of the Bronze Age boat and it's replica. In particular, I would like to thank Paul Bennett for firstly encouraging me to become part of the crew of the replica, and for being incredibly generous with his knowledge and time helping me to piece together the fragmented archive of the boat and gain first hand experience of the replica. Finally, I would like to thank Jon Iveson Curator of Dover Museum for encouraging and indulging my interest in the Bronze Age boat beyond the scope of my role within the museum.

7. AUTHORS BIOGRAPHY

Catherine Holt holds the current position of Collection's Manager at Dover Museum and Bronze Age boat Gallery. The author is responsible for the management and care of all the displayed and stored collections at Dover Museum, including the conservation of the internationally significant Bronze Age Boat. She is also part of a small working group that maintains and operates the replica of the Bronze Age Boat. The author's previous experience includes completing an apprenticeship in Historic Ship Conservation with National Historic Ships UK. As well a background in maritime archaeology and extensive experience working on maritime collections.

THE LAST FLOWER - PRESERVING HMCS SACKVILLE, LAST OF THE FLOWER CLASS CORVETTES

G P Cottingham, Director, Canadian Naval Memorial Trust, Canada, **J W Dalziel, MRINA**, Adjunct Professor Industrial Engineering, Dalhousie University, Halifax, Canada

SUMMARY

Shortly after the beginning of World War II, the first convoy of Merchant Ships slipped out of Halifax, Nova Scotia bound for Britain. It was followed by hundreds more convoys with thousands of ships. Had the trans-Atlantic supply chain been broken, Britain (and the world) may well have succumbed to tyranny.

At the beginning of World War II there was a desperate shortage of naval vessels suitable to protect the merchant ship convoys from submarine attack. This Paper briefly traces the Battle of the Atlantic, the history of the Flower Class Corvettes from inception in England, noting that over 100 were built in Canada, through wartime service, until disposal after war's end. Of these Flower Class Corvettes, only HMCS Sackville remains.

It discusses the preservation of HMCS Sackville after post war service and the on-going work necessary to preserve eighty-year-old ship to remind future generations of the wartime bravery.

NOMENCLATURE

Convoy designations:

- HX Convoy: Eastbound fast convoys (12 knots) from Halifax or New York to Liverpool or Londonderry
- SC Convoy: Eastbound slow convoys (8 knots) from Sydney, NS to Liverpool or Londonderry
- ON Convoy: Westbound fast convoys from Liverpool to Halifax.
- ONS Convoy: Westbound slow convoys from Liverpool to Halifax

1. INTRODUCTION

Winston Churchill once wrote that, '. . . the only thing that ever really frightened me during the war was the U-boat peril'.

Despite their modest size and humble beginnings, the Flower Class Corvettes were arguably the most important allied naval ships of World War II. Quickly, and efficiently built by shipyards which were not proficient in building more complex warships, and often crewed by relatively 'green' crews, they provided needed protection for merchant ships carrying vital supplies across the perilous North Atlantic.

Shortly after the beginning of World War II, the first convoy of merchant ships slipped out of Halifax, Nova Scotia bound for Britain. It was followed by hundreds more convoys with thousands of ships. Six years later and after the loss of 3,500 merchant ships, the Battle of the Atlantic was won. However, had the trans-Atlantic supply chain been broken, Britain (and the world) may have succumbed to tyranny.

The Presentation briefly traces the history of the Canadian Flower class corvettes, from inception, through wartime service, until disposal after war's end. It looks at why the class was initiated with Smith's Dock in England, and how so many were built so quickly, noting that 123 of the 294 corvettes were bult in Canada. Of these corvettes, only HMCS Sackville remains. In addition, it discusses the preservation of HMCS Sackville after post war service, and the on-going work necessary to preserve an eighty-year-old ship, so that future generations will be reminded of the historic efforts of naval and merchant seafarers.

2. BATTLE OF THE ATLANTIC – THE BEGINNING

The British Admiralty had learned its lessons about the importance of the Convoy system from World War I. In 1917, after the implementation of unrestricted submarine warfare, and prior to the implementation of Convoys, one in four ships bound for Britain was sunk.With only about six weeks of food reserves left, Britain was near starvation and in a dire situation. However, Convoys (groups of merchant ships sailing together, generally with naval escort) were introduced, and the situation dramatically improved with ships and supplies getting through.

On September 16, 1939, shortly after the outbreak of World War II, the first HX Convoy of merchant ships sailed from Halifax, Nova Scotia, Canada bound for Liverpool, England. There were not enough naval ships to provide escort across the Atlantic, escort was provided for a distance from Halifax and the convoy would be met by a naval escort somewhat west of Ireland. In the middle of the Atlantic it would be on its own,

With increasing numbers of escort ships, including corvettes and American lend-lease warships the situation stated to get better. Improvements in technology, detection, communication, air cover and military intelligence also played a role. However, the U-boat threat also evolved, in particular, new bases on the coast of France, the use of Wolf packs and improvements in technology. There were several turning points during the Battle of the Atlantic, often when escorts were diverted to other military operations. The peak was during the early months of 1943, in the month of March alone over 100 merchant ships, representing well over 500,000 tons, were sunk. However, soon the tide turned, and although losses to U-boats continued until the end of the war, convoys with allied naval escorts and air-cover, and massive shipbuilding production, ensured the delivery of essential supplies to Britain.

The Battle of the Atlantic was the single most important campaign of World War II. However, the cost was large, over 36,000 allied merchant seafarers and another 36,000 naval and military personnel lost their lives.

3. BIRTH OF THE FLOWER CLASS

Figure 1: Smith's Whaling Ship Southern Pride

The turmoil in Europe in the late 1930's alarmed the members of the British Admiralty Board, convincing them that war with a revitalized Germany was inevitable. A series of plans were drawn up to bring Britain and her armed forces to war readiness by 1940. One issue that quickly became apparent was that there would be a serious lack of small close escort vessels. All the existing shipyards capable of building naval vessels were already loaded to capacity to produce the complex cruisers and destroyers so essential to guarding the sea lanes. It was clear that something else was needed, capable of being built quickly, cheaply, and in less capable shipyards that normally used for naval construction.

The Smith's Dock Company at Southbank on the Tees submitted a proposal to the Admiralty Board, based on their recent design for a whaling ship. The proposed "Patrol Vessel, Whaler Type" was 205 feet overall, beam of 33 feet, drawing 15 feet. Its hull was designed to be very seaworthy, as well as manoeuvrable. The machinery plant was selected to be very common in merchant use, with two fire tube scotch boilers and a four-cylinder, triple expansion reciprocating steam engine driving a three bladed propeller. Range was to be 4000 miles on 200 tons of fuel at 12 knots, with a maximum speed of 16 knots. Armament was to be simple: a single 4" quick-firing gun, two Lewis guns, and 300 lb. depth charges. The cost was estimated at £90,000 each.

Designers from Smith's met with the Admiralty Board in January 1939, and after considering a number of submissions, the Board selected their design and let a contract to Smith's on 27 February. During the next 7 months the design was iterated and finally settled on 940 tons, 190 feet between perpendiculars, and 205 feet overall. The complement would be 47 officers and men. This would be the design built in mass to meet the initial demands for the Navy. On 25 July 1939 the first batch of 25 ships were ordered from Harland & Wolff, followed by a second order of 30 on 31 August. On 3 September 1939, the United Kingdom declared war on Germany. Another 20 Flowers were ordered from Harland & Wolff on 19 September, with orders to other shipyards for 10 ships on 21September, 2 more on the 23rd, and two more on 15 December. Additionally, 10 ships were ordered from Canadian shipyards in January 1940 bringing the total number of ships ordered to 116.

4. CANADA BUILDS

At the commencement of the war the Royal Canadian Navy (RCN) consisted of 6 River class destroyers, 5 minesweepers and 2 small sail training vessels [1]. Canadian planners were faced with a conundrum: they needed ships – lots of them - but all their usual suppliers were frantically building to meet their own needs and had no surplus capacity to assist Canada. No Canadian shipyard was capable of building ships to naval standards, which was necessary to build all current warship designs. The initial plan was to build minesweepers, but these required naval standards, ruling out construction in Canada. The RCN had received the plans for the 'Patrol Vessel, Whaler Type', and just days after the commencement of hostilities, a Canadian proposal was sent to the Admiralty that Canadian shipyards could build these ships. In January 1940 the Admiralty placed orders for 10 Flowers with Canadian shipyards. Canada also began its own program, ordering 70 Flower class from 12 different yards across the country.

Table 1. Canadian Shipyards Building Flower Class. [3]

Shipyard	Built
Burrard Dry Dock Co. Ltd., North Vancouver, British Columbia	4
Canadian Vickers Ltd., Montreal	8
Canadian Yarrow, Esquimalt, British Columbia	5
Collingwood Shipyards Ltd., Collingwood, Ontario	8
Davie Shipbuilding & Repairing Co. Ltd., Lauzon, Quebec	13
Kingston Shipbuilding Co. Ltd., Kingston, Ontario	3
Marine Industries Ltd., Sorel-Tracy, Quebec	7
Midland Shipyards Ltd., Midland, Ontario	2
Morton Engineering & Dry Dock Co., Quebec City	4
Port Arthur Shipbuilding Co., Port Arthur, Ontario	8
Saint John Dry Dock and Shipbuilding Co. Ltd., Saint John, New Brunswick	3
Victoria Machinery Depot Co. Ltd., Victoria, British Columbia	5
Total Flower class built in Canadian Shipyards	**70**

The Canadian built Flowers differed from their British built sisters in several ways. First, they weren't named after flowers, but rather Canadian small cities and towns. From a material point of view, there were a number of changes. Canada was unable to supply gyrocompasses for the first builds, so they were fitted with a magnetic compass only. Unfortunately, they were not fitted at build with the low-power electrical system that the gyrocompass would require. As a result, the only ASDIC set that could be fitted was the earliest: the 123A. No updates would be possible until the low power system was fitted, which required a more extensive refit. [2] This would lead to the early Flowers not being modified or updated in an expedient fashion.

Canada was also not able to outfit the early Flowers with 2-lb guns, Oerlikon cannons or .5" machine guns until later in the war. They made do with Lewis guns, which proved woefully inadequate against surfaced U-boats.

Canadian design changes included moving the galley from right aft to just behind the wheelhouse. This was a much more useful location, particularly when supplying meals to the seamen's messes in the forecastle. A bilge keel was added to reduce rolling. Canadian radios and direction-finding (DF) gear were used in lieu of the British kit.

The platform for the after gun was moved in Canadian built ships to right aft, the easiest distinguishing feature from the British built ships.

Warfare brings rapid change, and it did not take long until the flower class experienced modifications to improve performance and liveability for the crew. The last six of the Canadian Flower 1940-41 program were fitted with water-tube boilers instead of the fire-tube boilers in the initial 64. They also incorporated an extended forecastle.

The remaining ships built in the 1940-41 program were considered Revised Flower class. The rake of the bow was increased, and the draft was increased. The forecastle was extended, and the foremast moved aft of the bridge to improve visibility. The bridge wings were extended and on 20mm Oerlikon cannon was installed on each wing. There were 10 of these ships built.

In the 1941-42 program, additional fuel tanks were added, extending the range to 7400 miles at 10 knots. 16 of these ships were built.

The final program of the Revised Flower class in Canada (1942-43) built 12 ships. The only change to these was an increase of displacement from 940 tons to 970 tons.

Table 2. Canadian Shipyards Building Revised Flower Class for RCN. [3]

Shipyard	Built
Collingwood Shipyards Ltd., Collingwood, Ontario	6
Davie Shipbuilding & Repairing Co. Ltd., Lauzon, Quebec	1
Kingston Shipbuilding Co. Ltd., Kingston, Ontario	6
Marine Industries Ltd., Sorel-Tracy, Quebec	8
Midland Shipyards Ltd., Midland, Ontario	3
Morton Engineering & Dry Dock Co., Quebec City	13
Port Arthur Shipbuilding Co., Port Arthur, Ontario	1
Total Revised Flower class built in Canadian Shipyards for RCN	**38**

Table 3. Canadian Shipyards Building Revised Flower Class for USN. [3]

Shipyard	Built
Collingwood Shipyards Ltd., Collingwood, Ontario	6
Kingston Shipbuilding Co. Ltd., Kingston, Ontario	2
Midland Shipyards Ltd., Midland, Ontario	2
Morton Engineering & Dry Dock Co., Quebec City	5
Total Revised Flower class built in Canadian Shipyards for USN	**15**

5. HMCS SACKVILLE AT WAR

Sackville was the second corvette order placed by Canada, at St. John Dock and Shipbuilding Company, but was the thirty-fourth completed. St. John Dock was the closest building yard to Halifax and had been busy repairing war-damaged ships, slowing the new construction work. Sackville was laid down in 1940, launched 15 May 1941 and commissioned into the Royal Canadian Navy (RCN) on 30 December 1941.

Sackville's war-time experience was split into 4 periods:

1st commission - January 1942 to 13 January 1943

Refit: Liverpool, Nova Scotia – 14 January to 3 May 1943

2nd commission – 4 May 1943 to 27 February 1944

Refit: Galveston, Texas – 28 February to 7 May 1944

3rd commission – 8 May 1944 to 8 November 1944

Refit: 9 Nov 1944 to 21 May 1945 - Conversion to loop layer.

Final wartime commission: 22 May 1945 to 8 April 1946 – Decommissioned and Laid Up into Reserve.

5.1 FIRST COMMISSION

Sackville began her first commission escorting convoy HX175 from Newfoundland to Iceland in January 1942. This was followed by SC72 from Newfoundland to Iceland, with the return convoy ON70 from Iceland back to St. John's, Newfoundland On her path she diverted to recover survivors from the torpedoed Lily east of Sable Island.

In April Sackville received her first radar set, the Canadian made SW1C. By the time she sailed again, the Battle of the Atlantic had entered a new phase. The U-boats had pushed west to plunder the US eastern seaboard and the intensity of the battle intensified. No longer was Iceland being used as a relay point and convoys were now sailing directly from Newfoundland to northern Great Britain (Londonderry or Liverpool).

During the remainder of 1942, Sackville escorted 12 more convoys across the Atlantic. A total of 553 ships were escorted, with 5 torpedoed, leading to the loss of 3. While escorting convoy ON115 from east to west, the convoy was attacked by 2 wolf packs: group Wolf (7 U-boats) and group Pirat (6 U-boats). In heavy fog off the coast of Newfoundland, Sackville engaged 3 U-boats in less than 24 hours. U43 was sighted at less than 200 meters, forced to dive then attacked with a shallow pattern of depth charges. The submarine's bow rose from their wake at a steep angle before a depth charge exploded directly underneath, and it slid under. The ship claimed a kill, however, post war analysis determined that the boat returned to France with heavy damage. Several hours later they sighted another boat on the surface, on their port beam at less than 200 meters. They turned to engage but the submarine dove and no contact was established with ASDIC. Finally, just about noon using passive ASDIC bearings, they located a third U-boat (U552) on the surface, taking advantage of the heavy fog to load their last torpedo. Sackville tried to ram but couldn't turn tightly enough. The 4" gun got a direct hit at the base of the fin. The Commanding Officer, LCdr Allan Easton, wrote that he could see "a gaping hole at the base of the U-boat's conning tower". The submarine dove and Sackville and HMCS Agassiz failed to gain ASDIC contact. U552 did manage to return to France requiring significant repair. [2]

Figure 2: HMCS Sackville during 1st Commission

5.2 REFIT: LIVERPOOL, NOVA SCOTIA

By the time Sackville entered refit after her first hard year at war, all new corvette construction were of the modified Flower design, with an extended forecastle and improved and expanded accommodation spaces. Their range was also extended, and in the case of Canadian ships, twin Oerlikon mounts were added to the bridge wings and gyrocompasses were fitted. Unfortunately for Sackville, escorts were still too short in supply to allow the RCN to grant her the time and resources to receive this greatly desired change. It was a case of do the minimum work required to get her back out to sea. Her propulsion machinery was stripped down and rebuilt and her weapons were serviced. The bridge wings were strengthened and the Oerlikons added. Had Sackville carried the Oerlikons during her engagements with U43 and U552, much greater damage could have been inflicted on the submarines. Although no gyrocompass was fitted, a second magnetic compass was added on the bridge. The minesweeping winch at the stern was removed and the large minesweeping davits were replaced with a simple davit for loading depth charges.

5.3 SECOND COMMISSION

During her second commission, Sackville escorted 15 trans-Atlantic convoys, for a total of 717 ships, with 8 ships torpedoed and 7 sunk. By the end of the second commission the ship and crew were worn out. Luckily, the opportunity came to refit at the Todd Shipyard in Galveston, Texas, in late February 1944.

Figure 3: HMCS Sackville during 2nd Commission

5.4 REFIT: GALVESTON, TEXAS

When Sackville arrived in Galveston, she was so rusty that the locals referred to her as "the gravy boat" since her sides were so brown. During this much needed refit she received overdue improvements:

- Low power electrical system and gyrocompass.
- Extension of the forecastle.
- Second seamen's mess and offices, galley now fully enclosed.
- Improved ASDIC model 127 DV. This allowed the operators to control and better monitor bearing and depth information on the target, and it required a gyrocompass for proper control and accuracy.
- Hedgehog anti-submarine mortar.
- Radar upgraded to SW2CP.
- Two 16-foot rowboats replaced with one 27-foot whaler and enough Carley floats for 150% of ship's complement.

5.5 THIRD COMMISSION

In June 1944, Sackville escorted convoy HX297, consisting of 116 ships, from Halifax to Londonderry, without losses. As she sailed from Londonderry, her No.1 boiler developed a leak, causing her to return to port for repair. After attempted repairs failed, the boiler inspectors determined that during construction, the rivet holes had been over-drilled, leading to cracking in the rivet countersinks. This allowed leakage of water and boiler compound into the combustion chamber of No.1 boiler. The result was caustic embrittlement of the plates in the central combustion chamber, requiring replacement of the boiler. If this had occurred earlier in the war, the boiler would likely have been replaced. In mid-1944, the RCN felt the major repair was not cost effective and decided to convert Sackville to a training role. For one month she served as a training vessel, attached to HMCS Kings, in Halifax. The decision was then made to convert her to a maintenance vessel to support the magnetic range in the approaches to Halifax harbour. This range was used to calibrate the degaussing systems of ships when entering or leaving the harbour. [2]

Figure 4: HMCS Sackville during 3rd Commission

The No.1 boiler was removed from the ship and converted into a cable storage well, capable of holding 10 miles of cable. The armament was removed, and a steam winch was installed in place of the 4" gun. Two samson posts and derricks were installed forward of the bridge and two cable sheaves were fitted to the bow. The mainmast was fitted with a boom so that a sail could be fitted to steady the ship while handling cables.

Figure 5: HMCS Sackville During 4th Commission as a Loop Layer

Sackville's conversion to a loop layer completed on 21 May 1945. This proved to be excellent timing since the war in Europe ended on 10 May. Between the 10th of May and the 31st of July, 94 of the surviving 113 RCN corvettes were decommissioned and sent to the scrapyards. Sackville, in her new role, was spared this fate, and, after lifting the magnetic loops from Halifax and Sydney harbours, she finally was decommissioned and placed into reserve on 8 April 1946.

6. HMCS SACKVILLE – CIVILIAN NAVAL AUXILIARY

It was not long after the cessation of hostilities that tensions in the world again began to rise. The communist take-over of Czechoslovakia in February 1948 marked a new period of clashing national ideologies. The blockade of Berlin was followed by the birth of the North Atlantic Treaty Organization (NATO) and the Cold War had begun.

During the war the RCN, had developed a practical appreciation of the difficulty in conducting anti-submarine warfare. There were large knowledge gaps in the understanding of underwater acoustics and oceanography, and the Soviet Union's large-scale production of submarines led to NATO countries beginning research programs in acoustics, oceanography, and anti-submarine warfare.

In 1950, during the Korean war, Sackville was called out to be the depot ship for the reserve fleet, and to be ready "for loop laying if needed". In 1951, Sackville received an extensive work period designed to prepare her for reactivation. During this refit, scientists from the Canadian National Research Establishment (NRE) installed a new device they had developed: an impressed current, cathodic protection system. This addition probably assisted greatly in the extension of Sackville's hull life.

In December of 1952 she was reactivated as Canadian Navy Auxiliary Vessel (CNAV) Sackville. Her first 10 years as a civilian vessel were roughly split between civilian and military tasks, working for the Joint Committee on Oceanography. During this period, she retained the same appearance as before, other than paint colour.

Figure 6: CNAV Sackville in 1954

Sackville spent most of 1957 in refit and suffering problems with her remaining boiler. She came very close to being scrapped during this period, but the need for support vessels won out and her boiler was repaired. She had a very busy schedule from 1953 right to the end of 1961. In 1962 her work focus shifted as she began supporting the Institute of Oceanography at Dalhousie University in Halifax. The next ten years of her career were largely spent conducting civilian geologic and oceanographic research for the Bedford Institute of Oceanography (BIO), in Halifax.

After years of busy schedules and little funding for maintenance, Sackville had a refit in Lauzon, Quebec, in 1964 where a large laboratory was constructed across the after end of the engine room casing. Although the working conditions for scientists were now improved, there were loud complaints about the primitive accommodations – these had not been updated from their wartime status. Finally in 1967, after much haggling between different government departments over who would pay the bill, another refit was conducted in 1968.

In this refit, the wartime bridge was extensively reworked into a two-deck fully enclosed modern bridge. The main deck was enclosed from the midships position aft to the end of the laboratory block. The additional space was used for more accommodation and the accommodations, heads and washplaces were all modernised.

Figure 7: CNAV Sackville Flying Paying Off Pennant 16 Dec 1982

During the 1970s, Sackville was used mostly in support of the Defence Research Establishment Atlantic, conducting acoustic research. Her steam plant proved to be quiet, and well suited to the task. She continued working a busy schedule right to the end. Finally, her aged boiler being past the point of repair, she was decommissioned on 16 December 1982, almost exactly 41 years after her commissioning.

7. HMCS SACKVILLE – MUSEUM SHIP

The Sackville was sold to the Canadian Naval Corvette Trust 28 Oct 1983 for the princely sum of $1. The Trust, which changed to the Canadian Naval Memorial Trust, largely consisted of retired naval personnel. The Maritime Museum of the Atlantic (MMA) was very interested in obtaining the Sackville but was unable to obtain provincial funding. The RCN and the Trust worked out an agreement where the navy would provide some maintenance on an on-going basis. The MMA provided a venue for the ship during the summer months, while she wintered in the RCN dockyard.

A study was conducted to determine the feasibility of returning the ship to her 1941 state, but it was determined that this would be too large a work-package and too expensive. Instead, the decision was made to revert her to her condition on completion of her 1944 Galveston refit.

Stripping the "civilian bits" from the ship took little time, and the ship was docked in October 1983. Her propeller was removed, and all her underwater openings were plated over. She was sandblasted, painted and re-floated. By the spring of 1985, the exterior of the ship was largely restored. The ship was able to participate in the celebrations for the RCN's 75th birthday celebrations, although her interior had not yet been restored.

Many contributions were received to assist in the restoration. The Amherstburg Ontario Legion owned the last surviving 4" breech-loading Mk IX gun in Canada and were persuaded to donate it after the Canadian War Museum (CWM) offered to provide them with a World War 2 vintage 4" gun. The Irish Navy donated the 40 mm pom-pom and the control gear for the 144 ASDIC set. The CWM had one 20 mm Oerlikon and the last remaining Hedgehog mount in Canada, which they donated. Another Oerlikon was later acquired, as were depth charges, throwers and various miscellaneous pieces.

The ship currently sits directly behind the MMA, in a prime spot on the Halifax waterfront. She is painted in Western Approaches light disruptive camouflage as per 1944. She is definitely the cleanest corvette ever seen in Halifax harbour. She is now open to visitors from May until December. Admittance is free, with donations requested.

Figure 8: HMCS Sackville Alongside at the Maritime Museum of the Atlantic, Halifax

8. CANADIAN NAVAL MEMORIAL TRUST

The Mandate of the Canadian Naval Memorial Trust as stated in the 2023 Strategic Plan is:

The Trust affirms its enduring commitment to:

1. Ensuring that HMCS Sackville is preserved in perpetuity as a cherished Canadian naval icon and major artefact of one of the country's defining moments, for future generations of Canadians to experience and enjoy;

2. Ensuring that the service and sacrifice of Canadian sailors during the Second World War is commemorated;

3. Promoting education and awareness of the stories associated with the Canadian Navy's extraordinarily important contribution to the Allied effort to safeguard democracy, and secure peace during the Second World War;

4. Continuing to develop, and make accessible to the public, the historical archives held by the Trust (including both physical objects and stories); and

5. Ensuring that the Trust has sufficient resources and support for the fulfilment of its Mission.
   The Mission of the Trust is: To preserve the ship and share the stories.

9. SACKVILLE'S RELATIONSHIP WITH THE RCN

Canada's Department of National Defence (DND) has a Memorandum of Understanding (MOU) with the Department of Canadian Heritage (Canadian Conservation Institute – CCI) to provide funding for annual maintenance of the Sackville. Each year Sackville can submit work requests to CCI, which then forwards to Fleet Maintenance Facility (FMF) in Halifax Dockyard. The value of work not to exceed $100K (£ 60K), plus there is a pool of 1000 hours of dockyard labour.

The HMC Dockyard Halifax provides a serviced berth for Sackville during the winter months (December to May) and as required for weather events such as hurricanes. In the Dockyard, steam and shore power are provided at no cost.

The King's Harbour Master provides tug moves to and from her summer berth, as well as moving the ship to participate in the annual Battle of the Atlantic Ceremony at the entrance to the harbour. As well, at least twice a year, the ship gets a tug move to the harbour approaches where Committals of Ashes of RCN veterans are made.

Another key way in which the RCN supports the ship is through the provision of uniformed, active-duty naval personnel to assist staffing the ship when alongside downtown. This is a key feature that greatly assists the Trust in following its mandate while also providing the sailors a chance to know and appreciate their heritage.

In November 2023, the RCN announced their intention to ceremonially recommission the Sackville into the RCN. This is intended for 2024.

10. CURRENT STATE OF THE SACKVILLE

During her most recent dockings, in 2018/19 and 2020/21, some plating and framework were renewed due to corrosion. At the hull surveyor's recommendation, doubler plates were welded over the entire hull from one foot above the waterline. It is hoped that this will provide at least 10 years of safety; however, a new concern is the difficulty of knowing what is going on between the original hull and the doubler plates. The Hull Working Group (HWG) was formed in 2022 and has the mandate to preserve the ship in perpetuity.

To achieve that lofty aim the HWG's short term goals are to gain a better understanding of the state of the hull, determine how much time we have to replace the hull, and provide a rough order of magnitude estimate for the Trust Fundraising group. In the ship's next docking, hopefully 2024/25, it is intended to cut out sample coupons for detailed analysis to provide the necessary understanding of the condition and timeline for hull replacement.

11. CONCLUSIONS

HMCS Sackville is the sole survivor from the Flower class of 275 ships that served in 6 navies during World War 2. She is a visible artefact to a particularly important campaign of the war, but more importantly, is a memorial to all the people who toiled and sacrificed during the Battle of the Atlantic. For Canada, she is an important representative of the achievements Canada made in producing so many ships from a completely unprepared industrial base.

The story of the Sackville and her crews must be preserved and made available for future generations.

12. ACKNOWLEDGEMENTS

The excellent reference works "HMCS Sackville 1941 – 1985", 1998, by Dr Marc Milner, Professor of History at the University of New Brunswick, and "Canada's Flowers", 1981, by Thomas G. Lynch were instrumental in the preparation of this paper and presentation.

The late Arnold Hague's extensive database of wartime convoys is a principal source of data for anyone researching the Battle of the Atlantic. The most complete version of this database is not available on-line but may be viewed at Guildhall Library, Aldermanbury, London EC2V 7HH. An abbreviated version is available on-line at http://convoyweb.org.uk/hague/index.html.

John Dalziel, Adjunct Professor, Industrial Engineering at Dalhousie University was the catalyst who got this paper and presentation going for the conference. Thanks John.

13. REFERENCES

1. GERMAN, T., 'The Sea is at Our Gates: The History of the Canadian Navy', *McClelland and Stewart*, 1990

2. MILNER, M., 'HMCS Sackville 1941 – 1985', *The Canadian Naval Memorial Trust*, 1998

3. THOMAS, G L., 'Canada's Flowers', *Nimbus Publishing Limited*, 1981

4. HAGUE, A., 'Convoy Database', *http://convoyweb.org.uk/hague/index.html*

14. AUTHORS BIOGRAPHY

Greg Cottingham, MEng, PMP, CD holds the current position of Chair – Hull Working Group at Canadian Naval Memorial Trust and is also the Engineering Officer of the HMCS Sackville. He is responsible for planning, monitoring and executing the work required to preserve the HMCS Sackville's hull and structure. The Author's previous experience includes 28 years in Royal Canadian Navy (Canadian Forces Sea Element), as a Marine Systems Engineering Officer. This included service in steam frigates and submarines. He was Engineering Officer of 3 submarines, two of them Royal Navy while on exchange duties. He also managed projects and programs with General Dynamics Canada and Ultra Electronics Maritime Systems.

John W. Dalziel, P.Eng., MRINA, is a Naval Architect who has sailed with the Canadian Coast Guard and fishing industry, supervised ship construction, refits, and repairs, and been a safety inspector / Surveyor on behalf of Flag State and Class. He has presented internationally on maritime safety and on search and rescue technologies, including at several RINA Conferences, As an Adjunct Professor in the Department of Industrial Engineering, Dalhousie University, Halifax, Canada, his interests include Maritime Search and Rescue, Safety Culture and Maritime History.

POMMERN DOCK – BUILDING FOR THE FUTURE

H C Hagmark, Åland Maritime Museum Trust, Åland Islands

SUMMARY

In 2017-2018, a bespoke dock was built for the sole purpose of preserving the historic vessel Pommern of Mariehamn. As the dock was built in the water rather than dug into land, the construction posed several challenges that rarely feature in dock constructions. However, it was these challenges that made the project exciting, attracting international interest among engineers as well as heritage professionals. That Pommern now has its own custom-built dock provides new possibilities to pursue strategic preservation work. Furthermore, the ship and the dock together have the potential of becoming a hub and workshop for international co-operation regarding different methods for preserving historic ships.

1. INTRODUCTION

The historic vessel Pommern is a four-masted steel barque built on the Clyde, Scotland in 1903 for a German shipping company. From 1906 until the end of WWI, it belonged to the Hamburg-based Ferdinand Laeisz's fleet of "flying P-liners". In 1923, the vessel was bought by Åland shipowner Gustaf Erikson, who in the interwar period was the owner of the world's largest fleet of cargo-carrying windjammers. The vessels were registered in Mariehamn but deployed in world-wide shipping, mainly the Australia-UK grain trade. In 1939, Pommern made its last commercial voyage around the globe, and in 1952 Erikson's heirs donated the ship to the Town of Mariehamn on condition it was maintained as a museum ship. Today, Pommern is Åland's most popular landmark and visitor attraction, moored next to the Åland Maritime Museum in Mariehamn.

Neither while in active service nor as a museum ship, has Pommern undergone any major structural changes. This makes it a unique example of the last generation of ocean-going sailing vessels built and used for commercial shipping. That said, during its the seven decades as a museum ship, the ship has been through a few restoration projects for preservation and security purposes. It was docked in Turku, Finland in 1969 and in Stockholm, Sweden in 1987 and 1996. Due to Pommern's advancing age it was deemed too risky to tow it to shipyards outside Åland and from 2001, all dockings took place at the local shipyard in Mariehamn. In 2008, the shipyard was decommissioned, leaving the islands without a local site for future dockings. This forced Mariehamn's town council to look for a long-term solution for the preservation of Pommern. A working group, set up by the town council in 2009, unanimously recommended that a bespoke dry dock be built for the ship, in which it would be kept afloat, bar regular dockings. The recommendation was heeded by the council and in 2015 a committee was appointed to manage and oversee the realisation of the project. After some initial hurdles, the construction began in the autumn of 2017.

2. CONSTRUCTION

For seventy years, Pommern has been moored in the Western Harbour of Mariehamn, a stone's throw away from the Åland Maritime Museum. In that time, the ship has become an iconic landmark of the town, like the Eiffel Tower in Paris, the Opera House in Sydney, and the Statue of Liberty in New York. The idea of moving Pommern from that location to the old shipyard was briefly considered and then quickly discarded. It had to remain *in situ*. This decision turned the dock construction from a straightforward job into an engineering challenge, as building a dry dock in water calls for a very different kind of construction compared to digging it into land, which is the common method. In the tender process, the bidders were allowed to suggest their own solutions to the challenge. The winning tender was founded on traditional quay construction, complemented with different special features that this project required.

The building phase lasted 16 months from signing the contract in August 2017 to final inspection and hand-over in December 2018. The construction work began in earnest in October 2017. One of the first steps was to form the dock walls by driving down steel sheet piles in the seabed. A total of 204 sheet piles were used, measuring up to 23 metres in length and weighing on average 7.5 tonnes. The purpose of the sheet piles was to form the supporting structure for the concrete casting, which was done both above and below the waterline. Above the waterline, the head beam that was to become the dock promenade was cast. Below the surface, divers cast the corrosion protection on the outside walls of the dock. The sheet piles also formed a frame, which on the inside was filled with gravel to even out the heavily sloping seabed. The sheet piles were knitted together with a material that reacted with water to develop a watertight seal between the sheets. This left the construction workers with just one chance to get each sheet into its exact position. Considering the inhospitable seabed under the dock, with layers of soft clay, hard silt till and broken rock on top of the solid rock bed, it is remarkable that only one of the sheet piles got into the wrong position and had to be removed.

The next substantial stage of the building process was to pound 43 steel piles all the way down to the bedrock inside the steel frame. Each pile measured 0.7 metres in diameter and ranged from 18 to 26 metres in length. Through each pile, rock anchors were locked into at least 12 metres deep holes that had been drilled into the bedrock itself and injected with concrete. If one considers the steel sheet pile framing to be part of regular quay construction, the steel piles and rock anchors were among the solutions the construction engineers had developed especially for Pommern Dock. This was their solution to one of the greatest challenges in the whole design, which was to find a method that would simultaneously hold up and anchor down the construction. When it is empty, the dock has a lift force of 11,500 tonnes. In those circumstances, it is the rock anchors and the dock's own weight that stops it from floating up. When the dock is filled with water, on the other hand, it is the steel piles that support it and stop it from sinking. The heavy incline of the seabed from east to west inside the sheet pile wall posed another major challenge. Starting at -6.5 metres nearest land, the seabed drops down to -14 metres by the outer wall. In preparation for casting the dock's bottom slab, the seabed had to be evened out, which was done by covering it with a sinkable grade 1 fibre cloth before depositing 10,000 m3 of gravel on top of it. However, as most of the gravel ended up towards the deepest end by the outer wall, it created an internal side force that pushed the entire structure outwards. To counter this, 20 % of the 43 steel piles were driven down at a 15-degree easterly angle. On the other hand, to counter for the external side force from ice during wintertime, another 20 % of the piles were driven down at a 15-degree westerly angle, keeping the dock perfectly stable in its position.

In May 2018, after seven months of construction work, the time had come to pump out the water inside the dock walls. In preparation for this, 21 heavy steel piles were mounted across the dock's upper edge and the corresponding number of H-beams at -4 metres inside the dock. Their purpose was to support the structure against the external water pressure that occurred as the dock was emptied. The pumping out of water was done slowly to give the walls time to adjust to the outside pressure. It was a nerve-racking time; everyone involved followed the process closely, but few dared to believe that the dock would be watertight at the first attempt. However, after ten days of pumping, it became evident that the construction workers had done their job; no water leaked through either walls or gravel bed. At this point, the project moved on to a new stage, to cast the bottom slabs and inner walls. That the dock was watertight significantly improved the working conditions as there was no need to use divers for underwater casting work. Not only did it make the work much easier, but it also saved a lot of time. When the annual summer vacation started in July, the first layer of the bottom slab had been cast and work was well underway forming and reinforcing the 6-metre-high inner walls. By the end of August, the bottom section of the walls had been cast and the supporting H-beams could be removed. A few months later, all walls were cast.

The dock gate was another one of the unique solutions devised for the dock. In place of the simple but impractical gate that was included in the tender documents, the contractor proposed a significantly improved solution that eliminated the need for cranes and barges to open and close the dock gate. The solution was a floating dock gate, a hollow u-shaped steel construction designed with the same physical principles in mind that had been considered for the dock itself; lift force and own weight. To open it, the water ballast inside the gate is pumped out, causing it to float up from its frame. In that position, it only takes a couple of people to pull the gate open and moor it on the outside of the dock. The same procedure in reverse is used for closing the gate; it is pulled into position and filled with water. As it gets heavier, the gate sinks into its frame. The heavier the gate is, the tighter the seal is. Even though the principle was simple, the construction was a technical challenge. The gate had to stay floating upright when it was without ballast and lie perfectly in its frame when it was full. Failing that, the gate would not be watertight. The buoyancy issue was solved by fitting the gate with stabilizers. The fit between gate and frame was acquired by making a 1:1 drawing of the gate from which the casting mould was formed. The meticulousness in forming the casting mould, in addition to the precision of the casters, resulted in a perfect fit. Finally, the frame was fitted with a rubber seal especially developed to stand variations in water pressure. In October 2018, the 38-tonne floating dock gate was lifted in place, and once it was confirmed that the gate was watertight, the temporary wall that had been fitted at the "stern" of the dock was removed. The next step was to cast the second layer of the bottom slab, on top of which the keel blocks were placed along the dock's centre line. Once the four fenders had been fitted, the work was finished and on 19th December 2018 the project was officially completed. A month later, the process of filling the dock with water begun. The filling was deliberately done slowly to avoid undue stress on the structure. Within the space of a week, 12,000 m3 water was pumped into the dock and with that everything was ready for Pommern to be towed into the dock from her temporary berth, which happened on 27th March 2019. On that day, hundreds of people had gathered on the quayside to watch the event and welcome "their" ship home. The dock and its adjacent quay have since become a very popular area. From morning to evening people stroll along the dock promenade, they have spontaneous picnics on the quay, and they enjoy their afternoon coffee by the dock gate. The quay has also been used for concerts and festivals, and thanks to the footbridge between the Åland Maritime Museum and Pommern, the ship itself has become accessible to all. A new visitor experience, Pommern - 100 days under sail, has already offered nearly 200,000 visitors an immersive and innovative "close to history" experience.

3. CONCLUSION

Pommern is still owned by the Town of Mariehamn but since 2015, it is managed by the Åland Maritime Museum Trust. The trust's vision is to preserve and interpret the tangible and intangible maritime heritage the ship represents locally and globally. When it comes to hull preservation in particular, the dock gives us fantastic opportunities to pursue important conservation activities. Instead of carrying out only the most acute repairs, which was the case when the ship had to be taken to commercial shipyards, we can work strategically and systematically. We can dock the ship whenever we want to, we can work at our own pace and use the methods of our choice. A first docking is planned for the autumn of 2025. The primary aim is to conduct a thorough survey of the hull to create a long-term preservation programme. In this work, we want to involve conservators, naval architects, structural engineers, and corrosion specialists to investigate different approaches. We want to use the opportunity the dock offers to initiate international co-operation concerning methods of historic ship preservation, particularly regarding riveted steel hulls. In the long term, and further motivated by the inclusion of Gustaf Erikson's shipping company archives on UNESCO's Memory of the World Register, we would like to see Pommern and the dock become a hub and workshop for international co-operation regarding the preservation not only of underwater steel constructions, but of all aspects of the tangible and intangible heritage that has shaped our shared global maritime legacy.

4. AUTHORS BIOGRAPHY

Hanna Hagmark has been the Executive Director of the Åland Maritime Museum Trust since 2004, steering it through two major capital projects. She leads the daily operation and long-term strategic planning for the trust, which comprises the Åland Maritime Museum and the historic vessel Pommern as well as archives, library, and a world-class collection of artefacts. She has a PhD in maritime history from the University of Hull (UK)

FORENSIC NAVAL ARCHITECTURE – RECREATING *WATERBIRD*, THE WORLD'S OLDEST OPERATIONAL SEAPLANE

J Gifford, Jack Gifford Naval Architect Ltd., Member RINA and **P Wrobel,** FREng, Emeritus Professor University College London, Fellow and Honorary Vice President RINA

SUMMARY

On Lake Windermere in November 1911 *Waterbird* conducted the first successful flight in the world of a craft with a "stepped hull". Over a century later a replica was built using available historical information but the first water trials in 2020 showed what history failed to record, the craft was unstable in both roll and pitch. The hull and floats had to be re-designed from first principles. Although successful at low speeds, violent porpoising occurred at 25 knots. The design of modern seaplanes was of little help since *Waterbird*'s two-step configuration quickly became obsolete. Further changes were made based on empiricism, historical deduction and engineering intuition, leading to successful flight trials on Windermere in 2022. The question remains of exactly what worked in 1911?

1. INTRODUCTION

The Wright brothers first flew a land plane in December 1903. Due to the high mortality rate the idea of flying from water was conceived as being safer! The first successful flight in the world of a craft with a "stepped hull" was made by Edward Wakefield's *Waterbird* on Lake Windermere in November 1911. He came from a family of bankers and adventurers. Wakefield took ideas and successful elements from wherever in the world he could. For the floats he both designed his own and borrowed from others. He then had them made by a local boat builder Borwick & Sons. At that time twin floats were considered dangerous because of the risk of water ingress into either hull and subsequent inability to keep the plane level.

2. *WATERBIRD*

Figure 1: *Waterbird* flying 1911

A.V. Roe designed and built the original land aircraft based on a design by Curtiss in the USA. There was then a brief process of trial and error at Brooklands in June 1911 that resulted in the design of a new tailplane. Curiously remnants of the rejected tailplane have survived including the rudder. There are no records of the water trials, nor the reasons for the modifications between August 1911 when the craft arrived on Lake Windermere, and 25 November when it first flew – Figure 1. There are just a few long range photos when airborne, some inconsistent photos when waterborne and some press articles.

The majority of Wakefield's first float still exists from 1911 – Figure 2. It is 12 feet long overall, 2 feet wide and 1 foot deep. Its external shape was modelled on the Curtiss float used for his first successful flight from water in the Spring of 1911.

Figure 2: Internal framing

Wakefield's float had robust mahogany sides with fifteen latticed internal frames, no watertight sub-division and a one-piece canvas deck cover. Damage to the hull or cover and the whole craft would have gone straight to the bottom. There were no internal baffles, so even a small amount of water in the float would have led to the risk of fore and aft sloshing. Curtiss had initially done the same - he did get water inside the float, was unable to control the craft's pitch once in flight, and he crashed. That was R&D in the pioneering days.

Unlike Curtiss who used all wood for his main float, Wakefield employed 22 gauge (0.7mm) Aluminium for the floor, which was revolutionarily for an aircraft at that time. This would have presented the local boat builder with some new challenges.

Interestingly the float shows signs of being hastily modified:

- The addition of a bustle on the deck to provide buoyancy aft.
- The addition of one 4inch deep wedge shaped step - and then a second 3inch (aft) - providing both buoyancy and the potential to reduce the float's wetted surface area at high planing speeds. It should be noted that these steps were very crude and led to a very "aggressive" inflow angle that probably would have resulted in the porpoising behaviour referred to later (see Section 8).
- The float shows three sets of holes (the two aft sets are plugged) for the connection with the airframe. These indicate the need for appreciable trial and error in order to achieve a satisfactory longitudinal balance once proceeding at speed in the water.

3. *WATERBIRD* RE-BORN

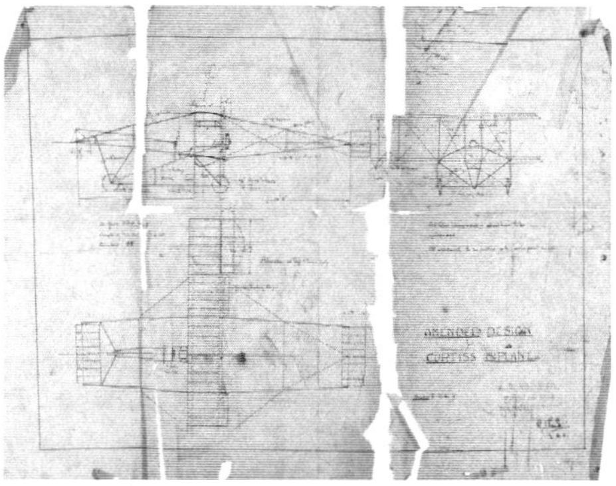

Figure 3: A.V. Roe's land plane

A hundred years later the replica *Waterbird* took the best available historical information starting with A.V.Roe's land plane – Figure 3. Then adding a single main float (including all of the modifications to the original listed above) and the two side floats – each as best recorded from 1911. The replica land plane flew without any problems.

Despite the main float incorporating the modifications from 1911, the first water trials of the replica at Nottingham in September 2020 very quickly established what history had failed to record. The craft was unstable in roll with a negative metacentric height - GM(T), and negative righting level arm - GZ(T) - at all angles. It had very little freeboard and was unstable in pitch when proceeding above 5 knots due to the high location of the engine thrust line and minimal longitudinal stability. There was only a modest righting lever arm - GM(L), combined with an extremely small angle of vanishing stability – Figure 4.

Figure 4: Nottingham 2020

The roll problem was addressed by temporarily lashing additional buoyancy (two 10cm diameter drain pipes from a local builder's merchant) to each side float. This resulted in a loll angle of ~10 degrees and then a healthy GZ(T) curve.

However the lack of overall buoyancy in the main float and rapidly vanishing longitudinal pitch stability, were not fixable. The pilot nearly caused the craft to capsize by the stern as he moved backwards to adjust the engine. As even low engine thrust was applied the foredeck went under water - a few empty barrels had been attached to the top of the forward deck as a safety measure!

The floats were evidently not fit for purpose even at low speed. We learnt the hard way just as they must have done in 1911. However no records of their efforts are available.

The trials were terminated since it was agreed that everything in the water had to be re-designed and made from scratch. This required going back to first principles, whilst also satisfying the constraints of modern regulations, remaining authentic and all for the same weight.

4. THE COMPETITION IN 1911

The historical record was examined in greater detail. This revealed a far from clear picture which is still emerging even today.

In early 1911 Glenn Curtiss in the USA tried a variety of float configurations and sizes [1]. His first experiment at an aeroplane on water involved the simple replacement of his land aircraft's undercarriage with a Canadian canoe and two small side floats. No calculations - no regulations. Exciting times but rapid learning!

After a short period of much trial and error. Curtiss finally adopted the configuration of a single main float with two small side floats. Curtiss did not use steps, relying instead on a more powerful – and heavier – engine than Wakefield was to use. Nor did he have the additional buoyancy that Wakefield was later forced to add (see Section 2). Although Curtiss was reported to have used a main float of 12 feet length, 2 feet beam, 1 foot depth – all the many drawings and photographs show a length of 16 feet. Lacking the additional buoyancy of Wakefield's bustle and steps – combined with a heavy engine - would have very likely led Curtiss to submerge with a 12x2x1 feet float. The actual length of Curtiss' float only came to light when a member of the *Waterbird* team visited the US Navy Museum in 2022 with a tape measure! It is quite conceivable that Curtiss' 16 feet float was simply described incorrectly by the non-technical press - just as say one could describe a craft's waterline length rather than its length overall.

The US Navy replica (now known to be 16x2x1 feet) shows the craft to be nearly awash at low speed. The shore team were essential to keep the aircraft safe during launch and recovery - so that at 12 feet long it would have been impractical.

Even at 16 feet long Curtiss was still concerned about pitching bow down as the speed picked up and then again on landing. He initially fitted a 3 feet wide hydrofoil – pitch angle up - forward of the bow, just above the static waterline. As confidence grew the hydrofoil was removed and a nose wheel fitted to allow the whole craft to operate amphibiously.

In the UK during the summer of 1911 Wakefield had begun *"a long and rather tedious, but interesting, series of experiments with floats of many shapes and sizes, weighted with sand-bags to represent the weight of my airplane, plus pilot, and towed at the top speed of our motor launch"* [2]. Frustratingly no records of what was tested, nor the results, can be found. However he appears to have cast aside his own ideas and based his first float for *Waterbird* on what he could glean from Curtiss' developments in the USA - only to discover it needed major modification and possible replacement. Curtiss was definitely using a 16 feet long float as early as April 1911. Wakefield's first float for *Waterbird* was built three months later in July. So why did Wakefield start at just 12 feet long – poor intelligence, the availability of Aluminium sheet in the required length, other?

In the UK and also Lake Windermere, Wakefield had another rival – Figure 5. Oscar Gnosspelius was a pioneering aeronautical engineer and ultimately the only person on Windermere to design and fly his own "hydro-aeroplane". He was a very competent engineer and had quite different ideas from both Curtiss and Wakefield regarding the configuration of the main float. His design was 12 feet long but with a **beam of 6 feet**. Whilst looking rather strange to the modern eye, this would have provided plenty of buoyancy and stability. He also had a more advanced underwater step configuration than Wakefield's crude wedges. Gnosspelius' step configuration was based on Knight's 1906 patent for high speed motor launches [3]. This offered the ability to take off at lower power and avoid the pitch instability that would probably have plagued Wakefield (see Section 8). Had it not been for Gnosspelius also piloting his own aircraft he might have beaten Wakefield by a few days. As it was he crashed.

Figure 5: Gnosspelius on Lake Windermere

It is not clear what Wakefield was doing with *Waterbird* in the time window between August - when *Waterbird* first arrived on the lake - and November 1911 when it flew successfully. He was in a race to succeed and there was certainly enough time and resources for several floats to be designed, built & trialled. In a shorter time window earlier in 1911 Curtiss had tested at least four different floats. Curtiss' efforts are well documented but no records of Wakefield's activities during this frenetic period can be found.

5. OTHER WAKEFIELD FLOATS – A HYPOTHESIS

Despite being competitors Wakefield and Gnosspelius became collaborators. The latter's 12x6x1 feet float - including the advanced step configuration - were adopted by Wakefield for his successor aircraft to *Waterbird*. This was a two seater called *Waterhen* and it successfully flew in April 1912. Each of the changes from *Waterbird* to *Waterhen* provides clues to the problems found with the former.

- The much larger float and step configuration for buoyancy & stability (including porpoising - see Section 8).
- The use of "Duralumin" and spruce in lieu of mahogany sides and lattice frames respectively. This was necessary to reduce the weight of a much larger float.
- The ailerons were extended in lieu of the "flaperons" or "elephant's ears" for improved roll authority in cross winds.
- The rear elevator was made controllable by the pilot - in conjunction with the canard - for greater pitch authority (noting the propeller being located well aft of the pilot).

We now know that *Waterbird* had at least one other main float built – possibly more – since the construction of one is described by a visitor from the Admiralty to Wakefield's site on Lake Windermere in January 1912. Reference is made to the use of "Duralumin". No record of this other float's design nor its dimensions can be found. The same writer also describes the main float fitted to *Waterbird* at that time – which he then flew with no previous seaplane experience. The description of the main float is both precise and at slight odds with the surviving float.

Wakefield was a magpie picking up whatever worked. If he required another main float for *Waterbird* to make it viable he would have been able to tap into the ideas of Gnosspeilius and others, such as the use of Aluminium alloys in experimental airships at Barrow-in-Furness. He evidently did just that for *Waterhen*. So we know of his floats in August 1911 and April 1912 – what others joined the dots between these radically different designs?

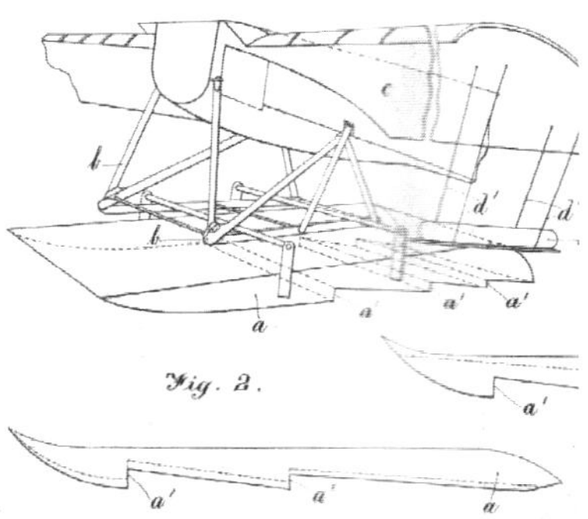

Figure 6: Wakefield's 1913 patent

The original *Waterbird* was unfortunately damaged beyond repair when the hangar collapsed in March 1912. It is noteworthy that the surviving first 12 feet long float that lay in the hangar shows no signs of any damage. Had it been supporting the airframe and engine the holes in the float would have taken the full weight of the roof structure. This supports the hypothesis that the first float had been discarded by that time and was no longer part of the aircraft when the roof collapsed on *Waterbird*.

Since the remaining artefacts from *Waterbird* include parts of the discarded tailplane, one could surmise that the surviving 12 feet long float is also from the "discard pile". What Wakefield later described in his memoirs, as the "successful float" could be of a different – subsequent – design that was the result of extensive trial and error between early August and late November 1911.

By 25 January 1912 *Waterbird* had made about 60 flights on 38 days. Its greatest distance had been about twenty miles and its best height 800 feet. Modern analysis and the recent trials have demonstrated that the original *Waterbird* with Wakefield's first float would have been very unlikely to operate successfully and safely many times as indeed *Waterbird* did.

Wakefield continued to make sketches for further floats with multiple steps, for example there is a 1913 patent – Figure 6. However no other design of float - or actual artefact – from 1911-13 can be found apart from the very first one. The search continues.

6. BACK TO FIRST PRINCIPLES

The historic picture at the time of the *Waterbird* replica's first unsuccessful water trials in September 2020 was very opaque and often mis-leading. A new direction was set based on first principles rather than solely on incomplete fragments.

The beam of the main float was retained to avoid changes to the airframe. The main float's length and depth were proportionally increased until analysis indicated that the craft would remain stable in heave and pitch across the speed range. This was estimated as 15 feet long and 15 inches deep. The two steps were scaled similarly thus retaining the same angles – Figure 7.

Figure 7: The re-designed float

In retrospect the flotation characteristics are remarkably similar to where we now know Curtiss ended up.

The resulting design even meets the current Civil Aviation Authority (CAA) regulations [4] for a seaplane's overall reserve of buoyancy, as well as intact and damaged stability in roll and pitch. The aircraft will survive upright with the loss of any one compartment in any of the floats and remain afloat with any two compartments lost.

The primary load case was considered to be slamming and rapid deceleration in the event of a forced landing, with the structure designed accordingly.

The first replica float was not designed to any rules or specific loads. It was built entirely of relatively thick plywood, with minimal stiffeners and watertight bulkheads. The structure was lacking in longitudinal structural continuity largely due to the inclusion of the steps into the hull, rather than installing them as "add-on" appendages. The need to completely re-design the float and build a new one post the trials at Nottingham may well have been fortuitous in avoiding what could have been a catastrophe at speed once the craft experienced significant dynamic loading.

Priorities for the replica float were weight and strength. A heavy structure would overburden the already built aircraft and compromise performance on take-off, flight and landing. The new structure had to be compliant with modern regulations and also fit for purpose in the eventuality of a stalled take-off or emergency landing. Flight speeds were taken into consideration when defining the load cases and a keen eye kept on weight when specifying the scantlings. The experience from the first replica and following risk assessment led to the approach of separating the steps from the main buoyant structure - future modifications would therefore be easier and the buoyancy of the craft would not be impacted in the event of a step being punctured. The target weight from estimates and records of the original was set at 38kg with the aim of retaining as similar a centre of gravity as possible.

The hull was built around four plywood bulkheads, forming five watertight compartments – shorter at each end and longer in the middle. 3mm Honduran Cedar was specified for the side panels and 0.7mm aluminium alloy for the bottom skin. Longitudinal stiffness was provided by continuous Sitka Spruce stringers at the chines and deck edges with lightweight extruded aluminium alloy 'L' sections used to stiffen the foil-like bottom skin. Wooden components were bonded using epoxy resin whilst the specialist, high bond strength - yet flexible - mast building epoxy 'SpaBond' was used to bond the aluminium alloy. The products and preparations for bonding were specified on advice from SP Gurit. The float was locally reinforced in-way-of the airframe fastenings, with Sitka Spruce backing pads joining the deck edge to the chines and padding out the thin cedar sides to take the airframe bolts.

As per the original the deck is nothing more than sections of canvas covering, re-created in a modern synthetic material; the only deck stiffness comes from a 75mm plywood margin that rings the top of the float and the tops of each bulkhead.

The side floats were completely re-designed with the same volume as the stand-ins used during the first water trials. Thin wall UPVC lightweight ventilation duct was used for the body. Structural foam for the ends as well as the two internal watertight bulkheads - all glued into place. Unlike the first set of side floats there were no fixing holes in the tubes, relying instead on glue and straps – the latter mostly for visual effect! The critical loads were hydrodynamic slamming on landing and buckling failure due to the water depth pressure in the event of rolling. The angle of the wooden skids to the water was retained from 1911. Losing a side float would be a critical hazard.

Figure 8: New side float on its skid

The CAA's understanding of boats and operation in water sometimes required an education into naval architectural principles and practice. For example the use of the optimum - rather then equal - longitudinal spacing of the watertight bulkheads, the use of epoxy rather than rivets etc.

The boat builder had never used such extremely light scantlings nor the combination of wood and aluminium alloy required for the main float, UPVC tube and structural foam for the side floats on a wooden skid - Figure 8. He needed constant re-assurance that whilst the component parts were extremely flexible individually, they would be fine once assembled. At his suggestion the very thin wooden components were glassed with a single layer of lightweight reinforcement both inside and out.

The overall weight of the new floats was within 2% of the target, which was more than offset by the selection of a lightweight pilot!

7. TO LAKE WINDERMERE

The replica *Waterbird* went to Lake Windermere for the first time in mid 2021. Just as in 1911 there was some experimentation in the water to find the best location of the fore/aft connection of the main float with the airframe. This was to ensure that the initial set up would manage the transition from hydrostatic, slow speed/high engine thrust, hydrodynamic hull-borne, hydrodynamic "on the steps" and aerodynamic behaviour. This process was complicated by a late request for the addition of more than 20kg of lead ballast at the forward canard. This would move the longitudinal centre of gravity forward to more closely match the centre of lift and thus ease the flying characteristics. It was frustrating noting the huge effort at keeping the new floats to the original weight. After correcting the arithmetic a compromise was reached of just 10kg. There is no record if any ballast was fitted in 1911.

Figure 9: Windermere 2021

The new floats enabled *Waterbird* to float with a large reserve of buoyancy – Figure 9, comfortably taxi and turn in light wind and waves, and be stable in roll and pitch throughout. She was then able to safely go through the transition onto the plane. The pilot had good rudder authority even at low speed. The lower rudder remained clear of the water at all times.

As the craft's speed increased the behaviour of the main float was carefully observed. As at Nottingham the year before there was initially a bow down pitching moment as the engine's thrust was increased, but then the bow started to rise as speed increased and hydrodynamic forces become significant, then the whole float rose in the water – initially planing hull-borne, then onto the steps alone - the float eventually started to level off. It was still several degrees bow up at 25 knots.

However another completely unforeseen problem occurred as the float started to porpoise approximately +/- 5 degrees at about 1 hertz when attaining 25 knots of water speed. No report existed of such a phenomenon from 1911/12 but the laws of physics would have been the same.

Several wind and wave conditions were tried as well as "surgery" - with a saw - to the aft step. Whilst very emotional for the boat builder, it was possible to implement without compromising the main float due to the steps being structural appendages to the main hull. The tests demonstrated that the problem was water driven i.e. hydrodynamic not aerodynamic. The changes were ineffective and the trials were curtailed. Nothing was broken and nobody injured. We had progressed from 5 knots waterborne at Nottingham to 25 knots on Lake Windermere.

8. PORPOISING

Various indirect "fixes" were proposed to solve the porpoising behaviour. Adding more lead ballast forward at the canard would have reduced the running trim angle but would also have messed up the running angles at lower speeds. Quick fixes aside it was decided to truly understand the fundamental causes of the problem.

The design of modern seaplanes was of only limited help since *Waterbird*'s two-step configuration quickly became obsolete during the rapid development of seaplanes post 1911. The two-step configuration on the main hull is unlike the evolved single step on most modern seaplanes that have a very shallow step angle to the water, together with a keel line aft that sweeps upwards allowing the aircraft to rotate and thence lift off.

Consideration was given to the use of CFD and model testing, but these expensive and time-consuming methods would raise as many questions as they would answer such as whether to model just the float or the whole aircraft including the wings and engine? As models they would not guarantee success. after all, a model is just that and computer output does not equate to truth.

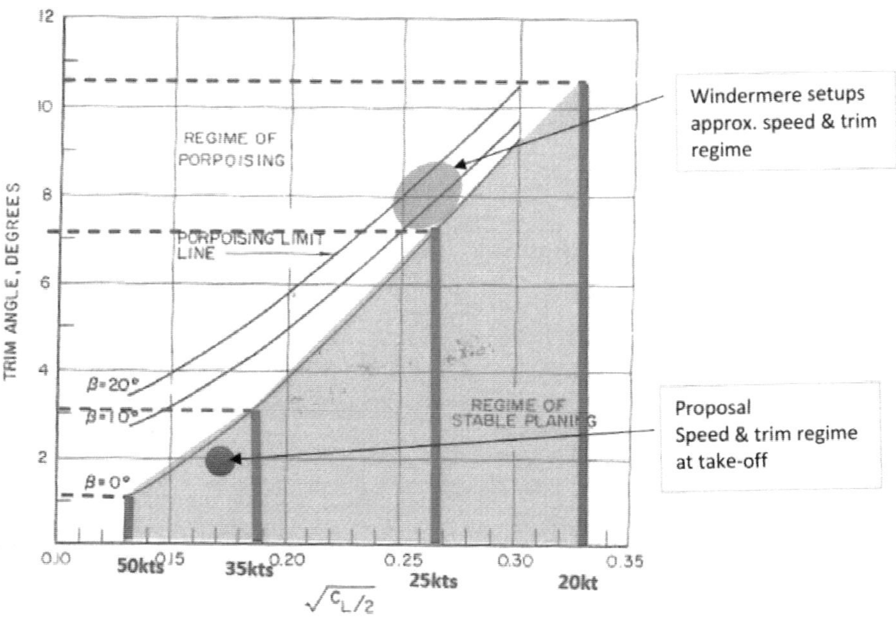

Figure 10: Savitsky's porpoising assessment - β is the deadrise angle

The empirical method for high-speed boats due to Savitsky [5] was chosen as the best method and adapted for a multiple step configuration. This approach confirmed what had been observed during trials which provided confidence – Figure 10. The method indicated the most probable cause as the step configuration – in particular the local angle of entry with the water. It also indicated that it was not possible to simply accelerate through the phenomenon as some had believed.

Wakefield's wedge-shaped steps had a particularly aggressive underwater profile compared to that patented by Knight in 1906, and the configuration used by Gnosspelius in 1911.

Seaplanes and racing boats ever since have used similar step configurations to the latter. The safest route would be to adopt a similar solution, but this would compromise the authenticity of the *Waterbird* replica. By re-designing the steps, the onset speed of porpoising could not be eliminated but pushed well above the anticipated water speed required to generate sufficient aerodynamic lift to take off.

Even if *Waterbird* had managed to get planing with Wakefield's first float in the Autumn of 1911, they would have almost certainly run into the same porpoising problem we found, although nothing was reported. Gnosspelius was either lucky or more knowledgeable than Wakefield, since his float design and step configuration has a predicted porpoising onset speed well in excess of the design take off speed for his aircraft. Wakefield's later adoption of Gnosspelius' float for *Waterhen* was very wise - as noted he was a magpie. Did Wakefield fit a second float to *Waterbird* with input from Gnosspelius prior to November 1911?

The video data showed the forward step emerging clear of the water – Figure 11. Three other possible causes of *Waterbird*'s porpoising were considered:

- Thrust line: There was a view that the set up of the main float and airframe lead to the the thrust line of the engine being a few degrees pitch down was leading to the nose digging in – generating a significant pitch up moment

Figure 11: Video frame showing porpoising

– the forward step leaving the water – the craft then pitching back down, et seq.

- Angle of attack: The trials also indicated that the angle of attack of the wings was far too low to allow the craft to achieve take off at the desired air speed of 35-40 knots. An increase in angle and hence lift would also reduce the hydrodynamic load on the steps and thus affect the porpoising behaviour.
- Rotation: The video shows the pilot often using high angles of the canard. Modern seaplanes get up to sufficient water speed to be "on the (single) step" that is located amidships. Then the pilot uses the (rear) elevator to rotate the aircraft to achieve the desired angle of attack over the wings and thus take off. A lack of a controllable elevator and the two-stepped hull make this all but impossible for *Waterbird*. The pilot's instinct to use high angles of canard remained and may well have stimulated the instability.

All these factors needed to be considered together since the angle of attack and the engine's thrust line are not just a function of the static set-up. They both vary as the float's pitch angle to the water changes across the speed range. Initially bow down compared to the craft at rest, then well up and then gradually levelling off as the steps dominate and take off speed approaches. Anticipating this final condition is critical – too low and the wings won't generate enough lift – too high and the aircraft could stall. These angles were modelled across the pitch range to ensure the desired performance and amongst other things, ensure the aircraft's trail did not drag in the water!

Lots of issues to consider and balance.

9. 2022 – THE LAST CHANCE

Trials on the lake are severely time restricted and expensive. The *Waterbird* replica project had been running for over a decade and there were people that believed the project had already ceased to be. If we failed to fly in 2022 that would probably be the end of the road. It was decided to adopt a package of changes that addressed <u>all</u> of the potential issues. Options that were not authentic such as having a controllable rear elevator were not adopted.

Changes were made as follows:

- The depth of the steps was halved making the local inflow angle much less aggressive. Note that this also has the detrimental effect of reducing the step volume/buoyancy and changing the longitudinal centre. Savitsky's empirical analysis indicated that this should push the water speed for onset of porpoisng to well over 40 knots.
- The relative angle of the main float to the airframe was increased by five degrees. The latter was also made adjustable between trial runs +3/-2 degrees. This required a new and stronger design for the forward connection of the airframe to the main float.
- The side float aft strut lengths were adjusted to retain the original freeboard and angle to the water.

With the following considerations:

- Too small an angle and the craft won't generate enough lift and might still porpoise if the water speed had to be increased to ~40 knots.
- Too large an angle could lead to the pilot stalling the wings on take off as the drag of the float reduces dramatically on leaving the water causing a pitch up moment. The result of such a low altitude stall would be large slamming loads and could be potentially catastrophic if the float's nose was to immerse.
- Increasing the set-up angle also directly increases the engine's thrust line to be positive.
- It also increases the possibility rudder becoming immersed whilst at rest and during landing has a large nose up attitude.

Finally:

- Instructions to the pilot to leave the stick alone until take off occurs by design rather than by pilot control!

Waterbird went back to Lake Windermere in June 2022. The set-up procedure was repeated – also the discussion about the amount of lead ballast on the canard, noting that this would change the waterborne trim and draft – and everything finally re-checked.

Waterbird transitioned through 25 knots of water speed with no porpoising, the steps behaved as desired reducing the wetted surface and hydrodynamic drag, the craft reached 32 knots water speed (~37 knots of air speed) with the main hull gradually both rising and reducing its absolute pitch angle to just a few degrees bow up, there was no aircraft rotation and *Waterbird* gently took off – Figure 12.

Figure 12: First take-off in June 2022

Waterbird's first water landing was wonderfully dull – the pilot adopted a slight pitch up attitude and ~35 knots air speed. There was no rudder immersion, the aft step engaged, the aircraft rotated, the forward step engaged. There was a gentle deceleration, no bouncing, and no nose-dive. Subsequent inspection found no water in the main float.

It all worked perfectly, first time. Which of the four main changes was the most effective? We cannot be sure. The pitch adjustment of the airframe to main float was not required. Unfortunately, this was not an engineering development programme – the objective was simply to succeed.

10. THE FUTURE

The craft has now successfully taken off from the water, flown and landed many times with the pilot progressively pushing the envelope during flying and landing.

The naysayers have changed their spots and become champions of *Waterbird.* The CAA has granted the replica *Waterbird* its certification. *Waterbird* has now carried out several public demonstration flights on Lake Windermere.

There is an original 1929 Hamilton Metalplane H-47 floatplane in the USA and a 1935 Caproni Ca.100 floatplane in Italy. Otherwise, there is no airworthy, original, replica or restoration seaplane prior to 1936 anywhere in the world. The replica *Waterbird* from 1911 is unique for representing the pioneering era of aircraft in general and seaplanes in particular.

As an indication of the esteem in which the replica *Waterbird* is held, the team has swept the board of awards receiving:-

- The Phoenix Group Diploma - The Federation Aeronautique Internationale: the World Air Sports Federation;
- The Bremont Special Recognition Award - The National Transport Trust;
- The Robert Pleming Memorial Award - Aviation Heritage UK;
- The Desmond Penrose Silver Salver – for vintage aeroplane of the year - The Vintage Aircraft Club;
- The Salomons Trophy - for an outstanding performance by a British aviator in a flying apparatus designed and built in the UK - The Royal Aero Club;

- A Red Wheel from the National Transport Trust unveiled at Hill of Oaks adjacent to the original *Waterbird* boathouse.

11. BUT......

We are left with the questions "what exactly was the design of float that worked successfully and repeatedly - in 1911/12". "Where are all the records and artefacts?"

12. REFERENCES

1. CASEY, L S., Curtiss The Hammondsport Era 1907-1915 - 1981.

2. WAKEFIELD, E., The Westmorland Gazette - 22 March 1941.

3. Albert Edward Knight – Improvements in or relating to the Construction of Mechanically Propelled Navigable Vessels – 22 August 1906.

4. Certification Specifications for Very Light Aeroplanes – European Union Aviation Safety Agency 2009.

5. SAVITSKY, D., – 'Hydrodynamic Design of Planing Hulls', *Marine Technology*, 1964.

13. AUTHORS' BIOGRAPHIES

J R Gifford, Jack Gifford Naval Architect Ltd., Member RINA

P G Wrobel, FREng, Emeritus Professor University College London, Fellow and Honorary Vice President RINA

STRUCTURAL DESIGN OF WOODEN BOATS

J-B R G Souppez, Aston University, UK

SUMMARY

The contemporary environmental challenges faced by the maritime industry have triggered a regain of interest in wood as a sustainable material. Modern classics and replicas of historical vessels have been increasingly popular. However, owing to their historical nature and small market share, wooden yachts and small ships do not benefit from a robust and detailed regulatory framework. Instead, their structural design remains driven by designer and builder experience, historical scantlings and rules of thumb. As such, the lack of detailed regulations for the scantlings of traditional timber vessels has been identified as a limitation to their wider implementation in the maritime industry. Consequently, this paper explores existing regulations and recent experimental research to provide novel insights into wooden structures and their associated scantling rules. First, a classification of wooden boatbuilding techniques is offered. Then, the lower carbon footprint of traditional carvel construction compared to modern cold moulding, strip planking, and plywood construction is quantified. Finally, recommendations are provided regarding the remaining outstanding research questions that need to be addressed to ensure reliable traditional wooden boat scantlings in modern regulations. It is anticipated these findings will facilitate the wider adoption of wooden construction for small crafts, support the use of more sustainable materials, and may further contribute to the development of future regulations.

NOMENCLATURE

c	Curvature coefficient [-]
C_f	Carbon footprint of hull [kg kg^{-1} m^{-2}]
C_m	Carbon footprint of boatbuilding method [kg kg^{-1}]
k	Aspect ratio coefficient [-]
L	Rule length [m]
L_{OA}	Overall length of hull [m]
L_{WT}	Waterline length [m]
m	Mass of hull [kg m^{-2}]
P	Design pressure [kPa]
s	Short side of panel [mm]
t	Planking thickness [mm]
ρ	Density [kg m^{-3}]
σ_d	Design stress [MPa]
σ_{uf}	Ultimate flexural strength [MPa]
ABS	American Bureau of Shipping
GL	Germanischer Loyd
ISO	International Organization for Standardization

1. INTRODUCTION

Wooden boats have been predominant throughout history [1, 2, 3], prior to the advent of metal [4, 5] and, more recently, composite construction [6, 7]. The significant progress in adhesives associated with the latter [8] has led to the development of modern timber construction, namely cold moulding, strip planking and plywood, all relying on epoxy and modern adhesives for timber encapsulation. This alleviates the expansion of timber with increasing moisture content [9, 10] as found in traditional wooden construction, namely carvel and clinker (also known as lapstrake), where planks are left exposed to the environment. Carvel relies on caulked seams to cope with the swelling of planks with increasing moisture content, while clinker construction employs overlapping planks.

The past decade has seen a regain of interest in historical wooden boats and their analysis using modern naval architecture techniques [11, 12, 13, 14, 15]. This is further evidenced in the development of modern replicas [16, 17, 18, 19, 20, 21], new builds [22, 23] and wooden cargo vessels [24, 25, 26]. Yet, there remains a lack of regulations to support the adoption of traditional wooden boatbuilding techniques for sustainability purposes [27, 28, 29], leading to new considerations for the regulatory implications of timber constructions [30, 31] and regulatory compliance [32, 33]. Pre-1950 designs, whether original historical crafts or replicas built predominantly with the original materials, are beyond the scope of legislation [34, 35] and associated structural regulations (e.g. ISO 12215-5:2019 [36]). In fact, regulatory frameworks primarily focus on modern wooden construction [37, 38]. While plank thicknesses are provided for a given estimated shell area by Germanischer Lloyd (GL) [39], only the American Bureau of Shipping (ABS) [40] features traditional construction with dedicated scantling calculations.

Because clinker is less common than carvel, it has received comparatively less attention, but for the recent work of Loscombe [12, 41]. As a result, clinker scantlings are adjusted from carvel ones, instead of tackled in their own right. While ABS [40] does not cover clinker, GL [39] states that, for a given planking thickness, stiffener spacing can be increased by 65% for clinker compared to carvel. The guidelines of Gerr [42] also support the underlying assumption that greater strength is achieved with clinked construction thanks to the overlapping planks, recommending that, for a given stiffener spacing, clinker planking thickness can be reduced to 85% of that of carvel. Finally, the MGN 628 [43] gives, on average, plating thicknesses 24% thinner for clinker compared to carvel for a given vessel size.

Consequently, this paper focuses on carvel construction and further limits its scope to the shell planking thickness. The aim is to characterise the environmental benefits of traditional compared to modern timber construction, as well as identify critical areas currently overlooked by existing regulations in order to support its future wider implementation in small craft regulations. The remainder of this paper is structured as follows. Section 2 offers a classification of modern versus traditional wooden construction. Then, Section 3 tackles the current regulatory scantlings and identifies the benefits of traditional wooden boatbuilding. Subsequently, structural challenges beyond the scope of current regulatory bodies are discussed in Section 4 to provide novel insights into traditional wooden construction. Finally, the main findings are summarised in Section 5.

2. MODERN VERSUS TRADITIONAL TIMBER CONSTRUCTION

A simplified classification of the most commonly employed wooden boat construction methods is presented in Figure 1. A primary distinction is made between modern and traditional wooden boatbuilding. The former features a predominant use of modern adhesives (e.g. epoxy) to fully seal the timber from the outside environment, thereby maintaining a low moisture content. In addition, modern construction relies on machined timber, namely thin veneers, edge-machined strips and plywood, and commonly features a fibreglass sheathing. Three subcategories are presented here: cold moulding, strip planking and plywood (or plywood on frames), matching the options available for wooden construction under ISO 12215-5:2019 [36]. On the other hand, traditional boatbuilding requires the planking to absorb moisture and swell to yield a watertight hull. Planks are made from solid timber and mechanically fastened to transverse stiffeners. Planks may either be laid edge-to-edge, with a small seam later caulked (carvel construction) or overlapping edges (clinker construction).

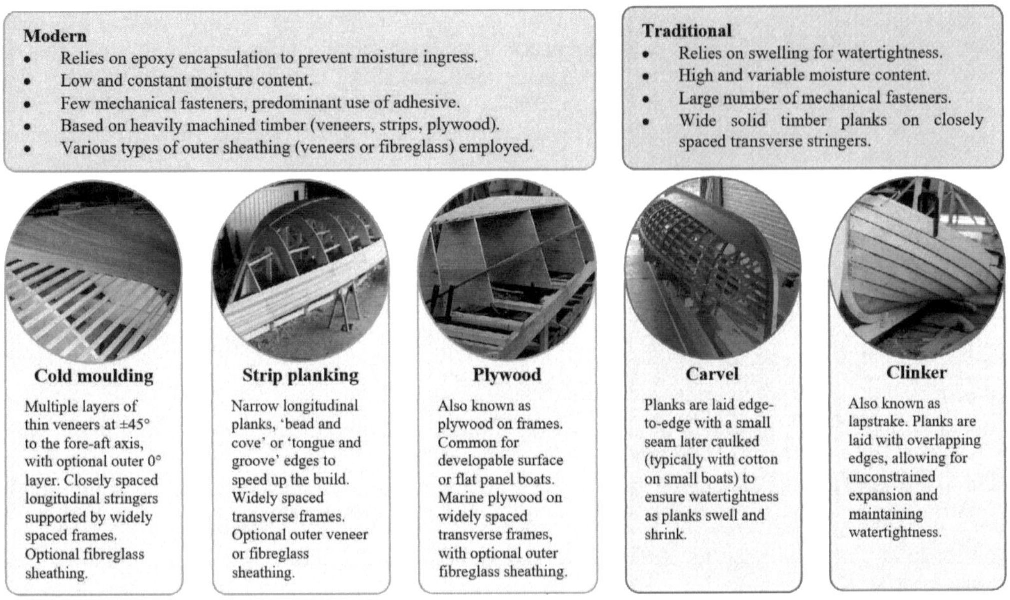

Figure 1: Classification of the main wooden boatbuilding methods for small crafts. Illustrations taken from [44] for cold moulding, [45] for strip planking, [46] for plywood on frames, [47] for carvel and [48] for clinker.

First, the application of existing regulations for small wooden crafts will be investigated in Section 3. Then, the specifics of traditional boatbuilding and associated challenges will be detailed in Section 4, with the intention to identify outstanding research questions and provide an overview of the limitations that may be associated with regulatory scantlings for carvel and clinker vessels.

3. CONTEMPORARY SCANTLINGS FOR STRUCTURAL DESIGN

The planking thickness t of modern wooden boats is given in both ISO [36] and ABS [40] by an equation of the form

$$t = c\, s \sqrt{\frac{P\, k}{\sigma_d}}, \tag{1}$$

where c is a curvature correction coefficient taken as 1 (i.e. no curvature), s is the short side of the panel, P is the design pressure, k is an aspect ratio coefficient taken as 0.5 for panel with an aspect ratio greater than or equal to 2, and σ_d is the design stress, based on the ultimate flexural strength, σ_{uf}. Equation (1) assumes a built-in beam a under uniformly distributed load [49].

For carvel, ABS [40] gives the planking thickness as

$$t = 1.09\left(4.2 - 0.84\sqrt[4]{L}\right) s \sqrt{\frac{P}{\sigma_d}}, \tag{2}$$

where L is

$$L = \frac{L_{\text{OA}} + L_{\text{WL}}}{2}, \tag{3}$$

and L_{OA} and L_{WL} are the overall hull length and waterline length, respectively.

Despite varying pressure and material properties, rules and regulations have been shown to yield consistent scantlings [50]. To avoid bias in the subsequent analysis, a pressure range 0 kPa $\leq P \leq$ 150 kPa is adopted, representing small craft pressures [51]. It is noted that longitudinal pressure distributions vary between ISO [36], ABS [40] and GL [39], and thus could impact scantlings for individual panels under different regulations. In this work, the mechanical properties are taken as defined by ISO 12215-5:2019 [36] and the short side of the panel is constrained to the practical ranges employed in small wooden boat building (see Table 1). Note that ISO [36] does not provide scantlings for carvel construction, while ABS [40] does not cover strip planking.

For the purpose of the paper, the assumptions inherent to the materials and associated properties for each boatbuilding technique are detailed in Table 1. Eco-properties, in the form of the carbon footprint C_f are provided [52], and account for the timber used, necessary machining, and epoxy (cold moulding and strip planking), phenol-formaldehyde (plywood) or copper nails (carvel and clinker), as relevant.

Table 1. Boatbuilding techniques, assumed construction material [10, 42, 53] and material properties [36, 52].

Construction Method	Cold moulding	String Planking	Plywood	Carvel
Stiffener spacing, s [mm]	$0 \leq s \leq 360$	$0 \leq s \leq 1200$	$0 \leq s \leq 900$	$0 \leq s \leq 360$
Wood species (common name)	African mahogany	Western red cedar	9 ply marine plywood	European Oak
Wood species (latin name)	*Khaya anthotheca*	*Thuja plicata*	*n/a*	*Quercus spp.*
Wood density, ρ [kg m^{-3}]	514	368	500	689
Ultimate flexural strength, σ_{uf} [MPa]	67	52	32	77
Design stress, σ_d [MPa]	10	26	16	38.5
Carbon footprint, C_m [kg kg^{-1}]	1.942	1.301	1.133	0.602

The regulatory thicknesses, calculated according to ISO 12215-5:2019 [36] for cold moulding, strip planking and plywood, and based on ABS [40] for carvel are presented in Figure 2. Cold moulding yields significantly lower thicknesses than the other boatbuilding methods under consideration. This may be expected given the original use of hot moulding for lightweight aircraft [54], before being adapted to yachts [55]. Note the difference in terminology between hot moulding and cold moulding, the former required the application of heat for the resin to cure, while advances in adhesives later enabled the curing process to take place at room temperature (cold moulding).

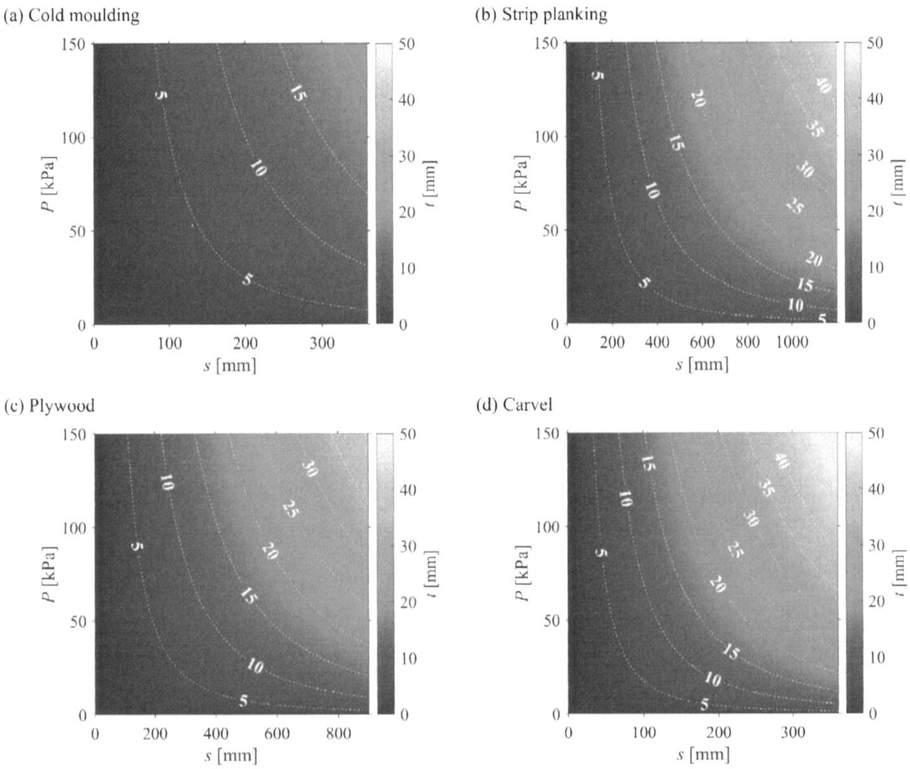

Figure 2: Required planking thickness for (a) cold moulding [36], (b) strip planking [36], (c) plywood [36], and (d) carvel (and clinker) [40].

In Figure 2, the carvel scantlings are given for $L = 15$ m, i.e. in the middle of the range where ABS [40] applies
($6 \text{ m} \leq L \leq 24 \text{ m}$). This is compared to the upper bound (minimum length $L = 6$ m, maximum pressure $P = 150$ kPa) and lower bound (maximum length $L = 24$ m, minimum pressure $P = 10$ kPa) of ABS [40] (Equation (2)), as well as the discrete values provided by GL [39], and the guidelines of Gerr [42] in Figure 3. There is a clear common trend across all three methods, despite their quantitative differences, which may be explained by their varied input parameters and associated assumptions.

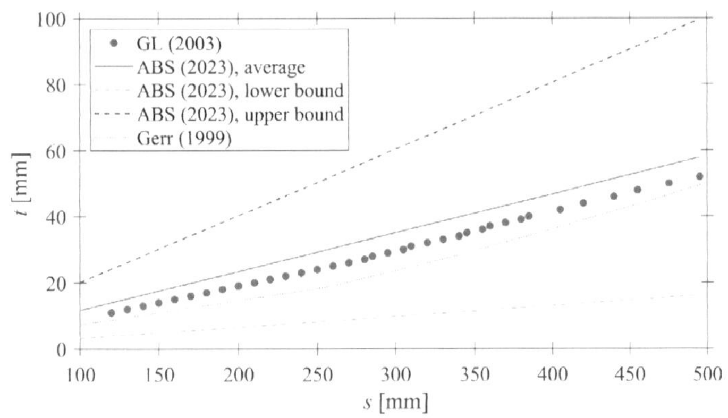

Figure 3: Comparison of the GL [39], ABS [40] and Gerr [42] carvel planking thickness.

Because different construction methods are associated with different timber species (see Table 1), with varying densities ρ, the resulting hull mass per square meter m can be computed as

$$m = t\rho. \tag{4}$$

The results are presented in Figure 4, further revealing the potential for cold moulding to achieve the lightest hull weights. Additionally, despite initially high thicknesses, the relatively low densities of western red cedar for strip planking (ρ = 368 kg m^{-3} [36]) and the marine plywood assumed (ρ = 500 kg m^{-3} [36, 53]) compared to that of oak for carvel (ρ = 689 kg m^{-3} [36]) makes the latter a much heavier option. Note that this finding remains valid for all species of timber commonly employed for carvel planking (e.g. larch, Douglas fir, mahogany, etc).

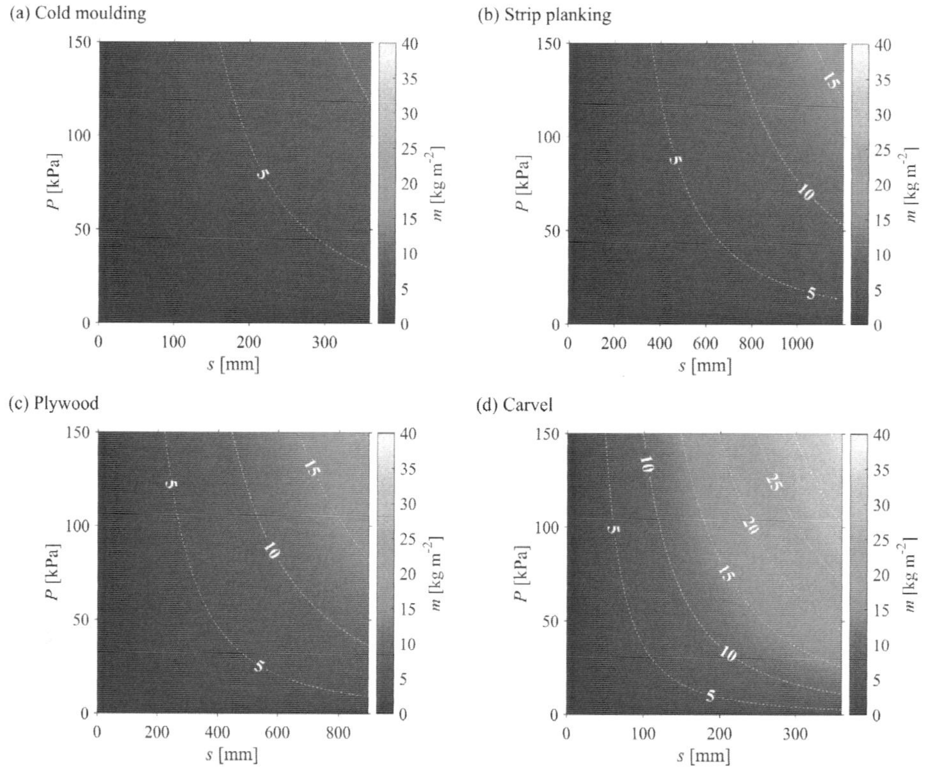

Figure 4: Planking mass based on required thickness for (a) cold moulding [36], (b) strip planking [36], (c) plywood [36], and (d) carvel (and clinker) [40].

Carvel planking may, therefore, not appear as an obvious choice for contemporary wooden boatbuilding. However, carvel differs from modern wooden boatbuilding methods in its absence of adhesives, which have a high environmental impact. Indeed, the regain of interest for wooden boats has been driven by sustainability considerations, which has yielded investigation into the life cycle assessment of vessels [56, 57], including wooden boats [58], albeit with high uncertainty as life cycle assessments for small crafts remain to be improved [59]. Nevertheless, assessing the carbon footprint per meter squared of hull shell C_f for the four construction methods under consideration using Equation (5) shows carvel to have the lowest carbon footprint, as evidenced in Figure 5.

$$C_f = m\, C_m \tag{5}$$

The benefits of carvel construction from an environmental point of view demonstrated in this section justify the increasing need for its inclusion in rules and regulations to support sustainable developments in small craft design and construction. Recent work [33], however, has highlighted the difficulty in achieving a reliable rule-based thickness equation for traditional wooden boatbuilding owing to some specific structural features and considerations not accounted for by regulatory bodies, and for which a thorough understanding remains lacking. Consequently, Section 4 tackles the challenges associated with traditional construction to identify the outstanding research questions to be answered in order to include traditional wooden construction methods in modern structural regulations.

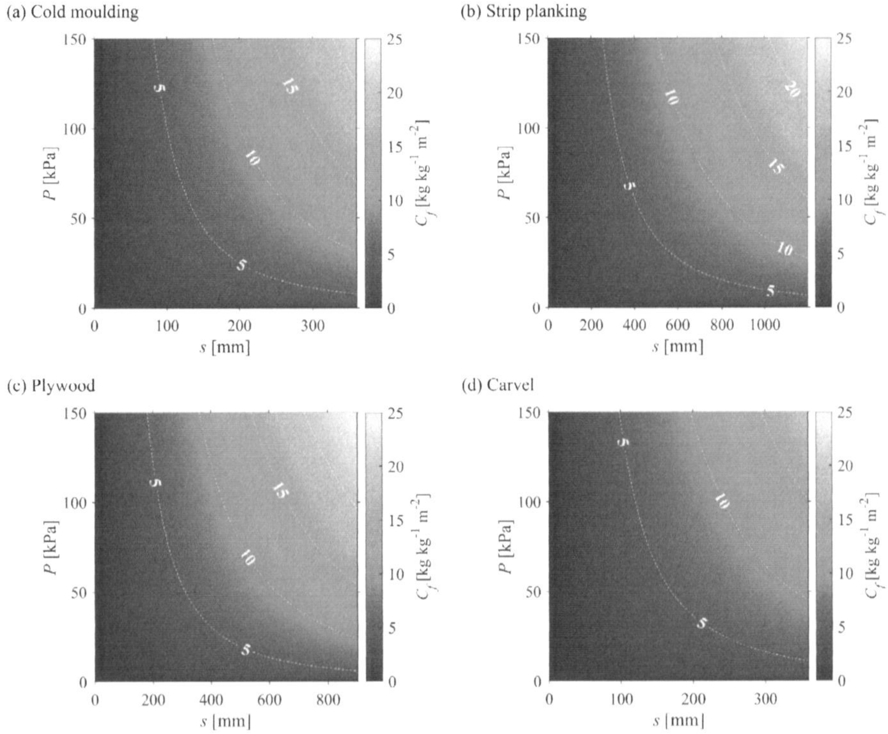

Figure 5: Carbon footprint for planking based on required thickness for (a) cold moulding [36], (b) strip planking [36], (c) plywood [36], and (d) carvel (and clinker) [40].

4. RECOMMENDATIONS FOR FUTURE CONSIDERATIONS

4.1 MOISTURE CONTENT

Moisture content is defined as the mass of water in timber compared to the dry mass of the timber. For boatbuilding applications, moisture content ranges from 7% (kiln-dried) to 14% (air-dried) [60] (7% to 16% allowed in ABS [40]; 8% to 14% in GL [39]). For modern boatbuilding, the epoxy encapsulation means the timber will remain within this moisture content throughout the operating life of the vessel. For traditional construction, however, the moisture content would increase to 28%-30%, also known as the fibre saturation point [60, 61], while in the water. This increase in moisture content is associated with a sharp reduction in material properties [62, 63, 64], as evidenced in Figure 6.

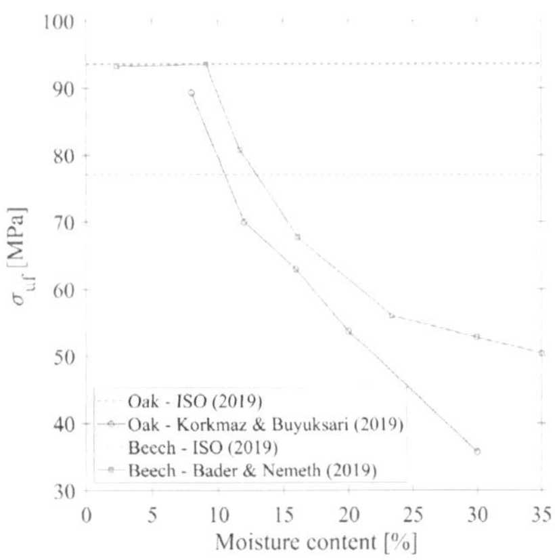

Figure 6: Variations in ultimate flexural strength with moisture content. Includes data from [62] for Oak and [63] for Beech, both compared to the default ISO properties [36].

Mechanical properties for timber are typically given at 12% moisture content [36, 39, 40]. Comparing σ_{uf} at 12% moisture content with 30% (fibre saturation point), the values are halved: 46% and 56% for oak and beech, respectively. This would imply that the factor of safety on the mechanical properties of timber for traditional construction needs to be at least doubled compared to current regulations.

4.2 STEAM BENDING

Steam bending allows to form timber into complex curved shapes, which is particularly appealing to boatbuilding. While suggestions for more effective application of steam bending in other fields have been made [65, 66], boatbuilding retains the rule of thumb that timber should be steamed for an hour per inch of thickness to bend [9]. Steam bending increases compressibility while having little effect on tensile properties [67]. However, under-streaming will not achieve the desired curvature, and over-streaming can lead to compressive fracture.

Studies have focused on the ability of timber species to bend following exposure to steam, investigating the effect of timber moisture content, steam time and breakage [68, 69, 70]. However, the material properties remain uncharacterised. The curvature of the timber thanks to stream bending provides better grain orientation. Whether the resulting mechanical properties deviate from that of solid timber assumed by regulatory bodies is unknown. As such, this may be an area of future research to quantify the effect of steam bending on traditional wooden boat scantlings.

4.3 MECHANICAL FASTENING

Both carvel and clinker planks rely on a large number of mechanical fasteners, typically riveted copper nails [9]. These require the use of a pilot hole and are counterbored, thereby introducing both a loss of mechanical properties, and a stress concentration, none of which is currently accounted for despite empirical methods available for their analysis [71]. Additionally, the pull force (or withdrawal force) exerted on a nail can be computed [60, 72], though it is unlikely to be of concern [73]. However, because the built-in end fixity of traditional construction may be questioned and simply supported seen as more suitable, as later detailed in Section 4.7, the prying action may be critical. Indeed, contrarily to built-in beams, simply supported ones experience a slope at their support. This would cause a prying moment on nails, the study of which requires further research [74].

4.4 SOLID TIMBER DEFECTS

Traditional construction methods are characterised by the use of thick solid timber, as previously evidenced in Figure 2. This leads to a high uncertainty associated with individual components, which may be affected by the grain orientation (further influenced by the slab or quarter-sawn nature of the timber [42]), deviation from straightness (cupping, bowing, twisting, diamonding, etc… [60]), or natural flaws (knots, sap pockets, etc…).

This renders default regulatory mechanical properties and the result of mechanical tests on small samples too optimistic. Indeed, while experimental testing of timber samples displays values close to that of ISO [23, 33, 75], lower values are achieved for larger timber components [76]. As such, higher factors of safety would be necessary. Conversely, modern construction techniques, such as cold-moulding, relying on thin, straight-grain veneers, where any defect would be visible, yield much more reliable structures. This justifies their lower thickness requirement (Figure 2), inclusion in scantling regulations [36], and lightweight, high-performance applications [10, 55, 77, 78]. While composite reinforcement of timber beams has been shown to alleviate the uncertainty inherent to timber defects [76], this would defy the environmental benefits associated with traditional construction.

4.5 PROPERTIES PERPENDICULAR TO THE GRAIN

Both carvel and clinker feature wide planks, meaning the properties perpendicular to the grain may become relevant. While properties perpendicular to the grain for plywood, and perpendicular to unidirectional fibre orientation for composite, are covered in ISO [36], this is not the case for solid timber. For the purpose of the application of ISO standards, flexural properties perpendicular to the grain may be taken as 7% and 9% of the flexural properties parallel to the grain for softwood and hardwood, respectively. There remains a lack of experimental data for timber perpendicular to the grain for boatbuilding applications. Moreover, any mechanical test would be affected by the limitations detailed in Section 4.4, and would further be affected by joint efficiency, tackled in Section 4.6, which has solely been investigated parallel to the grain.

4.6 JOINT EFFICIENCY

The finite size of timber planks requires the use of scarf joints. For plain scarfs, the length-to-thickness ratio drives the resulting strength, leading to higher scarf ratios being recommended for structurally critical applications. For instance, historical rules by Lloyds register of shipping [79] recommended ratios 4:1 for planks, 6:1 (possibly 8:1) for keels, and 12:1 for spars. In contemporary regulations, ratios of at least 10:1 for $t < 10$ mm and at least 8:1 for $t > 10$ mm are required by GL [39]. While both Birmingham [9] and Gerr [42] suggested scarfs are always weaker than solid timber, recent work by Souppez [33] has demonstrated that, thanks to modern adhesives, joints stronger than solid timber may be achieved for scarf ratios of 16:1 and higher. This is depicted in Figure 7.

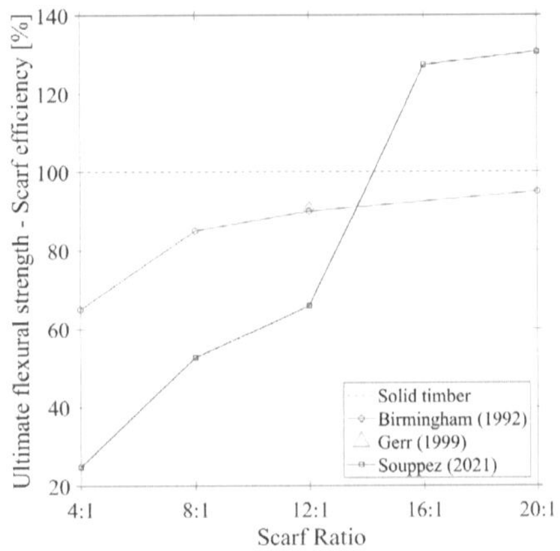

Figure 7: Scarf efficiency (ultimate flexural strength) compared to solid timber, represented by 100% efficiency. Includes data from [9], [33] and [42].

There is, therefore, a concern that, for commonly employed and recommended scarf ratios [9, 39, 42], flexural properties significantly lower than those assumed for solid timber may be achieved. Consequently, the scantlings of traditional crafts may be driven by the effectiveness of joints, as opposed to the mechanical properties of solid timber, an approach currently overlooked by rules and regulations.

4.7 STRUCTURAL ANALYSIS AND FAILURE MODE

Structural regulations are based on two assumptions: panels are treated as built-in beams (100% end fixity), and the load is considered uniformly distributed [80]. The first assumption may not be relevant to traditional construction, which features closely spaced small stringers, and thus a lower end fixity. Interestingly, recent developments in carbon fibre racing yachts have developed similar structures: small, closely spaces stringer, where panels have been assumed to be simply supported [81, 82]. Secondly, with short panels, a robustness criterion may be relevant [73], which would lead to a point load instead of a uniformly distributed load. The combination of both changes to the regulatory assumption would result in a change in the maximum bending moment from $Ps^2/12$ for a built-in beam under uniformly distributed load, to $Ps^2/4$ for a simply supported beam subject to a central point load assumed a Ps. The new plating thickness would, therefore, become

$$t = s\sqrt{\frac{1.5P}{\sigma_d}}, \tag{6}$$

in lieu of the current

$$t = s\sqrt{\frac{0.5P}{\sigma_d}}. \tag{7}$$

A thickness increase of $\sqrt{3}$ would therefore result from a change in underpinning assumptions. However, an additional failure mode may also need to be considered, namely shear. Indeed, given the high thickness and short span, panels with a thickness-to-span ratio up to 0.097 would be achieved for carvel, based on the parameter space considered in Figure 2. This compares to 0.031 for cold moulding, 0.013 for strip planking and 0.022 for plywood. The mechanical testing of samples to determine shear properties under ISO 14130:1998 [83] is to be performed for $t/s = 0.100$. Conversely, flexural tests under ISO 178:2019 [84] are to be performed at $t/s = 0.050$. Carvel planking is, therefore, most likely to fail under shear, which is overlooked by current regulations.

5. CONCLUSIONS

Modern timber construction is thoroughly addressed in contemporary structural regulations. However, traditional construction is only rarely accounted for, despite benefits that have led to a recent regain of interest in their design and construction for small crafts and wind-assisted ships alike. The comparative plating scantlings have been assessed for cold moulding, strip planking, plywood on frame and carvel planking. The latter, despite its higher thickness and mass, was shown to yield a much lower carbon footprint, demonstrating its relevance to sustainable boatbuilding.

To facilitate the development of structural regulations for traditional wooden construction, the key specificities of this construction method have been reviewed. These include high moisture content, the use of steam bending and mechanical fastener, the effect of natural defects and orientation perpendicular to the grain on the mechanical properties, the efficiency of scarf joint, and limitations to the regulatory assumptions for traditional construction. Areas of future research have been identified where knowledge currently is not available to affect scantling calculations.

This paper focused on carvel construction; however, clinker may be of further interest. Indeed, despite a lack of research, clinker planking is consistently taken as lesser than that of carvel owing to the overlapping planks. This would enable to achieve lighter hull shells than carvel and further reduce the environmental impact, which would warrant future work tackling clinker. Additionally, while planking has been detailed, the internal structure of traditional wooden boats would also benefit from further research to support the future development of thorough scantling rules for traditional wooden boats. Nevertheless, it is acknowledged that there remain numerous challenges associated with regulatory scantling rules for traditional boats, which may also prove overly conservative and discourage its adoption.

6. ACKNOWLEDGEMENTS

The valuable discussions and advice provided by Robin Loscombe is greatly acknowledge by the author.

7. REFERENCES

1. PARK, G., KIM, J C., YOUN, M., YUN, C., KANG, J., SONG, Y M., SONG, S J., NOH, H J., KIM, D K., and IM, H J., 'Dating the Bibong-ri Neolithic site in Korea: Excavating the Oldest Ancient Boat', *Nuclear Instruments and Methods in Physics Research Section B: Beam Interactions with Materials and Atoms*, vol. 268, no. 7-8, pp. 1003-1007, 2010.

2. WARD, C., 'Boat-building and its Social Context in Early Egypt: Interpretations from the First Dynasty Boat-grave Cemetery at Abydos', *Antiquity*, vol. 80, no. 307, pp. 118-129, 2006.

3. MCGRAIL, S., 'Ancient Boats in North-West Europe: The Archaeology of Water Transport to AD 1500', *Routledge*, Milton Park, 2014.

4. FAIRBAIRN, W., 'Treatise on Iron Ship Building: Its History and Progress as Comprised in a Series of Experimental Researches on the Laws of Strain', *Longmans, Green & Company*, 1865.

5. BAXTER, J P., 'The Introduction of the Iron-clad Warship', *Harvard University Press*: Harvard, US, 1933.

6. GREENE, E., 'Use of Fiber Reinforced Plastics in the Marine Industry', *Ship Structure Committee, National Academy of Science*, Washington, DC United States, 1990.

7. SOUPPEZ, J B R G., 'Structural Assessment and Scantlings of Traditional Small Crafts', *RINA Historic Ships*, London, UK, 2020.

8. BECK, R., BOOTE, D., DAVIES, P., HAGE, A., HUDSON, D., KAGEYAMA, K., KEUNING, J.A. and MILLER, P., 'Sailing Yacht Design', *17th International Ship and Offshore Structures Congress*, Seoul, Korea, 2009.

9. BIRMINGHAM, R., 'Boat Building Techniques Illustrated', London, UK: *Adlard Coles Nautical*; 2nd edition, 1992.

10. GOUGEON, M., 'The Gougeon Brothers on Boat Construction', Bay City, Michigan, US: *Gougeon Brothers Incorporated*, 2005.

11. SOUPPEZ, J B R G., 'On the Applications of Modern Naval Architecture Techniques to Historical Crafts', *RINA Historic Ships*, London, UK, 2016.

12. LOSCOMBE, P R., 'Analysing Ancient (Viking) Longship Structure', *International Journal of Maritime Engineering*, vol. 164, no. A2, pp. 207-220, 2022.

13. ROSE, K J., 'The Naval Architecture of Vasa, a 17th-Century Swedish Warship', *Doctoral Thesis, Texas A & M University*, Colle Station, Texas, US, 2014.

14. CANNON, S., BOYD, S., and WHITEWRIGHT, J., 'Development of a Quantitative Method for the Assessment of Historic Ship Performance', *14th International Symposium*, PRADS 2019, Yokohama, Japan, 2021.

15. THOMAS, J. and SOUPPEZ, J B R G., 'Comparative Performance Prediction of Historical Thames A Rater Class Designs', *RINA Historic Ships*, London, UK, 2018.

16. ALESSIO, L., SOUPPEZ, J B R G., and HAGE, A., 'Design Evaluation and Alteration of the Dark Harbor 17.5: Case Study of a Modern Replica', *RINA Historic Ships*, London, UK, 2016.

17. ALESSIO, L G., 'Redesign of a Classic Sailboat with FEA Investigation of the Plate Curvature', *MSc Thesis, University of Liege*, Liege, Belgium, 2017.

18. SOUPPEZ, J B R G., 'Design and Production of a Wooden Thames A Rater Class Sailing Yacht', *MEng Thesis, The University of Auckland*, Auckland, New Zealand, 2015.

19. MARTUS, V., 'Major Refit of a Sailing Replica of the 18th Century Frigate Experience and Traditional Skills Obtained by a Team of Volunteers', *RINA Historic Ships*, London, UK, 2018.

20. CASTRO RUIZ, M., and PEREZ FERNANDEZ, R., 'Galatea II: Reborn of a Classic', *RINA Historic Ships*, London, UK, 2020.

21. WHITE, R G., and PEREIRA, R S., 'Lilian: From Gentleman's Yacht to Scrap and Back', *RINA Historic Ships*, London, UK, 2016.

22. GUELL, A. and SOUPPEZ, J B R G., 'Combining Modern Hydrofoils with Wooden Classic', *British Conference of Undergraduate Research*, Sheffield, UK, 2018.

23. SCEKIC, S., 'Design of a 23 m Modern-Classic Wooden Sailing Yacht with Timber Investigation', *MSc Thesis, University of Liege*, Liege, Belgium, 2018.

24. LINDEN, V. and SOUPPEZ, J B R G., 'Sailing Towards Sustainable Trading with Wooden Cargo Schooner', *British Conference of Undergraduate Research*, Sheffield, UK, 2018.

25. DE BEUKELAER, C., 'Plain Sailing: How Traditional Methods could Deliver Zero-Emission Shipping', *The Conversation,* 2018.

26. ARMANTO, J., 'The Future of Sailing Cargo', *Thesis, University of Turku*, Turku, Finland, 2019.

27. WANG, X. and PEGG, N., 'Proceedings of the 21st International Ship and Offshore Structures Congress VOLUME 2 Specialist Committee Reports', *21st International Ship and Offshore Structures Congress*, Volume 2, Vancouver, Canada, 2022.

28. TRUELOCK, D., LAVROFF, J., PEARSON, D., CZABAN, Z., LUO, H., WANG, F., CATIPOVIC, I., BEGOVIC, E., TAKAOKA, Y., LOUREIRO, C., SONG, C. Y., GARCIA, E., EGOROV, A., SOUPPEZ, J B R G., SENSHARMA, P., and NICHOLS-LEE, R., 'Committee V. 5: Special Vessels', *21st International Ship and Offshore Structures Congress*, Vancouver, Canada, 2022.

29. SOUPPEZ, J B R G., 'Structural Challenges of Low-Emission Vessels: A Review', *International Journal of Maritime Engineering*, vol. 165, no. A2, 2023.

30. MEULEMEESTER, D., 'The Classification of Historic(al) Vessels and their Replicas', *RINA Historic Ships*, London, UK, 2018.

31. SOUPPEZ, J B R G., 'Timber Construction: An Experimental Assessment of the Strength of Scarf Joints and the Effectiveness of Various Adhesives for Laminated Wood', *RINA Historic Ships*, London, UK, 2020.

32. BUCCI, V., CORIGALIANO, P., EPASTO, G., GUGLIELMINO E. and MARINO, A., 'Experimental Investigation on Iroko Wood used in Shipbuilding', *Institution of Mechanical Engineers, Part C, Journal of Mechanical Engineering Science*, vol. 231, no. 1, pp. 128-139, 2017.

33. SOUPPEZ, J B R G., 'Experimental Testing of Scarf Joints and Laminated Timber for Wooden Boatbuilding Applications', *International Journal of Maritime Engineering*, vol. 163, no. A3, 2021.

34. EUROPEAN PARLIAMENT, 'Directive 2013/53/EU on Recreational Craft and Personal Watercraft', *Official Journal of the European Union*, Luxembourg, Luxembourg, 2013.

35. UK GOVERNMENT, 'The Recreational Craft Regulations 2017', *UK Government*, London, UK, 2017.

36. INTERNATIONAL ORGANIZATION FOR STANDARDIZATION, 'ISO 12215-5:2019 Small Craft - Hull Construction and Scantlings - Part 5: Design Pressures for Monohulls, Design Stresses, Scantlings Determination', *International Organization for Standardization*, Geneva, Switzerland, 2019.

37. LOSCOMBE, P R., 'Structural Design Considerations for Laminated Wood Yachts', *International Conference on the Modern Yacht*, Portsmouth, UK, 1998.

38. LOSCOMBE, P R., 'Developing an ISO Scantling Standard for Recreational Craft of Laminated Wood Construction', *The Modern Yacht Conference*, Southampton, UK, 2003.

39. GERMANISCHER LLOYD, 'Rules for Classification and Construction - I Ship Technology - 3 Special Craft - 3 Yachts and boats up to 24 m', *Germanischer Lloyd*, Hamburg, Germany, 2003.

40. AMERICAN BUREAU OF SHIPPING, 'Rules for Building and Classing - Yachts - Part 3 Hull Construction and Equipment', *American Bureau of Shipping*, Spring, Texas, US, 2023.

41. LOSCOMBE, P R., 'Some Observations on the Strength of Viking Ship Rudders', *International Journal of Maritime Engineering*, vol. 165, no. A2, 2023.

42. GERR, D., 'The Elements of Boat Strength: for Builders, Designers, and Owners', *International Marine/McGraw-Hill*, 2000.

43. MARITIME AND COASTGUARD AGENCY, 'MGN 628 (M+F) Construction and outfit standards for fishing vessels of less than 15m length overall', *Maritime and Coastguard Agency*, Southampton, UK, 2020.

44. ROXANE & ROMILLY, '*Cold molded Romilly*,' 23 08 2023. [Online]. *Available: https://www.roxane-romilly.co.uk/romilly-spv/cold-molded-romilly*, 2023.

45. VAN DE STADT DESIGN, 'Glass-Fibre Wood Core', 23 08 2023. [Online]. *Available: https://www.stadtdesign.com/pages/woodcore*, 2023.

46. NEXUS MARINE CORPORATION, 'Building a Custom Wood Fly Fishing Boat', 23 08 2023. [Online]. *Available: https://www.nexusmarine.com/flyfisher_construction.html,* 2023.

47. SMALL BOAT NATION, 'Parker Dinghy', 23 08 2023. [Online]. *Available: https://smallboatsmonthly.com/article-tags/plank-on-frame/,* 2023.

48. SUTTON TIMBER, 'Clinker Boat for Burnham Overy Staithe', 23 08 2023. [Online]. *Available: https://www.suttontimber.co.uk/blog/clinker-boat-burnham-overy-staithe/*, 2023.

49. SOUPPEZ, J B R G., 'Structural Design of High Performance Composite Sailing Yachts under The New BS EN ISO 12215-5', *Journal of Sailing Technology*, vol. 3, no. 1, pp. 1-18, 2018.

50. SOUPPEZ, J B R G., BEGOVIC, E., SENSHARMA, P., WANG F., and ROSÉN, A., 'Comparative Assessment of Rule-Based Design on the Pressures and Resulting Scantlings of High Speed Powercrafts', *HSMV - High Speed Marine Vehicles*, Naples, Italy, 2020.

51. NIKFARJAM, M., YAAKOB, O., SEIF, M S., and KOTO, J., 'Investigation of Wedge Water-Entry Under Symmetric Impact Loads by Experimental Tests', *Latin American Journal of Solids and Structures*, vol. 14, no. 5, pp. 861-873, 2017.

52. ANSYS, 'Ansys Granta EduPack', *Ansys*, Canonsburg, Pennsylvania, US: Ansys, 2023.

53. BRITISH STANDARDS INSTITUTION, 'BS 1088:2018 Marine Plywood. Requirements', *British Standards Institution*, Chiswick, UK, 2018.

54. BLUNDEL, R., 'Little Ships: The Co-evolution of Technological Capabilities and Industrial Dynamics in Competing Innovation Networks', *Industry and Innovation*, vol. 13, no. 3, pp. 313-334, 2006.

55. BARRY, C D., 'CNC Enabled Wood/Metal Composite Construction of (Relatively) High Performance Sailing Yachts', *Journal of Sailing Technology*, vol. 7, no. 1, pp. 152-185, 2022.

56. MIO, A., FERMEGLIA, M. and FAVI, C., 'A Critical Review and Normalization of the Life Cycle Assessment Outcomes in the Naval Sector', *Journal of Cleaner Production*, p. 133476, 2022.

57. TUAN, D D., and WEI, C., 'Cradle-to-Gate Life Cycle Assessment of Ships: A Case Study of Panamax Bulk Carrier', *Institution of Mechanical Engineers, Part M: Journal of Engineering for the Maritime Environment*, vol. 2, pp. 670-683, 2019.

58. POMMIER, R., GRIMAUD, G., PRINÇAUD, M., PERRY N., and SONNEMANN, G., 'Comparative Environmental Life Cycle Assessment of Materials in Wooden Boat Ecodesign', *The International Journal of Life Cycle Assessment*, vol. 21, pp. 265-275, 2016.

59. JACQUET, L., DUIGOU, A L., and KERBRAT, O., 'Holistic Lifecycle Assessments Made in Maritime Industry: A Systematic Review and Proposals for Future Research Directions in Racing Boats Eco-Design', *6th International Conference on Innovation in High Performance Sailing Yachts and Wind-Assisted Ships*, Lorient, France, 2023.

60. HOADLEY, R B., 'Understanding Wood: A Craftsman's Guide to Wood Technology', Newtown, Connecticut, United States: *Taunton Press*, 2000.

61. BARKAS, W W., 'Fibre Saturation Point of Wood', *Nature*, vol. 135, p. 545, 1935.

62. KORKMAZ, O., and BÜYÜKSARI, Ü., 'Effects of Moisture Content on Mechanical Properties of Micro-Size Oak Wood', *BioResources*, vol. 14, no. 4, pp. 7655-7663, 2019.

63. BÁDER, M., and NÉMETH, R., 'Moisture-Dependent Mechanical Properties of Longitudinally Compressed Wood', *European Journal of Wood and Wood Products*, vol. 77, pp. 1009-1019, 2019.

64. FU, W L., GUAN, H Y. and KEI, S., 'Effects of Moisture Content and Grain Direction on the Elastic Properties of Beech Wood Based on Experiment and Finite Element Method', *Forests,* vol. 5, p. 610, 2021.

65. WRIGHT, R S., BOND, B H., and CHEN, Z., 'Steam Bending of Wood; Embellishments to an Ancient Technique', *BioResources*, vol. 8, no. 4, pp. 4793-4796, 2013.

66. STEVENS, W C., and TURNER, N., 'A Method of Improving the Steam Bending Properties of Certain Timbers', *Wood*, vol. 15, no. 3, pp. 79-84, 1950.

67. WANGAARD, F F., 'The Steam Bending of Beech', *Journal of the Forest Products*, vol. 2, pp. 35-41, 1952.

68. RATNASINGAM, J., LATIB, H A., LIAT, L C., and FARROKHPAYAM, S R., 'Comparative Steam Bending Characteristics of some Planted Forest Wood Species in Malaysia', *BioResources*, vol. 17, no. 3, p. 4937, 2022.

69. KULJICH, S., CÁCERES, C B., and HERNÁNDEZ, R E., 'Steam-Bending Properties of Seven Poplar Hybrid Clones', *International Journal of Material Forming*, vol. 8, pp. 67-72, 2015.

70. NIEMIEC, S S., and BROWN, T D., 'Steam Bending Red Alder', *Western Hardwoods*, vol. 12, 1995.

71. YOUNG, W C., BUDYNAS, R G., and SADEGH, A M., 'Roark's Formulas for Stress and Strain', New York, US: *McGraw-Hill Education*, 2012.

72. JONES, T H., 'The Encyclopedia of Wood', New York, US: *Sterling Publishing Co INC International Concepts,* 1989.

73. LOSCOMBE, P R., 'Remarks Related to Scantling Formulae for Wooden Craft', *Personal Conversation*, 2023.

74. VAN DE LINDT, J W. and DAO, T N., 'Performance-Based Wind Engineering for Wood-Frame Buildings', *Journal of Structural Engineering*, vol. 135, no. 2, pp. 169-177, 2009.

75. LINDEN, V., 'Wooden Cargo Schooner for Sustainable Trading Initiatives', *Dissertation, Solent University,* Southampton, UK, 2018.

76. CORRADI, M., BORRI, A., RIGHETTI, L. and SPERANZINI, E. 'Uncertainty Analysis of FRP Reinforced Timber Beams', *Composites Part B: Engineering*, vol. 113, pp. 174-184, 2017.

77. MILLER, P H., and DILLON, D L., 'The International Sailing Canoe: A Technical Review', *Marine Technology and SNAME News*, vol. 31, no. 4, pp. 296-304, 1994.

78. PARKER, R B., 'The New Cold-Molded Boatbuilding: from Lofting to Launching', Brooklin, Maine, US: *WoodenBoat Books,* 2005.

79. LLOYD'S REGISTER OF SHIPPING, 'Rules and Regulations for the Classification of Yachts and Small Craft, Part 2, Hull Construction', *Lloyd's Register of Shipping*, London, UK, 1970.

80. SOUPPEZ, J B R G., 'Ships and Maritime Transportation', in *Springer Handbook of Mechanical Engineering*, Springer, Berlin, Germany, pp. 1139-1164, 2021.

81. LORIMER, T. and ALLEN, T., 'Concurrent Multi-Component Optimization of Stiffened-Plate Yacht Structures', *Journal of Sailing Technology*, vol. 7, no. 1, pp. 203-227, 2022.

82. HARRIS, S P H., 'Structural Design, Sizing and Optimisation for a Foiling Monohull', *Thesis, The University of Auckland*, Auckland, New Zealand, 2020.

83. INTERNATIONAL ORGANIZATION FOR STANDARDIZATION, 'ISO 14130:1998 - Fibre-reinforced Plastic Composites - Determination of Apparent Interlaminar Shear Strength by Short-Beam Method', *International Organization for Standardization*, Geneva, Switzerland, 1998.

84 INTERNATIONAL ORGANIZATION FOR STANDARDIZATION, 'ISO 178:2019 - Plastics - Determination of Flexural Properties', *International Organization for Standardization*, Geneva, Switzerland, 2019.

8. AUTHORS BIOGRAPHY

Jean-Baptiste R. G. Souppez is a Senior Teaching Fellow in Mechanical, Biomedical and Design Engineering at Aston University, and contributes to the Erasmus Mundus Masters in Advanced Design of Ship and Offshore Structures (EMship+) and Sustainable Ships and Shipping 4.0 (Seas 4.0) as a Visiting Professor at the University of Liege and Naples Federico II, respectively. He was awarded a National Teaching Fellowship for outstanding impact on student outcomes and the teaching profession in higher education. He is also the Deputy Editor-in-Chief of the Society of Naval Architects & Marine Engineers (SNAME) Journal of Sailing Technology, the world's leading peer-reviewed journal for sailing research. His work in fluid dynamics features a wide range of experimental techniques, supporting his expertise in hydrofoils and award-winning research on the aerodynamics of yacht sails. As the UK Principal Expert in Small Craft Structures, he represents the interests of the British Marine Industry in the development of international structural regulations (ISO 12215) thanks to his extensive knowledge of composites and structural design and analysis, for which he was awarded a Medal of Distinction by the Royal Institution of Naval Architects (RINA). Recognised as an international leader in yacht design and marine engineering, he sits on several prestigious and influential international committees, and oversees all small craft regulatory developments on behalf of the Institute of Marine Science, Engineering and Technology (IMarEST).

THE EDWIN FOX: THE SECRET SHIP, YOU HAVE NEVER HEARD OF, AT THE OTHER END OF THE WORLD, PRESERVED AGAINST ALL ODDS BY A HANDFUL OF ENTHUSIASTIC TOWN LOCALS…

K McLeod, The Edwin Fox Ship and Museum, New Zealand

Figure 1: Edwin Fox by Gainor Jackson
Courtesy of Edwin Fox Museum

SUMMARY

The Edwin Fox ship, built in India in 1853, (Figure 1) has many stories to tell, from its adventures on the high seas to how it came to rest in a small, amazing little town at the other end of the world from where it began its life.

The Edwin Fox is not well known, yet its story is not only important to Picton, Marlborough, a Provence with a rich maritime heritage from the early Polynesian explorers to Captain Cook in 1770, [6] but to Aotearoa, (New Zealand) as a country, and indeed the world, there is much to be learnt from the ship.

The Edwin Fox is unlike any other historic ship due to the way the preservation of the ship was undertaken. It is one of the few historic ships in the world that people have been able to physically interact with. In this paper I hope to inform the audience of the history of the Edwin Fox with reference to the various RINA conference topics. I will showcase the importance and tenacity of volunteers, whom with passion and very little else, achieved remarkable outcomes while also highlighting some of the negative aspects.

I will also highlight our present plight today with regards the ship preservation project and the need for expertise, particularly in the field of engineering given the complication of the ship being docked in an Earthquake prone area, making this a truly unique opportunity.

1. INTRODUCTION

If not for the tenacity of these men and many others there would be no story to tell today:

Figure 2: Three of the volunteer's instrumental in saving the Edwin Fox Ship, from left, Mike Sextus, Norm Brayshaw (deceased) and Harry Stace (deceased).
Courtesy of The Edwin Fox Museum.

Figure 3: Chris Brown, longest serving volunteer.
Courtesy of Boating NZ and late Roger Kingsbury

Research suggests Edwin Fox is:

- The only surviving wooden New Zealand immigrant ship.
- The only example of a wooden refrigeration ship in New Zealand.
- The last of this design anywhere in the world.
- The last convict ship out of over 1000 that took convicts to Australia;
- The last remaining ship of the Dunbar Shipping fleet;
- The last 'Coolie' ship and
- The last surviving wooden transport ship which served in the Crimean War (Transport 109).

2. THE MANY LIVES OF THE EDWIN FOX

2.1 (a) Construction

The Edwin Fox was constructed of Indian Teak and other Indian hard woods by shipwright William Foster at the Thomas Reeves shipyard, near Kolkata (Calcutta). With signal letters JDMN she is loosely based on a Blackwall Frigate design and when launched had a registered tonnage of 836 tonnes, was 49.4 metres long, and 9 metres wide. Although originally configured as a fully rigged sailing ship later in her career she was converted to a barque.

The ship was built at a time when the East India Company's need for ships was declining and was built by the shipyard owner himself. Certified by Lloyds as A1 12, the ship went up for auction and was purchased in 1854 although quickly re sold to Sir Duncan Dunbar, of Dunbar Shipping in 1854. [1]

2.1 (b) Crimean War

The ship was chartered by the British Government as a troop and supplies carrier in 1854 and renamed temporarily to Transport 109. She carried British and French troops to war and evacuated the injured to different ports for treatment. On one occasion September 9th, 1854, it was noted there was a woman on board the ship tending to injured soldiers, which could have been Florence Nightingale as she wrote about the Edwin Fox. [1]

2.1 (c) Chinese Workers

The shift to end slavery at the beginning of the nineteenth century caused a shift in the global labour supply. In 1833 the British Government passed the *Slavery Abolition Act* and freed over 800,000 African Slaves [2]. This left a labour shortage and in the late 1840s 'Coolie' labour or indentured labour was introduced. Chinese and Indian workers entered voluntary negotiations for work, although fraud and kidnapping were also used. The conditions on the vessels in which these workers travelled to their contracted work were often poor and many suffered and died on the journeys. [3]

The Edwin Fox was used in 1857 for one voyage taking 309 Chinese workers to Cuba to work in the sugar plantations. The ships involvement is this abhorrent trade was mentioned during a British Parliamentary debate on Monday 21 June 1858. The Earl of Carnarvon noted that of the 309 Coolies who boarded the ship 40 died during the voyage, the Earl argued strongly against the trade. [4]

Later, on one occasion in 1871 the Edwin Fox Ship was also tasked with retuning Indian 'Coolie' workers home after their contracts had expired.

By 1874 international pressure ended the Coolie trade.

2.1 (d) Convict Charter

During the eighteenth-century Britian had sent its convicts to the American continent. However, after the American Revolution, convict transportation ceased and in England prisons and prison hulks were overflowing. This caused disease and high mortality rates. In 1768 Botany Bay in Australia was identified as a potential place to send convicts and in 1788 the first fleet of convicts arrived.

On 11th August 1858, the Edwin Fox departed Gravesend, once again chartered by the British Government to take 279 convicts to the Swan River Colony, Western Australia. The voyage was without incident and no deaths reported; However, three babies were born to the guard's families [1]. Western Australia had resisted the importing of convicts, but it was struggling to get up and running as a colony, so the region petitioned the British Government to send them convicts. This was designed to help with skill shortages. When they stopped convict transportation in 1868 around 10,000 convicts had passed through the Fremantle prison. [5]

The Edwin Fox delivered several interesting people while a convict ship, including the famous Great Gold Robbery assailants of 1852, Tester, and Burgess [Website 1] and Leopold Redpath known as the Kings Cross Fraudster who had embezzled colossal sums of money from the Great Northern Railway Company. [Website 2]

The crimes of those transported varied from pickpockets to murder. Being habitually drunk was the downfall of William Weir who received 7 years. Many of the convicts did very well for themselves and did not complete the sentences they were given before being released into the colony to use their skills. One of the most fortuitus convicts was Timothy Kelly, an 18-year-old convicted of double murder who already had luck on his side when he was transported rather than hanged. Timothy only completed five years in prison before being released on a ticket of leave. After a total of 11 years, even with 14 further minor convictions, he was declared a free man. [1]

2.1 (e) Transport Service

The Edwin Fox as with many ships of that time, worked in whatever capacity they were needed. During its career it was often chartered by the British Government, between 1865 and 1867 it was contracted to move soldiers between England and India after the British takeover in 1858. [1]

2.1 (f) Immigrant Ship

One of the themes of nineteenth century history was the movement of people from Europe to 'new world' countries. Although New Zealand had been under British rule since 1840, the small, isolated set of islands found it difficult to attract migrants when faced with competition from larger countries closer to Europe. In 1870 the New Zealand government began an ambitious project to borrow money to fund assisted passage for migrants to New Zealand. From 1873-1876 New Zealand saw some on its highest levels of migration, and the Edwin Fox was part of this movement of people. It was chartered in 1873 to transport immigrants to New Zealand, then again in 1875, 1878 and 1880, bringing over 774 new settlers to New Zealand visiting the ports of Wellington, Nelson and Lyttleton. [1]

2.1 (g) Merchant ship

The ship also traded all over the world, carrying a selection of goods such as, animal skins, rice, safflower used for cooking and rouge for pale cheeks of English maidens, reindeer tongues in jelly, Stilton cheese, Geneva Gin, Hennessy Brandy, Saltpetre (potassium nitrate) used for gunpowder, whale oil, wheat, sugar and of course tea, beer, and IPA to name but a few. [1]

2.1 (h) Pioneering Refrigeration Ship

With the age of steam, sailing ships were made redundant and many ships, became coal hulks. Yet this was not the fate of the Edwin Fox. Instead, she left London in with her own refrigeration equipment and coal ready to be fitted out as a refrigeration ship on her arrival to Dunedin in 1885. These ships were to be used to freeze the carcases after processing. With much of her masts and rigging removed she was towed around various New Zealand ports until eventually ending up in Picton on the main Town wharf in 1897.

In December 1898, the ship held the Picton record for freezing stiffens (nickname for frozen carcasses) with the tally being 12,000 in 43 days. By 1900 the new Picton Freezing works was operating and then the Edwin Fox became a coal hulk for the next sixty years. [1] (Figure 4)

It was the ships arrival into Picton, a small Port Town with important maritime histories that locals first began their relationship with the 'Edwin Fox' ship, good or bad over the next 126 years, many have adored and appreciated the ship but equally it has been controversial, particularly around the saving of the ship.

Figure 4: Picton Freezing Works 1902 (Edwin Fox in the background)
Photo courtesy of Picton Museum

2.1 (i) The Edwin Fox Shipwreck

In the early 1960's 'The Edwin Fox' came to the notice of Norman Brayshaw, a local historian who recognised the importance of the ship, he was integral in the setting up of the first Edwin Fox Society aiming to save the ship and restore it. He purchased it from the refrigeration company for the grand total of a shilling, which arguably was the easiest part of the project. The ship was moved to a nearby bay, but there was limited support for the project and the ship sat there in the mud for the next 20 years.

Contrary to belief the Edwin Fox never sank but filled up with silt and mud and in time weather, tide and locals with chainsaws inflicted much damage. This did not deter a new group of enthusiastic locals to give the project another go, so in 1980s work began again. Their tenacity and passion for the ship did not waiver and in fact some 'expert' advice was intentionally ignored. Many said she would never be refloated, but with the aid of pumps and continual work she floated for 15 years until she was dry docked.

2.1 (j) The Conservation

Earthwatch International from Australia, headed by Paul Hundley and accompanied by Jay Lawry, surveyed the ship in 1987 at the early stages of the preservation project with the view of restoring but due to funding bodies not willing to support restoration but rather preservation, the project eventually conceded to preservation.

In 1999 the dry dock was built and ready for the Edwin Fox. Then in 2002 the next phase of the preservation process began consisting of the Maritime Archaeology Association of New Zealand, (MAANZ) based in Wellington, developing a plan to preserve the timbers led by Jack Fry.

Firstly, a slow reduction of moisture content in the timbers was needed, followed by application of Busan an organic biocide, effective against fungus and a range of wood-destroying insects. Gel was pumped into other timbers, and this was followed by an application of consolidant to further protect the timbers. The iron work was preserved by using a rust inhibitor and more recently a fish oil, white spirit solution has been applied. [1]

A sprinkler system was installed for fire protection and a roof over the dock built, since then little else has been done.

3. HERITAGE OVERSEAS – THE NEW ZEALAND PERSPECTIVE

New Zealand is a relatively new country from a colonial perspective and has had a very different approach and appreciation in preserving heritage items, compared to other countries like the United Kingdom. To date New Zealand has only 3 UNESCO listed sites, although there are many sites that should qualify. [9]

The New Zealand Government oversees the Ministry of Culture and Heritage. The purpose of the ministry is to "Enrich the Lives of all New Zealanders" the belief being the more, New Zealanders value their culture and heritage the more they will take part in producing and consuming culture. The Ministry primarily provides advice on arts, culture, heritage, and broadcasting with consultation to review legislation and policy. [7]

Under the Ministry of Culture and Heritage sits Heritage New Zealand, Pouhere Taonga, funded and supported by the Government, who's role is to "identify, protect and promote New Zealand Heritage with the belief of honouring the past and inspiring future generations while ensuring bicultural history is understood by New Zealanders and all people." Heritage New Zealand has a portfolio of heritage assets it administers itself but also offers support for other heritage assets deemed important and has a listing criterion, The Edwin Fox is listed as Category 1 status, under Heritage New Zealand criteria, the highest status possible meaning it is seen as a place with special or outstanding historical or cultural significance. [8]

New Zealand heritage has some big hurdles:

- Public perception and appreciation of heritage assets and recognising the importance of safeguarding these assets for the future. Many assets are seen to not be 'that old' when compared to countries overseas like the UK. New Zealanders will travel overseas to find worthy heritage items. It has only in recent years that the New Zealand school curriculum is encouraging the teaching of Aotearoa/New Zealand history, which is compulsory from 2023, although there is still work to do it is an exciting time for the heritage sector.

- Underfunding. There are 1064 category one listings and Heritage New Zealand financially look after 40. The balance of which are mostly run by small charitable organisations who are often competing for funding and are run by volunteers with passion but not necessarily expertise.

New Zealand Heritage seems to be largely at the bottom of funding and priority lists; however, many district museums are fully supported by the local district councils. Before July 1st, 2023, the Edwin Fox was run by a Trust, which had some funding from the Marlborough District Council. However, we are now full funded and run by the District Council.

4. HISTORIC SHIPS AS A BASIS FOR EDUCATION

The ship has been the basis for education since the 1960's when Norman Brayshaw (Figure 2) came to its rescue and tried to convince local people that the ship was worth saving. Unfortunately, at this time his enthusiasm was lost on locals and the project essentially was abandoned and not picked up again until the mid-1980s, when the second wave of educating locals to the historic ship began. The ship has been used as an educational tool since its opening to the public in 1986, seeing many school groups coming through as well as the public. According to Museums Aotearoa survey in 2022 65% of visitor's come to museums for learning and 46% for understanding. [10]

Work is now underway to move the collection of artefacts, descendant records, and research online which will open information about the ship to the wider New Zealand and global communities. [Website 3]

4.1 ARCHEOLOGISTS

The ship has been an incredible educational tool, not just for school groups but for modern day Maritime Archaeologists like Dr Kurt Bennet, of Flinders University, South Australia and Sara Gainsford, Director of Red Ox. Kurt believes the ship is the last of its design anywhere in the world and will be the source of continued learning for the maritime archaeology field, as did Dr Ian MacLeod previous Director of Western Australian Maritime Museum who stated, 'her career has hardly begun in terms of contribution of eloquent scientific data.'

Kurt completed his PhD on the Edwin Fox in 2020, his research focussed on the design and construction development of English East Indiaman throughout the eighteenth and nineteenth centuries. He concluded that shipwrights quickly adapted to working with foreign timbers for ship building components and with the gradual acceptance of opening the domestic ship building industry to include foreign skills and ideas, British colonial ships began a new era of innovation, adoption, and refinement. [11]

Kurt also took the learnings of 'the Edwin Fox Museum' to the 'Conserving Cultural Heritage Conference' in Spain 2017, with reference to preserving heritage during natural disasters, namely earthquakes which our region is susceptible too. Canterbury Museum had passed their disaster response learnings on from the 2011 Christchurch quake to other museums including 'the Edwin Fox Museum' in the form of emergency response training. We worked with Kurt to present a disaster response plan at the conference. [11]

4.2 DIGITAL ARTISTS

Rina Ash, a 3D Scan artist based at Auckland University has used the Edwin Fox as the subject of her PhD. The ship served as an example for developing a digital workflow for digitising historical artefacts of such a scale (more on digital technology later). [12] (Figure 5)

Figure 5: 3D Scan images courtesy of Rina Ash

4.3 CONTRACTORS

In more recent times the ship has been a learning experience for local authorities and contractors with the major construction of a new ferry terminal which connects the North and South Islands of New Zealand. This is immediately adjacent to where the ship lies in dry dock.

The project, named IREX, is a truly unique situation. The project is expected to be completed in 2025, and to date there has been great collaboration and communication between the IREX project team, who enlisted the help of WSP (Williams Sale Partnership) archaeologists to monitor the ship for vibrations and dust.

As this project progress's its under constant review as to best practices given the complexity and sensitivities of the ship. [Website 4]

4.4 AUTHORS/ HISTORIANS

The first ever comprehensive book on the Edwin Fox was commissioned and financed by our local Marlborough District Council in 2014, named Teak and Tide which tells the story of the Edwin Fox and its travels and indeed the story to date, including the saving of the Fox [1]. A second book has just been released, 'The Edwin Fox,' how an ordinary sailing ship connected to the world in the age of globalization,' by Boyd Cothran and Adrian Shubert (Figure 6).

"It began as a small, slow, and unadorned sailing vessel – in a word, ordinary. Later, it was a weary workhorse in the age of steam. But the story of The Edwin Fox reveals how an everyday merchant ship drew together a changing world and its people in an extraordinary age of rising empires, sweeping economic transformation, and social change." [13]

Published by York University in Toronto it will educate globally the next generation with the most up to date research of the Edwin Fox story.

In 2015, we were lucky enough to be approached by the renowned 'BBC Coast' series filming the first 'New Zealand Coast' show. This first aired in New Zealand and then went on to be aired around the world on the history channel.

"It is unbelievable. It is the last surviving ship that conveyed soldiers to the Crimean war. It's the last surviving ship that carried convicts to Australia and the last surviving ship that carried emigrants to New Zealand for me it's like an Egyptian Pyramid, Greek temple, or a stone circle. That it is a survivor from another time and it's a monument to people who have made extraordinary journeys and in due course changed the World and yet it sits largely overlooked". Neil Oliver [15]

4.5 DESCENDANTS

An incredibly special part of the ship as an educational tool, is what descendants of original passengers learn and experience when visiting the ship Our data base now holds names of over 2000 people who have travelled on board the ship, many families especially descendants of the New Zealand immigrant passages are aware of their connection to the ship and indeed many visit regularly and consider themselves very privileged to be connected and see and learn first-hand what their great grandparents experienced. In the past, events to bring together those descendants have been organised which has truly been a special occasion for everyone, many descendants share their family stories and photos, and we even have bequests left to the project.

Figure 6: New book on the globalisation era of history
Courtesy of Adrian Shubert and Boyd Cothran

5. WHERE TO FOR THE EDWIN FOX

Although the Edwin Fox ship could be said to have similar risks and threats as other historic ships, there are additional concerns and risks the ship faces.

5.1 ENVIRONMENT

The ship has been in dry dock since 1999, with a roof over the dock installed in 2002, however the ship is still exposed to the elements and local wildlife (birds). In recent times we have had above average wet weather that has influenced the ship, but our region Marlborough is also a very dry climate and fire risks during summer months are high. Effects of global warming and higher sea levels that would breach dry dock wall is also a real threat. It is predicted the site without mitigation will start to see effects by 2050 and potentially be under water by 2130. [Website 8 and 9]

5.2 FUNDING CONSTRAINTS

Lack of funding and use of volunteers has also potentially posed a risk to the ship over the years, but the more recently it has been lacking volunteers and passionate people. The change of governance structure to Marlborough Heritage Trust in 2016 and then the Marlborough District Council from July this year, a contributor.

Going forward with the extensive 'to do' list for the project we will face more funding barriers, given that Marlborough (our local Provence) is one of the smallest regions with around 50,000 in population and New Zealand a country of just over five million.

5.3 VISITORS NUMBERS

Picton is an important port for the New Zealand cruise ship industry with an ever-growing number of ships visiting each year, in the 2023/24 season there are currently 63 ships booked to visit between October and April. [13]. Increasing visitor numbers last season, meant the ship saw over 15,000 visitors. This has grown from around 6000 a year in 2015 and 8312 in 2018.

These numbers have taken many years to achieve on small marketing budgets and much work has been done to achieve Qualmark Tourism accreditations and Tripadvisor awards to highlight this special museum as a quality tourist attraction to aid its financial sustainability. [18, 19, 20]

Although these numbers represent good revenue for the Edwin Fox, the extra visitors have come with a downside. Previously visitors could walk on the ship, entering on the main deck, which is replica, they could descend into the hold where they were walking on the actual fabric of the ship and wear is now visible.

As of September 2023, the difficult decision was made to restrict access to the ship due to the combination of rising visitor numbers, work needed on the supporting structure and museum best practice/ advice from Maritime Archaeologist Sara Gainsford after her condition report on the ship. At present, we are undecided as to whether we will allow people the experience of going inside the ship again. We face a tension between wanting many people to experience the Edwin Fox, whilst preserving the ship.

5.4 EARTHQUAKES

Earthquakes and rising sea levels are the two main and potentially disastrous risks we face, the most recent being the magnitude 7.8, 2016 Kaikoura earthquake [Website 7]. Museum staff have been trained by Canterbury Museum in how to best manage museums and artefacts in a natural disaster but ultimately even with the best planning could be catastrophic with the loss of the collection and the ship.

The Marlborough region is the most at-risk area in New Zealand from Earthquakes and has many fault lines throughout that are connected to the major Alpine fault line, this is considered one of the world's most major geographical features. Historically producing quakes in excess of magnitude 8, it is said that the fault line is 50 years overdue for a rupture. [14 & 15]

It is very difficult to find specialists who can assess the earthquake risk as well as best practice for preservation of the ship.

5.5 SUPPORT STRUCTURE

One of our most urgent risks for the ship is the support structure, the ship was placed in the dry dock in 1999 on wooden tongs and concrete keel blocks, 23 years later these are now deteriorating, and we need to re-evaluate the best way forward regarding the stabilisation of the ship whilst having minimum impact on the fabric and allowing for the issue of pending earthquakes.

5.6 CONSERVATION

We have recently undergone a condition report by archaeologist Sara Gainsford, Director of RedOx who trained in Sweden and was part of the team at Stockholm University that helped excavate the Garrison (Garnisonen). The purpose of the report being to identify and describe changes in the vessel's archaeological fabric and produce a condition assessment. The report summaries and states that 'compared to surveys in 2016 and 2020 minimal changes in archaeological records have been seen. The vessel is continually degrading over time and slowly losing its fabric as seen by falling metal sheathing and tacks. Additional conclusions are that the ship shows components not always preserved in archaeological records on shipwreck sites i.e., masts and continues to have large research potential'. [17]

Although the ship appears to be relatively stable it is of course continuing to deteriorate, and planning is now needed to slow this process and make sure all protections are in place. The next stage is to carry out a full survey of the ship and develop a comprehensive conservation plan. With special consideration to the dry dock and the support structure itself which has a limited life span.

5.7 VOLUNTEERS

Volunteers are the only reason this ship survives today.

Since the initial inception of The Edwin Fox project volunteers have come and gone and contributed greatly, there are a notable few that deserve special mentions.

Norman Brayshaw was the first to see the value in the ship and rewarded with a BEM from the Queen in 1975 for services to Marlborough Heritage, followed by Harry Stace who received a MNZM in 2001 (Figure 2). Mike Sextus and Chris Brown received Civic Honors just this year for their contributions (Figure 3).

In 2014 The World Ship Trust (based in London) honoured all that had worked on the project with an award for the successful preservation project, this was the last award to officially be handed out by the London based trust before folding. The patron of the society Sir James Dunbar-Nasmith spoke on the volunteer effort. "The conservation of the Fox is a superb example of individuals who appreciate the importance of a nautical historic relic in their locality and have done something about it. I know of no project in Britain that is comparable in terms of volunteer effort." [1]

New Zealand has a huge reliance on volunteers throughout many sectors and with a changing society and pressures of everyday life, numbers are stretched, we currently have one volunteer at the museum.

Today we face a raft of challenges some of which are directly connected to the reliance of volunteers and funding constraints over many years.

6. FUTURE

New Zealand is starting to appreciate its heritage, but we do need to continue to move forward keeping the ever-changing expectations of visitors in mind. We need to remember we do not own these objects, but we are merely Kaitiaki (guardians) of them until the next generation takes on the responsibility. Figure 7 shows the initial dream of 'The Edwin Fox Restoration Society' and figure 8, shows the reality of 'The Edwin Fox Preservation Society' The volunteers must be credited with their open minds and problem-solving abilities in regards being able to adapt and move forward with the challenges the project has faced over the years.

The coming years ahead we need to plan for the next stage of the Edwin Fox preservation, which will need to include re-supporting of the vessel, work to the dry dock structure, fully enclosing the ship to the elements, conservation work as needed and with consideration of visitor experience, earthquakes and rising sea levels.

The ship has been laser scanned by Rina Ash and 3 D Scans. There is now a digital model to within 1mm of the ships actual size and shape which ensures it is recorded should disaster occur, but also be used as a tool to monitor changes in the ship fabric. We are also exploring this technology as a replacement for the physical visits on the ship in the form of a VR experience too. A very exciting opportunity of having the digital model of the ship has been the possibility to restore the ship but in a digital format. [12]

Figure 7: The dream, Courtesy of Edwin Fox Museum

Figure 8: The reality (The Edwin Fox Museum) courtesy of Richard Briggs Photography

7. CONCLUSIONS

The Edwin Fox is a story that will forever keep evolving, from its sailing career to refrigeration ship to the preservation project to modern day learnings. It highlights people with a passion and a kiwi 'can do' attitude also known as 'number 8 wire' who although were not experts in the complex field of maritime archaeology saw the historical value in the ship that, just so ended up in Picton, New Zealand. It can be argued they became experts. They achieved remarkable outcomes given budget constraints and intense public scrutiny, without them an important Taonga would be lost.

Forty years on from the initial arduous work re. the inception of the preservation project and 23 years since the museum opened to the public proving itself, as a sustainable quality tourism asset, Marlborough District

Council our local authority has had the confidence to take us on as a rate payer funded project, something the volunteers from the 1960's and 1980's could only dream of.

Hopefully, New Zealand will continue to better understand the importance of preserving such heritage objects like the Edwin Fox, not just for Marlborough or New Zealand but in the case of the ship, The World.

It is now time to revisit this project to ensure its here for future generations.

To conclude, we find ourselves with an extensive list of needs in part, due to the funding constraints over the years and there is a definitive need to enlist expertise in the area of historic ship preservation. Being that New Zealand has limited resources in this area, we hope through telling our story, new connections can be made to plan for the future of this very unique and special historic ship at the other end of the world.

I hope the Edwin Fox story has informed and inspired readers to keep searching for little known treasures such as our wonderful ship and admire the passionate people who have the foresight to get these projects off the ground.

To follow our project please refer to websites five and six.

8. ADDITIONAL INFORMATION

8.1 WHO WAS EDWIN FOX?

For many years there was a mystery around why the ship was named Edwin Fox and who he was, research had led to a short list of three possible answers. It was not until 2018 that we definitively knew who 'our' Edwin was.

The ship was named after Edwin Goodhugh Fox (Figure 9), an auctioneer and surveyor from London, a member of the Royal Artillery Company, a Freemason and early member of the Carlton Conservative Club in London.

He was friends with Thomas Reeves, the shipyard owner that built the Edwin Fox, so much so that Reeves left his children to Edwin should his wife ever remarry in the event of his death. It is unclear if Edwin was a financial partner in the build of the ship. Edwin went on to have four sons, the eldest was also named Edwin.

Edwin Goodhugh Fox died after a short illness at his son Charles's home in Twickenham, little did he know how his name would live on, a ship, a wine, a beer, a wood and reservoir in London all named after him, not to mention several of Edwin's and Edwina's, that were babies born on the ship during voyages.

Figure 9: Edwin Goodhugh Fox
Courtesy of The Honourable Artillery Company

8.2 USEFUL WEB LINKS

1. https://www.btp.police.uk/police-forces/british-transport-police/areas/about-us/about-us/our-history/crime-history/gold-robbery/
2. https://www.oldbaileyonline.org/
3. https://digitalnz.org/records?text=edwin+fox
4. https://www.irex.co.nz/picton
5. https://www.edwinfoxship.nz/
6. https://www.facebook.com/EdwinFoxMaritimeMuseum/
7. https://www.kaikoura.co.nz/stories/kaikoura-earthquake-story/
8. https://www.marlborough.govt.nz/repository/libraries/id:2ifzri1o01cxbymxkvwz/hierarchy/documents/your-council/meetings/2023/environment-planning-2023/5_October_2023-Item_11-Marlborough_District_Sea_Level_Rise_Assessment_NIWA_final.pdf
9. https://experience.arcgis.com/experience/caa461e173ab41d588f78acd8e85ccb4/page/Map/

8.3 VOCABLARY

Kaitiaki: Caretaker, guardian
Taonga: treasure, prized.
Aotearoa: Māori for New Zealand and means land of the long white cloud.
Tiaki: To care, protect, preserve.
Manaakitanga: To show respect, care, generosity.
Kia Ora: Greeting, wishing good health.

8.4 LIABILITY

As a museum we follow the ICMM (International Congress of Maritime Museums) code of ethics and the Museums Aotearoa code of ethics and take guidance from our National Museum in Wellington, Te Papa.

This is a disclaimer stating that the opinions expressed are solely those of the author, however references are made to other experts works.

9. ACKNOWLEDGEMENTS

I would like to acknowledge, Marlborough District Council, Heather Fryer, Liz Ward, Dr Kurt Bennet, Professors, Boyd Cothran and Adrian Shubert, Rina Ash and Sara Gainsford and Kirsty Sykes for their help and continued expertise re. The Edwin Fox Ship.

Port Marlborough, the IREX team and WSP for their continued care, protection, and support during the Ferry Terminal Project.

The Edwin Fox Trust, who have funded me to deliver this presentation at the RINA Historic Ship Conference 2023 at Greenwich Maritime Museum.

10. REFERENCES

1. '[Teak and Tide by Nigel Costley],' [*book*], [2014].

2. '[Britannica]', [*website https://www.britannica.com/topic/Slavery-Abolition-Act*].

3. '[Prologue Magazine, by Robert Plowman vol 33 no 2 2001]', [*website https://www.archives.gov/publications/prologue/2001/summer/coolie-ship-kate-hooper-1*].

4. '[Labourers' Union Chronicle, p.1],' [*document*], [29.11.1873].

5. '[Freemantle Prison]', [*website https://fremantleprison.com.au/history-heritage/history/a-brief-history/*].

6. '[NZ History]', [*website https://nzhistory.govt.nz/culture/encounters/early-meetings*].

7. '[Ministry of Culture and Heritage],' [*website https://mch.govt.nz/what-we-do*].

8. '[Heritage New Zealand],' [*website https://www.heritage.org.nz/about*].

9. '[UNESCO],' [website *https://whc.unesco.org/en/list/&order=year*].

10. '[Museums Aotearoa],' [*website https://museumsaotearoa.org.nz/*].

11. '[Flinders University, papers by Kurt Bennett Thesis 2021 & 05/2017]', [documents via *website https://flinders.academia.edu/KurtBennett*].

12. '[Auckland University, papers by Rina Ash via Wordpress]', [documents via *website, https://katarinamarkovicportfolio.wordpress.com/2018/09/11/edwin-fox-heritage-preservation/ and https://openrepository.aut.ac.nz/items/382e4db4-66a0-48e7-a43a-8e29347e531e/full*].

13. '[The Edwin Fox, by Adrian Shubert and Boyd Cothran on Amazon]', [*website, https://uncpress.org/book/9781469676555/the-edwin-fox/*].

14. '[Marlborough NZ]', [*website,https://www.marlborough.govt.nz/environment/natural-hazards/earthquake*].

15. '[TV Guide NZ],' [*magazine*], [7.4.2016].

16. '[Port Marlborough]', [*website https://www.portmarlborough.co.nz/cruise-ship-schedule/*].

17. '[GNS]', [*website, https://www.gns.cri.nz/our-science/land-and-marine-geoscience/our-plate-boundary/alpine-fault/*].

18. '[Edwin Fox Condition Report by Sara Gainsford and Red Ox],' [*document*], [01/02/2023].

19. '[Tourism New Zealand]', [*website* https://www.tourismnewzealand.com/work-with-us/tiaki/].

20. '[Qualmark],' [*website* https://www.qualmark.co.nz/how-qualmark-works/].

21. '[Tourism Industry Aotearoa],' [*website* https://sustainabletourism.nz/].

11. AUTHORS BIOGRAPHY

Karen McLeod (Figure 10) originally from the UK (born in Dorset and grew up in Cornwall) but a New Zealand resident from 2000, holds the current position of Manager at The Edwin Fox Museum in Picton, and has held this position for the past 11 years. Karen is very passionate about her role and has facilitated several events and milestones for the museum, including a visit from the Governor General of New Zealand, Sir Jerry Mataparae, hosting Neil Oliver from the BBC Coast show and arranging reunions of descendants of the original passengers of the ship.

Karen under her new employers (Marlborough District Council) will continue with her work to educate and promote the Edwin Fox Story along with ensuring the security and condition of this unique artefact is kept as a priority for our future generations.

Figure 10: Karen McLeod, talking with Neil Oliver in 2015, courtesy of Edwin Fox Museum

PRESERVING THE LEGACY OF GREAT SHIPS THAT HAVE SERVED…AND CONTINUE TO SERVE TODAY!

L D Santiago, USS Midway Museum (CV-41), Commander (retired) US Navy, Chief Engineer

SUMMARY

USS Midway, in San Diego, California, was once one of the longest serving aircraft carriers sailing the Atlantic and Pacific oceans from 1945 to 1992 but lives today promoting naval service and unique traditional engineering. Named after the most critical 6-minute maritime battle in the beginning of World War 2 in the Pacific, a combination of pure luck and incredible naval tactical prowess, USS Midway is 78 years "young" today, restored in 2004 as a museum to share the exemplary engineering feat in structural modifications, preservation techniques, and educational inspiration while promoting the tradition of naval heritage. From thousands of aircraft small runway launches and landings, propellor to jet aircraft, as seen in the movie "Top Gun", our 800 volunteers are our new "main engines" as USS Midway serves the new mission: "**Preserve the historic USS Midway and the legacy of those who serve, inspire and educate future generations, and entertain our museum guests.**"

1. INTRODUCTION

The "Bravo Zulus" (naval flag signal for "Job Well Done"), history and credit for the construction of the first aircraft carriers, also known as "flat tops," that will launch fixed winged aircraft, belongs to the United Kingdom with the building of HMS Furious, HMS Argus and HMS Eagle starting 1915 and into World War 1 and World War 2. Battleships reigned over the ship design and construction world pre-World War 2 but control of the skies and long-range detection/combat across the oceans became a requirement with growing technology of bigger guns, farther range needing better accuracy. USS Midway's hull is based on the Montana Class battleship which is why her first real hull designation was a CVB-41 vice CV-41 later. The U.S. never built the Montana Class as the USS Iowa Class was the last. Today, those ships also continue as museum ships featuring and sharing their impressive naval engineering, firepower and armor. (USS Iowa (BB-61) Museum-Long Beach, California, USS New Jersey (BB-62)-Camden, New Jersey, USS Missouri (BB-63)-Pearl Harbor, Hawaii, USS Wisconsin (BB-64)-Norfolk, Virginia)

Aviation development began shifting from the heavier-than-air aircraft developed in the early 20th century with the growing interest in their potential use as scouts for their big gun warship fleets at the time. In 1909 the French inventor Clément Ader published in his book *L'Aviation Militaire* the description of a ship to operate airplanes at sea, with a flat flight deck, an island superstructure, deck elevators and a hangar bay. A US Naval Attaché in Paris, Lieutenant Benjamin Foulois, shared his observations back home: "An airplane-carrying vessel is indispensable. These vessels will be constructed on a plan very different from what is currently used. First of all the deck will be cleared of all obstacles. It will be flat, as wide as possible without jeopardizing the nautical lines of the hull, and it will look like a landing field."

The United States heard and watched across the Pacific at these early flying machines like the "Eole" and the "Avion". So, with the combination of these early efforts in France and United Kingdom, several experimental flights were made to test the concept in the United States. Eugene Ely was the first pilot to launch from a stationary ship in November 1910. He took off from a structure fixed over the forecastle of the US armored cruiser USS Birmingham at Hampton Roads, Virginia. He flew again in 1911 to be the first to land on a stationary ship on the USS Pennsylvania in San Francisco with the first version of sandbags with ropes which was the beginning of the tailhook and arrestor wire innovation.

But the true first global attempt of naval aviation completing the launch and recovery cycle we call today was with the United Kingdom again and Commander Charles Rumney Samson, Royal Navy, with his take off from a moving warship on 9 May 1912 from the HMS Hibernia deserves that "Bravo Zulu." USS Midway calls San Diego home where the United States calls this the "birth of Naval Aviation" when Lieutenant Commander Edward "Butch" O'Hare landed a Curtiss F5L Flying Boat on the USS Langley (CV-1), the first United States aircraft carrier which engineers with pencils and slide rulers, no Iphones or computers, was a converted coal transport ship (USS Jupiter) and recommissioned on March 20, 1922. Right in the bay of San Diego which the bow of USS Midway enjoys the best view of the west coast.

2. CONSTRUCTION AND LARGE MULTIPLE REMODIFICATIONS. SAME HULL. DIFFERENT "ROOFS"

In the middle of World War 2, the United States had the goal to build 690 ships to support the Allied efforts in the Atlantic and the Pacific. The main manufacture and ship building facilities were comprised of mostly the female population that was coined "Rosie the Riveter." It was their hands, dedication and love of country that was key in the industrial machine of the United States as 3 Midway Class ships were planned to be built: USS Midway, USS Leyte Gulf and USS Coral Sea (3 critical battles in the Pacific). Although the first ship, Midway was built in 17 months, with no computers, but instead followed handwritten number 2 pencil drawings (some we have onboard MIDWAY today) as the ship was built and commissioned September 10, 1945, 8 days after the end of World War 2 in the Pacific. USS Leyte Gulf was renamed/christened the USS Franklin Delano Roosevelt (CVB-42) on October 27, 1945. USS Coral Sea (CVB-43) was christened October 1, 1947. Today, we have some of the original crews of these ships, known as the "3 Sisters", who volunteer onboard the MIDWAY every day contributing thousands of hours (some up to 10,000 hours plus) who have been supporting the museum and have their "3 Sister Reunion" onboard Midway. Each ship was originally built with a 3-acre deck comprised of 190,000 pieces of steel. Each of the pieces was put together using a revolutionary arc-welding technique for the day. A floating 20 story building, 3 stories underwater, 1750 compartmented spaces almost 1000 feet long—twice as tall as any building in the city of San Diego--17 months construction with a cost $87 Million US 1945 ($4.3 Billion today). The United States spent 13 years and $13 Billion US dollars to build the latest USS Gerald Ford and Great Britain spent 8 years and £6.2 Billion GBP for HMS Queen Elizabeth. Both ships are now conducting their first deployments today in 2023.

Although Midway missed World War 2, we entered a new chapter in the world called the "Cold War." With the new 3.5-inch-thick steel flight deck, Midway tested captured German V-2 rockets that wreaked havoc during the Battle of Britain, the first attempts we call today Ballistic Missile attacks. The flight deck was active with 200 sailors in different color jerseys inches away from rotating propellers and jet engines. USS Midway was designed to hold 130 fighter aircraft at the time and conduct launches every 45 seconds on 2 catapults and landings which were called "controlled crashes" which we still do today on the larger 10 Nimitz class and 1 Ford Class carrier. 4500 sailors on the Midway made the 12 boilers run to all steam driven equipment to store, make, move, cool and heat:

- 240,000 gallons of fresh water daily.
- 100,000 gallons of fuel daily
- 40,000 pounds of laundry weekly
- 70 tons of food weekly,
- 1500 haircuts weekly.
- 1000 loaves of bread baked daily.
- 3000 pounds of potatoes every meal
- 10,500 cups of coffee in a moment.
- 4752 pounds of clothing washed every 12 hours.
- 6 galleys cooking 4 meals equaling 10 tons daily.

Generators were turning for electricity to see, fix, turns 4 shafts for propulsion at a max speed of 34 knots, lift 170,000 lb aircraft elevators with a load of 100,000 pounds and launching planes night or day. Every sailor is a fire fighter so just in case that disastrous event occurred, which happened multiple times sadly with death and destruction in peace. Midway provided combat support in the Korean War from 1950-1953 as the jet age started coming of age.

Also during this same time smart English naval engineers were working on the first angled deck aircraft carriers in response to these early large jet aircraft which required longer takeoff and landing distances compared to propellor piston aircraft of the time. Bravo Zulu again to the British naval officer and engineer Captain Dennis Campbell as his brilliant design ushered in a design which adjusted the landing area off by 8 to 14 degrees which created a safer landing zone and improved sortie rate, launch and landing aircraft capability into the first true aircraft carrier with this angle into the HMS Ark Royal (R09) which supported the United Kingdom 24 distinguished years from 1955 to 1979.

Midway went into a very extensive "remodeling" from 1955-1957 costing nearly $202 Million US (today $7.3 Billion US, £5.3 Billion GBP). Although very extensive and expensive, which prevented the other 2 sister ships from getting, a new angled deck at 9 degrees off-set from centerline provided simultaneous takeoffs and landings, improving efficiency and safety for the flight deck crew. Midway also received steam catapults, a hurricane bow, upgraded aircraft handling/arresting gear, electrical system upgrades, more armor due to increased weight and stress, deck edge (change from center line) aircraft elevators, communication systems and comprehensive overhaul. It was also redesignated from a CVB (Carrier Battleship) to CVA (Carrier Attack). The average airwing that was approximately 100 before to about 70 to 90 due to the various compositions of larger fighters, attack, reconnaissance, airborne early warning, anti-submarine and helicopters.

Midway went through another extensive modernization from 1966 to 1970 which caused the ship to be "decommissioned" until this extensive work was completed which was the founding blocks for contemporary naval operations today. This extensive "remodeling" the second increased angled deck of 13.5 degrees off-set from center line, upgrades to the island structure, another upgrade to arresting gear, steam catapults and aviation fuel systems. The flight deck grew by widening making it 4 acres (5.28 British Football Fields) to support more aircraft maneuvering. Although technology improved the tools to the design engineers, this last upgrade made USS Midway a slightly unstable platform and one of the most difficult aircraft carriers to land and launch. If you survived a tour/deployment onboard, you were a respected U.S. Naval Aviator. After these upgrades at a cost of $88 Million US (today $2.4 Billion US, £1.75 GBP), USS Midway was recommissioned and redesignated from CVA-41 to CV-41 which we use today. The average air wing size was between 65 to 85 for fighters, attack, reconnaissance, early warning, anti-submarine and utility. Today, as part of the Midway Museum, we have 32 aircraft spanning the vintage years of Midway's service and only place aircraft of the type and series that launched and landed on the USS Midway. Many other aircraft carrier museums have a mixture of organic aircraft, space shuttles, spy aircraft, and foreign aircraft. We believe the Midway museum approach contributes to the authenticity for our guests to only use these assigned aircraft and allow them to come up in person to touch these aircraft and even sit in the cockpits of an F-14 like Maverick and Goose in "Top Gun".

Following multiple flight deck configurations, added topside weight, additional equipment and changing aircraft, USS Midway experienced the infamous "Shake, Rattle N- Roll" of October 18, 1986. While navigating in the rough seas of the South China Sea under adverse weather conditions which included a typhoon, the USS Midway rolled to a scary 24 degrees known as the "Rolling 24". As any good naval architect/engineer knows, one never desires the center of gravity to approach the center of buoyancy but in this case, it was close. An already challenging aircraft carrier to land for naval aviators in jets, "ballast blisters" were added to Midway to improve stability and performance. These were external compartments and structures added to the hull of a ship, low and external, to help weight distribution and stability. The Chief Engineer and aircraft "Handler" who spotted aircraft had to work closely in reconfiguring the high weight called aircraft. Better training and procedures when moving fluids and equipment throughout the ship was critical for the Damage Control Assistant to the Chief Engineer to calculate this daily. Liquid loading on top of aircraft "spotting" was a daily evolution. This brought to light the importance of incline experiments/stability measurements to determine the metacentric height of the ship to be conducted regularly and measured with improved accuracy.

Critical world events before decommissioning and put to sleep for a little while.

- Operation Frost Bite (1946). Testing aircraft carrier operations and flight operations in the Labrador Sea and Davis Strait (Labrador, Newfoundland, Greenland). 19-to-48-degree rolls, ice, snow.
- Korean War (1950-1953) Critical air support and combat operations against North Korea. A war that is still in place and armistice is maintaining peace.
- Operation Mainbrace (1952). First mission to defend NATO in the vicinity of Norway.
- Operation Tachen/Operation Withdrawal (1955). While conducting Midway's first crossing of the "Line"/Equator 1955 for a port shift from Atlantic to the Pacific, Midway responded to assist the Republic of China (Taiwan) of 30,000 nationalists from Tachen islands near mainland China across the Taiwan strait.
- Vietnam War (1965-1973). Critical combat air support throughout the conflict to include the first shoot down of Vietnamese combat aircraft, 11,900 combat missions.

- Operation Frequent Wind (1975). Fall of Saigon, 3000 evacuees by helo onboard, Midway crew pushing the helicopters over the sand to land a small fixed wing aircraft for a South Vietnamese family. That family still visits very year as they celebrate this April anniversary as they say "Midway's flight deck represents Freedom."
- Operation Blue Nose (1978) Testing U.S. Navy operations in Arctic waters of the North Atlantic
- Operation Desert Storm (1991) Flag Ship (the oldest but most senior with a 2-star Admiral embarked onboard) with command and control as Commander Battle Group Five of aircraft carriers USS Ranger (CV-61), USS Theodore Roosevelt (CVN-71), assigned 180 combat aircraft, 3 AEGIS Cruisers, 7 Destroyers, 2 Frigates and 3 Support Ship. Launching aircraft with 7 other US and international coalition aircraft carriers: USS Saratoga (CV-60), USS America (CV-66), USS John F. Kennedy (CV-67), HMS Invincible (R05), HMS Ark Royal (R07), FS Foch (R99), ITS Giuseppe Garibaldi (C551). One of the largest gatherings of international aircraft carriers active in combat post World War 2.
- Operation Fiery Vigil (1992). The evacuation of 33,000 refugees from the Philippines due to the volcanic explosion of Mount Pinatubo. Midway's last operation was a peaceful rescue helping mankind.
- Decommissioned in San Diego, CA April 11, 1992, after 47 years of service. Stricken from the U.S. Naval Register 1997.
- Sleeping until the start of the next Midway battle.

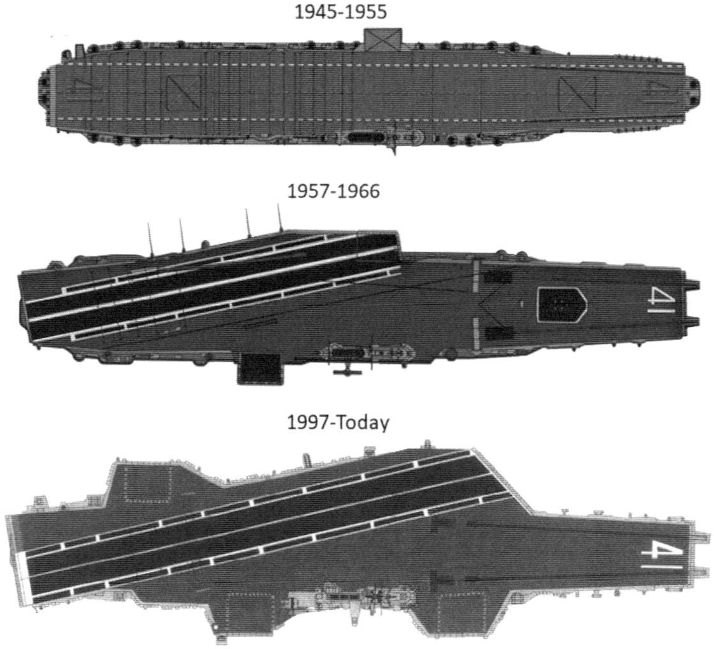

Figure 1: Comparison of USS Midway in its 3 different flight deck arrangements

3. THE SECOND BATTLE OF MIDWAY: REACTIVATION AND OPERATING AS A MUSEUM SHIP

Immediately after decommissioning, a group of local San Diego citizens sought to bring an aircraft carrier as a museum to downtown San Diego. San Diego was the birthplace of naval aviation and is home to 3 other Nimitz class aircraft carriers, each capable holding 90 modern warfare aircraft. Approximately one third of the U.S. Navy Pacific Fleet is homeported just a few miles south of USS Midway. There are multiple major operational commands and bases around USS Midway: Naval Station San Diego for surface ships, Naval Air Station San Diego for fixed wing and rotary aircraft, Naval Base Point Loma with fast attack submarines, Marine Corps Air Station for tactical fighter aircraft F/A-18 and F-35. Why not an aircraft carrier museum in downtown San Diego?

Alan Uke, David Flohr and a group called the San Diego Aircraft Carrier Museum pulled together their own personal financial resources, which included life savings and mortgaging their personal homes, to invest in the cleaning, painting, small staff salaries and electrical bills. A recently retired 2-star Navy Admiral named Rear Admiral Mac Mclaughlin, accepted the difficult job as many groups in San Diego were actively pushing against the USS Midway being downtown as it would be an ugly rusting ship. Surrounding food establishments were worried that the Midway would destroy their businesses. However, no one expected the current success nor the financial success of their own business while sitting next to USS Midway, earning millions annually today. A tow emergency fund of hundreds of thousands of dollars was mandated and is maintained today, 20 years later, in the event USS Midway needs to be towed out of San Diego. So far, so good, no need!

Midway's small staff, small group of volunteers were expected to open a ship June 6, 2004, just a few months after arriving in San Diego from Bremerton, Washington where the ship called home in suspended animation. No active sewage systems, no air conditioning, questionable electrical and limited fire protection only allowed the hangar bay, flight deck and a few spaces on the second and third deck. In addition to just opening as a museum, the ship ventured into the event world and the ship was booking events through 2004 as the event industry found the uniqueness to book 2004 to 2006. USS Midway showed up as the new overwhelming special place to have a special occasion after the museum closed at 5:00 pm and reopened 1 hour later as an entertainment venue.

Mitigation and abatement was a big challenge during the first days and remains a challenge today. Only a few spaces in the early days were open, cleared with old material from the 1940's that was ok then and later learned to be hazardous for people today. This requires the removal of old paint, fuel, lubricants, insulation material, fire hazards. So as newer projects are activated or started, the costs are usually 50% of a new project to be done properly for the long-term use of Midway as a public location.

Figure 2: USS Midway Island Superstructure 2004 vs 2023

Figure 3: USS Midway Flight Deck 2004 vs 2023

Figure 4: USS Midway hangar bay 2004 vs 2023

4. THE 3RD BATTLE OF MIDWAY UNDERWAY: PRESERVE THE USS MIDWAY AGAINST SEAWATER, METAL CORROSION AND TIME.

As the ship was originally 1945 steel rolled, welded and riveted together throughout the ship, bi-metallic and sea corrosion will occur no matter what. The modifications that followed in 1950 and 1970 in addition to the multiple shipyard maintenance periods created new sources of bi-metallic corrosion of various metal compositions 1945 to 1992. Sitting in the salty San Diego Bay is not conducive to the longevity of this precious Icon. From the bottom of the ship sitting 35 feet below the water surface to the top of the mast at 200 feet and 1000 feet bow to stern, preservation is the leading priority of "Preserve the USS Midway and the Legacy of those who have served." In order to support 12 600 lb boilers that created steam to operate steam electric generators, 4 shafts that pushed 64,000 tons of ship at 33 plus knots, launch aircraft and provide heating to cook, clean, wash and drink, all this needed to come through hundreds of sea suction and overboard discharge holes scattered through Midway's hull. During its operational days, this was a must. Now, as a museum ship, this is a curse as these hundreds of holes in the hull were sources of sea water corrosion, the curse of a museum ship. So, in the early years as a fledgling museum ship, USS Midway brought in Phoenix Marine to seal up hundreds of the potential rust cancer locations that would enter the ship.

Figure 5: Phoenix International sealing 320 hull penetrations/sea suction/discharge.

Another requirement for hull requirement would be the traditional drydocking, hull cleaning, hull inspection and hull repainting while "high and dry" out of the water. However, there are no local drydocks near USS Midway that could hold the ship to do this. The closest ones were back at Puget Sound Naval Shipyard where USS Midway was decommissioned from 1992 to 2004, 1100 nautical miles away. This would require tugs as Midway's engineering propulsion are shut down and there are no crew that had the experience to operate Midway's engineering system. So, USS Midway Museum today utilizes, in place without leaving San Diego and keeping the museum open and operating, the expertise and equipment of Seaward Marine Services that utilize a self-propelled, diver driven SCAMP cleaning machine as an effective and economical underwater hull cleaning technique. As Midway's hull is cleaned every summer due to the warm Pacific Ocean and abundant sea growth along the California coast, the divers conduct an underwater annual hull dive and inspection to verify hull holes that were sealed remain intact and examination from bow to stern running gear, propellers and rudders, is conduct. Through instrumentation as well as visual inspection, areas of hull delamination are addressed utilizing the HYCOTE underwater paint formulation which has a unique 2-part epoxy coating which is specifically designed to be applied in wet environments.

The ship's original internal cathodic protection was not able to be reactivated so a commercial system by Anode Solutions was acquired and installed onboard the ship which has been an outstanding layer of defense reversing the direction of corrosion. Annual inspections and measurements are conducted to ensure the electrical current is going in the right way at proper voltage to slow the metal electrolysis. They say one should never own a boat personally due to the expense, however, one should have a friend with a boat if you want to enjoy the world of sailing.

Above the waterline, external hull preservation is an ongoing fight. Once you finish painting an aircraft carrier sized ship, it's time to do it again so it's an ongoing evolution with no end in sight. The flight deck provides sources of large water lakes with crevices that were built in (e.g., catapults, arresting wire spools) and those that were not built in (e.g., flight deck cracks, corrosion holes, repairs on top of repairs) that need to be sealed with various techniques (epoxy to rewelding large areas).

Figure 6: Port catapult tracks before and after closing to prevent water intrusion.

4.1 ENVIRONMENTAL CONTROLS

The museum guest experience is the number one priority from the moment the guests walked onboard to the last exhibit they see. This only included the hangar bay and flight deck in the beginning, but this has now expanded to multiple locations up the Midway Island/superstructure from Primary flight control to the ship's navigational bridge area at the highest points which were critical to safely driving the 60,000-ton warship to launching and recovering fighter aircraft. The "tower" as it is called was critical during Operation Frequent Wind when many helicopters were pushed off the flight deck in support of refugee recovery and landing a South Vietnamese aircraft. Today, Midway has opened 40% of the nearly 2000 spaces of the ship which are a mixture of small 12-inch fans that the ship restoration volunteers put together and maintain in addition to a commercial air conditioning systems that can be controlled remotely by an iPhone. These are a mixture of the 50 plus exhibit locations that tell the story to our tourists along the routes or locations that are "behind the scenes" for our members to enjoy. Some of these locations are active workshops for our 800 volunteers for 14 different departments or office space locations for our staff of approximately 150.

5. "INSPIRE AND EDUCATE OUR FUTURE GENERATIONS, ENTERTAIN OUR GUESTS."

Along with our 50 plus exhibit locations which have been added along with workspaces for both staff and volunteers, we have added 10 classrooms that our education staff have provided STEM (Science, Technology, Engineering and Mathematics) who teach per the California Department of Education standards and training to our 50,000 plus kindergarten to 12th grade students that travel far and near to visit the USS Midway. Landing a jet aircraft on aircraft carrier sea requires the math of the Pythagorean theorem for a right triangle (height=altitude, length=distance to the ship, hypotenuse= approach angle to the landing zone, don't crash). Our instructors teach the basics of electricity utilizing the various inactivated damage control electrical casualty control stations of 3-Phase power and use models of the ship providing lessons in level flight deck balance as the ship is buoyant and weight movements cause moment arms as well as angles that do not support landing fighter aircraft. Even during the Covid shut down and challenges that impacted the world and schools, Midway's education staff developed 2020-2021 a virtual distance-learning program that reached 3,500 students. Since Midway University inception supported by Cox Corporation, 465 classroom sessions have been conducted, 283 live broadcasts visiting 26 states outside California reaching from the far northwest of Alaska to far southeast U.S. of South Carolinas. The STEM training is not done at 4:00 PM but continues sometimes overnight with families and chaperons sleeping onboard like sailors once did. This overnight program has supported over 6000 boys, girls, civic and church youth groups and adult leadership programs that inspire and educate. Summers are vacations for many schools but Midwaycontinues adult education with "Midway Institute for Teachers" to provide professional development resources for middle and high school teachers. Teachers from across the U.S. attend these programs in-person and virtually.

USS Midway Museum averages over 1,000,000 guests every year and since opening in 2004, approaching 19,000,000 total guests. The "main engines" of USS Midway are no longer the 12 boilers creating 600 psi steam, it's our 800 volunteers who contribute 235,000 hours of their personal time to inspire our guests during normal museum operations or during our 350 private events that can range from 100 guests to 4000 onboard at one time. Since Midway has opened, our volunteers have provided out of their own time, unpaid, 3.5 million hours which have contributed to our success as a non-profit and has helped us in the area of preservation by saving those labor costs and re-invest into the hull, electrical bill and cleaning staff.

Figure 7: 4000 guests on the flight deck for July 4th celebration

USS Midway is not only a museum for our guests (approximately 90% who have little or no military experience) but the conversion of the flight deck, hangar bay and many of the ship spaces on and off tour routes support the local military of multiple services, local law enforcement, fire department, scientists and small business researching new technology. Midway looks like a ship, smells like a ship and has the capability of providing a realistic environment of a maritime ship structure without the stringent requirements associated with military sensitivity or bureaucracy. The multiple small engineering spaces provided a good challenge for unmanned vehicles such as the SARCOS snake to investigate capabilities where humans cannot reach in tight places. (https://youtu.be/FX1au5S0z6s). Topside corrosion on a naval ship can be difficult to truly assess while at sea and usually is done when pier side in home port requiring scaffolding. However, allowing scientists to test the capability of unmanned aerial drones has opened new doors and safer approaches (https://youtu.be/nmeWj4h8FmA). Internal tank inspections on aging ships have become a growing issue for both military and commercial ships that are critical to a country's defense as well as commercial shipping. Midway has provided the platform for scientists and commercial companies to test different techniques with drones or improved scanning devices to assess the true condition of tanks for proper preservation. The vastness from the flight deck to the engineering spaces even challenged robotic dogs that learn to maneuver but can carry critical Forward Looking Infrared cameras or Gas Free Engineering testing equipment that can assess quickly and minimize risk to personnel.

Figure 8: Boston Robotics Dog testing capability throughout Midway with camera installed.

Fighting a shipboard fire or conducting military operations on a large ship is very challenging as each ship is a new maze of unknown spaces that are dark and can pose dangers to any individual who does not know the platform. Midway provides that safe controlled environment while providing the classroom spaces to conduct lectures, hands on training, and debriefing while also providing the ship environment while the museum is open with up to 8000 guests who may be onboard for a tour. Because of the vastness and hundreds of compartments, both evolutions can be conducted without any group realizing that they are occurring during their visit. The training also includes hands on mechanical, electrical and auxiliary systems training for local maritime schools preparing adult students for the maritime industry.

Figure 9: San Diego Fire Department training on Midway flight deck rescue operations.

Preserving the USS Midway mission also includes preserving the legacy of those who have served and utilizing the Midway to support the local military and community, investing in our future. Over 350 military events occur on the ship ranging from a simple promotion for a young sailor to remembering the key events of December 7, 1941, 9-11 and partnerships with allied nations such as Australia and New Zealand during ANZUS day. The ships community outreach volunteerism supports local high school students in study projects, vocational opportunities for challenged students. The USS Midway Foundation provides annual grants to organizations that support active-duty military and their families, veterans, and first responder organizations. Over $627,000 US Pillars of Freedom grants have been awarded in 2023, $3.2 Million US total to date that support services for veterans, law enforcement equipment and training, military to civilian life transition assistance and veterans college scholarships. General public education scholarships have surpassed $1,000, 000 US to San Diego schools and community organizations. Over $4,000,000 in "No Child Left Ashore" and "Bravo Zulu" scholarships have been awarded to more than 80,000 students through support from the Midway American Patriot Award Gala held annually on the flight deck. An independent economic analysis by National University has estimated Midway contributes $70,000,000 to the local San Diego, California economy.

Figure 10: Australia/New Zealand/United States (ANZUS) Celebration on the flight deck.

6. CONCLUSION

The USS Midway Museum is approaching its 20th anniversary and although it's 12 boilers powered the largest ship in the world at one time are now quiet as it traveled the world experiencing various chapters of conflict and human struggle, it stands today with a new mission as the #1 of "466 things to Do" in San Diego according to Trip Advisor. As the #5 most popular museum in the United States and the #22 most popular museum of any type in the world according to Trip Advisor, it serves a greater purpose to its visitors, students, local first responders, military, teachers, veterans and the general community of San Diego. Battling the effects of corrosion, sea, age and time, USS Midway's 800 volunteers are the ship's new "main engines" along with the small paid staff delivering the message to over 1 million guests from around the world. **"Preserve the legacy of the USS Midway and those who have served. Inspire and Educate our future generations. Entertain our guests."**

7. REFERENCES

1. HONE, T., "Innovation in Carrier Aviation", *Naval War College Newport Papers*, 2011

2. MCGAUGH, S., "From Mothballs to Magic", 2018

3. MCGAUGH, S., "Midway Magic", 2004

4. MCGAUGH, S., "Midway Memories", 2006

8. AUTHORS BIOGRAPHY

Leonard Santiago, Commander, United States Navy (retired)
USS Midway Museum Chief Engineer, Director of Engineering and Restoration

Leonard "Len" Santiago checked onboard USS MIDWAY April 11, 2017 as the ship's new Director of Engineering and Restoration, Chief Engineer (CHENG). A 23 year Navy veteran, Len graduated and was commissioned through the University of California, Berkeley NROTC program (the first one established by Admiral Chester Nimitz) in 1988 with a major in Biophysics. Qualifying as a Surface Warfare Officer early in his career, he has served as an engineer, operations and combat systems officer onboard USS LAKE CHAMPLAIN (CG-57), USS RUSSELL (DDG-59) and USS PELELIU (LHA-5) with six deployments throughout the Western Pacific and the Middle East with various Navy and Marine Corps forces (Operation Southern Watch, Operation Desert Storm, Operation Iraqi Freedom). In 1991, one deployment included supporting USS MIDWAY and a fleet of ships during one of the largest evacuations of all American personnel from the Philippines due to a major volcano eruption (Operation Fiery Vigil). Ashore, Len was a master training specialist in the AEGIS combat systems and engineer plant teaching the officers and enlisted crews of commissioned and new construction AEGIS cruisers and destroyers entering the Pacific Fleet. In addition, he was a requirements program officer for Commander in Chief Pacific Fleet in Pearl Harbor, Hawaii in the area of subsurface, surface, strike and air to air ordnance procurements, providing 4 star admiral level recommendations for future multi-billion dollar acquisitions in type and quantity while supporting post 9-11 operations defending the western United States and Pacific bases. A 1997 graduate of Naval Postgraduate School, Monterey California with a master's degree in Mechanical Engineering, Len utilized this education with his experience to help in future ship designs of the DDG-51 Arleigh Burke destroyers in the area of underwater shock blast analysis. He later served as the senior engineer and material officer for Commander Carrier Strike Group 15 and 7 while embarked onboard the newly commissioned aircraft carrier USS RONALD REAGAN (CVN-76), ensuring the material readiness and combat support of all ships and aircraft assigned to the RONALD REAGAN Carrier Strike Group during the ship's first deployments (Operation Enduring Freedom). Len completed his final active duty Navy tour as the Assistant Chief of Staff (N4) Material and Logistics at Commander Third Fleet here in San Diego. His primary duties involved the engineering readiness, operational training, logistics requirements and deployment certifications of all west coast aircraft carriers, surface ships, carrier air wings and military sea lift command ships for deployment overseas, homeland defense, Defense Support For Civil Authorities (2007 California Wild Fires) and integrating the multiple of international ships and aircrafts during Pacific Rim (RIMPAC) large scale exercises in 2008 and 2010. Len retired in 2011 as a commander and started a new adventure as an Engineering Program Manager for American Systems establishing the first Engineering Readiness Assistance Team (ERAT) for Commander Naval Surfaces Forces Pacific through 2016, providing senior engineering shipboard training (to junior enlisted to Commanding Officers) on all US Navy steam, diesel and gas turbine surface ships in the areas of main propulsion, damage control, electrical systems, auxiliaries, preventative maintenance programs up to and including USS ZUMWALT(DDG-1000). Len now enjoys the unique privilege and honor of maintaining, modernizing and preserving the USS Midway Museum-60,000 tons of America's living symbol of FREEDOM!

ROBERT SEPPINGS, A VISIONARY STRUCTURALIST

M Corradi and **C Tacchella**, University of Genova, IT

SUMMARY

The prolegomena of graphical statics in shipbuilding are uncertain and can only be deduced from a thorough reading of some authors' writings. Among them, Robert Seppings excelled for his innovative ideas especially concerning the increase of resistance of vessels, in particular with respect to the hogging phenomenon, by using reticular structures. The intent of this article is to tell a short story of the innovations and improvements that have taken place in wooden shipbuilding starting from the ideas of Seppings, regarding naval design engineering as well as the architecture of shipyard buildings. The importance of these ideas is proved by the fact that once entered the shipyard they were in use until the second half of the 19th century when they were finally replaced by the use of iron and steel. For these reasons, Seppings' ideas and drawings are a precious part of the maritime heritage which deserves to be remembered.

1. INTRODUCTION

Robert Seppings (1767-1840) was a naval architect without mathematical education. During his career, he studied and tried to find a solution for the hull deformations. Although not a mathematician, his empirical approach led him to express his intuition through basic graphic method already in 1814, forerunning the ideas of several scholars.

The German structural engineer Karl Culmann (1821-1881) wrote in the introduction to his treatise on graphical statics: «*At the School of Artillery and Engineers in Metz, a subsidiary of the Paris Poly-technical school, Poncelet primarily used graphic methods for the first time. Poncelet's constructions, however, were always just translations of analytic expressions that had been developed earlier. He had first used the funicular polygon, a fundamental aid in graphical statics to determine focal points, at the Metz artillery school, before 1837, when he was still working in Paris*» (translation by the authors) [1]. According to these words, Jean-Victor Poncelet (1788-1867), French engineer and mathematician forefather of the projective geometry, started to use graphic methods around 1825, date in which he became professor at the École d'Application in Metz. That mean that, Robert Seppings had to deal with a dearth of solid background knowledge concerning graphic methods due to a deficiency of studies, which went to zero in shipbuilding area. Previous studies had been anticipated by Pierre Varignon (1654-1722) in *Projet d'une Nouvelle Mécanique* [2] and then in *Nouvelle Mécanique ou Statique* [3], with the introduction of the polygonal of the forces and the funicular polygon, but these ideas had not yet had a diffusion in some areas of construction, including shipbuilding.

The applications of graphical statics suggested, quoting Galileo, «to the studious mind a large field for investigation» [4], but even if Galileo was referring to the ingenious shipwrights of the Venice arsenal, suggestions offered by this discipline stimulated scholars from a completely different field than shipbuilding. Indeed, the development of graphical statics took place at the École Polytechnique in Paris, and it was focused on the teaching dedicated to training civil engineering students, where the new discipline could find a wide range of applications. Civil engineering allowed a more than sufficient approximation in construction practices, as Poncelet recognized [*Cours de mécanique appliquée aux machines*, vol. 1, Paris, Gauthier-Villars, 1874; vol. 2, Paris, Gauthier-Villars, 1876]. In drafting a project drawn on paper, it would never have been possible to obtain an accuracy higher than that offered by the graphic method for determining the static behaviour of the structures from the point of view of resistance and deformability. For what concerns civil engineering, graphic methods have a long tradition [5] starting from the "mysterious rule of the parallelogram" [1985], as Edoardo Benvenuto (1940-1998) mentions in his essay on the parallelogram of forces [6]. Gaspard Monge (1746-1818), who achieved great results in the descriptive geometry, was one of the first to use graphical methods for the verification of retaining walls [7], and in the mid-nineteenth century, Barthélémy Édouard Cousinery (1790-1851) developed the «calcul par le trait» (or graphic calculus) [8]. In 1851, George Finden Warr (c. 1826-1858) was studying the equilibrium of systems of wooden and iron rods in his *Dynamics, Construction of Machinery, Equilibrium of Structures, and the Strength of Materials* [9], anticipating the solutions proposed by Jean-Barthélémy Camille Polonceau (1813-1859), Albert Fink (1827-1897), William Howe (1803-1852), James Warren (1806-1908), Squire Whipple (1804-1888), and many others. Gabriel Lamé (1795-1870) and Émile Clapeyron (1799-1864) clarified some properties of the funicular polygon [10], while the concepts of reciprocity between the polygon of forces and the funicular polygon were studied by James Clerk Maxwell (1831-1879) in 1864 [11] and in 1870-72 [12-13], although already in

1858, William John Macquorn Rankine (1820-1872) had proved a reciprocity theorem for trusses [14-15]. These studies were a reinterpretation in mechanical terms of the ancient tradition of the study of structures based on the reticular beams that traced back even to Andrea Palladio (1508-1580). All these innovative ideas in the design of trusses, linear or arched wooden bridges, and wooden roof structures, the so-called Palladian trusses, continued to evolve in more recent history. Indeed, in the second half of the 19th century, the German engineers Johann Wilhelm Schwedler (1823-1894) and Karl Wilhelm Ritter (1847-1906) were the first to propose again simple and highly effective methods for the analysis and calculation of lattice structures. However, it was in the mid-nineteenth century that Seppings faced and solved the problem of stiffening the skeleton of vessels through his new designed method of diagonal rids. Thus, despite the absence of the graphic tool, Seppings' intuitions allowed him to anticipate the methods later developed by the great scholars of projective geometry, such as Poncelet, Karl Georg Christian von Staudt (1798-1867), Karl Culmann, and Luigi Cremona (1830-1903) [16-17].

2. A LIFE AS INNOVATOR IN DOCKYARDS

Robert Seppings was born in the town of Fakenham, Norfolk, on the 11th of December 1767. His family was of humble origins. His father, Robert Seppings (c. 1730-1781), was a cattle dealer and his mother, Lydia Milligen (1734-1821), gave birth to seven children, of whom only five survived infancy, three girls and two boys. His father's business was not profitable enough to support the entire family, and thus the son Robert Seppings started working at the age of twelve, transporting the post from one city to another. The father died when Seppings was fourteen, and from then on, the mother's brother, John Milligen (1730-1788), took care of the family. Milligen had been a captain in the English Navy at Plymouth, and so in 1782, at the age of fifteen, Robert Seppings was sent to work at Devonport dockyard as master shipwright assistant to John Henslow (1730-1815) [18].

Although he had little mathematical knowledge [19], and his job was to manage the operations for supporting and lifting ships in the dry dock, his inventiveness was not long in showing itself. Indeed, not satisfied with the traditional method that involved lifting ships and then shoring them, the young Seppings designed a new system. He proposed hanging vessels partially rather than lifting them completely. This operation required particular purpose-designed wedges, later called "Seppings Blocks" [20], that had to be placed under the keel when a ship was in dry dock [21]. His new system reduced the time for repairs by two-thirds and required only twenty men, which was decidedly advantageous compared to the five hundred men required by the traditional method. They tested the new system on the first-rate 114-gun *San Josef* (formerly *San José*); the result was so satisfactory that Seppings received £ 1,000 from the Admiralty, and the system was quickly adopted by all British shipyards. Furthermore, Seppings devised a new technique for building masts that required smaller timber [22]. This method allowed making a first-class vessel's mast in which the largest logs measured 40 feet in length, instead of 84 feet. Moreover, Seppings proposed to use a single wood species with the advantage of avoiding decay resulting from the improper mixing of materials [23].

In 1804, he was promoted master shipwright at Chatham Dockyard, North Kent, and in a decade, achieved a prestigious position in English society. Indeed, during his career, he was involved in the design and construction of numerous vessels of different types, including frigates, sloops, corvettes, liners, and even steamships. His commitment to the British Navy focused more on technological research, developing innovative systems for shipbuilding. In particular, its main objective was to make British ships more solid, thus making them more efficient, durable, and above all, resistant during naval battles. In 1813, he was appointed Surveyor of the Navy (1813-1832), and on the 10th of March 1814, he became a member of the Royal Society, presenting himself with his essay *On a new principle of constructing His Majesty's ships of war* (1814). In a decade, he reached the pinnacle of his acknowledgements. In 1818, he received the Copley Medal, a prize awarded by the Royal Society to scientists who excelled in scientific research. Finally, on the 17th of August 1819, he became Appointed Knight Bachelor, and the House of Commons commented that Seppings introduced the "most valuable improvements in naval architecture" of England. In 1832, retiring from his career, Robert Seppings moved with his wife Charlotte to Taunton, Somerset, where he died on the 25th of April 1849 at 72.

2.1 THE ARCHITECTURE OF ROBERT SEPPINGS

When he was Surveyor of the Navy, Seppings faced a thorny problem in shipyards. At the time, they used dry docks for shipbuilding and restoration. However, these dry docks were open, and ships were dangerously subject to bad weather. Ships often remained on dry docks for a long time, some even for years, and often, work needed to be suspended due to bad weather conditions. Thus, ships underwent a deterioration of the structures because of material degradation and deformations due to the action of atmospheric agents, especially sun, water and wind. Seppings proposed to solve also this problem through the use of reticular structures, using them to support the dry docks roofing. In this way, the construction site would continue to work indoors even in adverse weather and climatic conditions. So, this practice, which was already in use in the arsenal of Venice, also spread in the British area starting from 1792.

Seppings's scheme is that of the truss, as can still be seen today in dock no. 3 of Chatham. Built in 1838, it was, at the time, the largest single-span wooden structure in Europe. Inside the space is covered by a three-dimensional reticular system. However, there is no information about any calculations made for its realization. The first developments of graphical statics were limited to simple operations on the forces and to the determination of the funicular polygon as illustrated by Varignon; nevertheless, they were sufficient to define the states of stress and to allow the sizing of the construction elements of the roofs of the Seppings dry docks. It is a complex reticular truss, which recalls the roofs of Gothic cathedrals, formed by a system of tie rods and struts connected to a "Palladian truss". The system could have been calculated only with the tool of graphical statics, then still the subject of study and research. It was in the second half of the nineteenth century that Culmann's theories allowed graphical statics to become an autonomous discipline and subsequently taught in the main European engineering schools.

Therefore, Seppings was an innovator in the field of shipbuilding, but also in the field of roofing of large spans. He managed to create complex structures with rectilinear elements even of modest size and suitably connected in a structural system capable of withstanding its own weight and external actions, but above all to allow the indoor construction of large vessels which, in those years, had reached considerable size. The structural scheme used was a reticular triangular mesh that could be verified through the graphical statics. It was not possible to operate with the analytical calculation as it was still far from the general methods for the structure analysis. These were proposed at the end of the nineteenth century by James Henry Cotterill (1836-1922), member of the Royal Naval College [24], Alberto Castigliano (1847-1884), Francesco Crotti (1839-1896) and finally, they became of common use in structural calculation with the force method by Heinrich F.B. Müller-Breslau (1851-1925).

3. SEPPINGS' SHIPBUILDING INNOVATIONS

3.1 ENGLISH CONTEXT

At the end of the eighteenth century, when Seppings started working at the Devonport dockyard, according to the mathematician and physicist George Atwood (1745-1807), even if English shipbuilding had achieved a high level, there were still many important factors, such as the shape of hulls, ballast, position and size of masts and sails, that had not been studied yet, even if they had a strong influenced on the seaworthiness. In Atwood's opinion, the design of the main parts of ships needed to be based on definite principles and no longer on the know-how of shipbuilders. He also noted that: «*It is frequently remarked by navigators, as well as by naval architects, that alterations apparently the most trivial, in the form of a vessel, in the distribution of the ballast, or in the position and extent of the masts and sails, will wholly change the qualities of a ship from bad to good, or the reverse. As these changes cannot be attributed to fortuitous causes, it is necessary to allow that they are consequences of principles certain and definite, though in many cases unknown, or imperfectly estimated by conjecture. The proportions and disposition of parts, which operate to produce good or bad effects on the sailing of ships, are probably in these instances so intricately combined as to make it scarcely possible from mere observation, however extended and diversified, to account satisfactorily for changes so remarkable: it must also be acknowledged, that some of the data on which the theory of naval architecture, is founded, being imperfectly known, particularly the laws of the different resistances to the ship's motion, it would be unsafe to rely entirely on deductions a priori for explaining this subject*» [25]. And: «*The practice of navigation requires the co-operation of many qualities in vessels, the laws and powers of which, considered as acting either separately or conjointly, it is the employment of theory to investigate. In respect to the construction of ships, it is obvious that no one of the component qualities can be regulated, without paying attention to all the others; because, by increasing or diminishing any of the powers of action, the others are commonly more or less influenced.*» [26]. In this way, Atwood wanted to show the state of the profound malaise of the shipbuilding industry of the state that, for an apparent paradox, was the sovereign of the seas.

Indeed, in the early 19th century, the Royal Navy shipbuilding still followed the practical and proportional rules consolidated for decades that almost ignored mathematics and physics. Thus, shipbuilding was still mainly an empirical process, relying on the experience of shipwrights. However, creating hulls in this way was art more than science. And art died with the individual, while science, if handed down, could survive in the memory of all and bring to a communal knowledge. Nevertheless, a continuous, even though slow, improvement in shipbuilding was pursued independently of the studies of ship theory, thanks to the progressive introduction of ingenious and practical advancement in shipyards. Moreover, the theoretical studies that begun at the end of the 17th century allowed a greater dissemination of knowledge in shipbuilding.

On the other hand, only a few shipbuilders were able to understand the complex mathematical formulas that the design and drawing of vessels required. Therefore, it is not a coincidence that one of the English shipbuilding reference texts remained for more than a century the work of William Sutherland (1668-1740), descendant of a family of shipwrights and carpenters in the service of His Majesty the King of England. Sutherland's work [27] is a practical treatise based on simple analytical and geometric calculations. Instead, the French equivalent in the half of the Eighteenth century, namely the treatise of Henri Louis Duhamel du Monceau (1700-1782), proved to be an excellent synthesis of mathematical premises and practical notions [28]. Indeed, France had become the reference point for European shipbuilding since it applied theoretical sciences to practical techniques. The regard with which England and the Dutch Republic looked at France is evident from their attempts to study French techniques. One of these attempts occurred in 1782-4, when the Dutch, on the occasion of their alliance with France during the Fourth Anglo-Dutch War (1780-1784), contacted the shipwrights Antoine Groignard (1727-1798) and Joseph Marie Blaise Coulomb (1728-1803) to ask for advice to solve some shipbuilding problems [29].

Seppings was aware that shipbuilding needed significant innovations but was equally aware that many shipwrights and people within the government of the Royal Navy, who were "not acquainted with that art" [30], had little attention to possible improvements. Still, the national maritime power depended precisely on its naval superiority and Seppings argued that the vast British empire needed new ships to control and manage the sea trade routes under its domain. In his opinion, the habits and traditions of shipwrights were a cause of backwardness in the search for new solutions that would make ships more functional from the point of view of resistance. Seppings emphasized how important the choice of materials for shipbuilding was and how a particular attention to its consumption and optimal use was necessary to avoid wastefulness. He also highlighted the need to bring technological innovations to vessels to improve their durability, resistance, and also safety of the people on board.

3.2 HULL STIFFNESS

Concerning the preservation of ship stability during the time, Seppings focused his attention on different aspects. He worked on the choice of the wood quality and how to protect the wood from its deterioration due to pathogens and humidity, up to the point of proposing the coverage of the dockyards to safeguard ships during construction [31]. He made studies on the shape of the transom, re-proposing a curved shape consisting of vertical stiffening parts, which increased the rigidity of the structure and allowed it to have superior impact resistance of projectiles, having to meet a receding curved surface, while a flat surface is more easily pierced [32]. Beside the choice of wood, and before designed a different aft shape, he tried to improve the constructive method of ships, implementing the wooden frame. Doing this, Seppings believed that it was essential not to disregard the two fundamental axioms at the basis of shipbuilding:

- "That the strength of a fabric consists not so much in the quantity of the materials of which it is composed, as in the disposition, the connection, and the security of its several parts."
- "That the strength of a ship, let its construction be what it may, can never exceed that of its weakest parts," and consequently "that partial strength produces general weakness" [33].

The traditional frame of wooden vessels was a rigid structure. Every frame-timber, or rib, was formed by a floor timber and futtocks and was rigidly connected to the keel. The frame was reinforced in the longitudinal direction by parts extending in the length, like the keelson and the shelves, and by a double planking outside and inside (ceiling) the frame. Deck beams reinforced the frame transversally and supported the deck planking. But as Seppings knew, that frame had critical issues related to the stiffness and deformability of vessels, starting from the launch operations onwards. Indeed, a ship just entered the water was subject to a deformation phenomenon at the extremities due to a substantial concentration of the weights amidships. That happened because the immersion of the hull was superior in the series of central frames, so the bow and stern were substantially cantilevered. Moreover, over time, the hogging phenomenon increased by the ageing and deformation of the wood, being a problematic aspect of the structure of ships. In the work of 1814 [34], Seppings highlighted how the general frame, as well as the secondary framework of bulwarks, and the connecting panels between gunports presented a structure that seemed bound to rigid ties, such as a hyper-static system. He was aware that because of the complex system of joints and nails between wooden parts, the structure had internal weaknesses that could arise in structural failures, for example, due to local stresses during navigation. That could even produce the ovalization of nail holes and the weakening of interlocking joints, transforming the rigid knots into elastically yielding parts. Seppings tried to overcome this problem by working on the individual frames connecting beams to stringers through specially designed brackets, first in cast iron and then in iron, with various shapes, called knees. These knees gave greater stiffness to frames and ties and guaranteed almost total non-deformability in the main section concerning the transversal stresses. In 1780, William Falconer (1732- c. 1770) had already dealt with this problem [35], and later Gabriel Snodgrass (c. 1759-1796), Surveyor of the East India Company, did so at the end of the century. Moreover, as noted by Fincham in *The History of Naval Architecture* (1851), these techniques were in use as early as 1810, and the use of iron in shipbuilding has much older origins, as Peter Goodwin pointed out [36]. However, the systematic use of these stiffeners is attributed to Seppings [37].

Continuing working on the structural weakness of frames, Seppings developed a system of diagonal rods. Even this system had been already studied by Snodgrass between the end of the eighteenth and the beginning of the nineteenth century (Figure 1). As expressed by Fincham «*Mr. Snodgrass proposed also, the use of diagonal braces, which should extend from the keelson to the gun-deck beams. (Plate 19, fig. 3.) This plan was tried in some of his majesty's ships, and in the Orion of seventy-four guns, in the year 1805. One end of the braces was attached to the keelson - an almost immoveable point, and the other end was attached to the side; and hence, at every alteration in the transverse form, especially when the ship was rolling, a motion took place tending to separate the side from its connection with the beams. As these braces were not found to answer the purpose for which they had been proposed, they were discontinued in ships of the royal navy.*» [38] In 1805, the Navy Board asked Seppings if he agreed to apply Snodgrass's method to the 74-gun *Kent*. Although Seppings was against the proposal, it seems that it was precisely on this ship that he began his experiments using the so-called diagonal rides [39]. Moreover, the architect John Walters (1782-1821) developed Snodgrass's ideas, proposing a similar system of diagonal struts amidships associated with a system of chains that, from the main section, traversed the entire hull longitudinally and in 1814, Walters obtained a patent for his system [40].

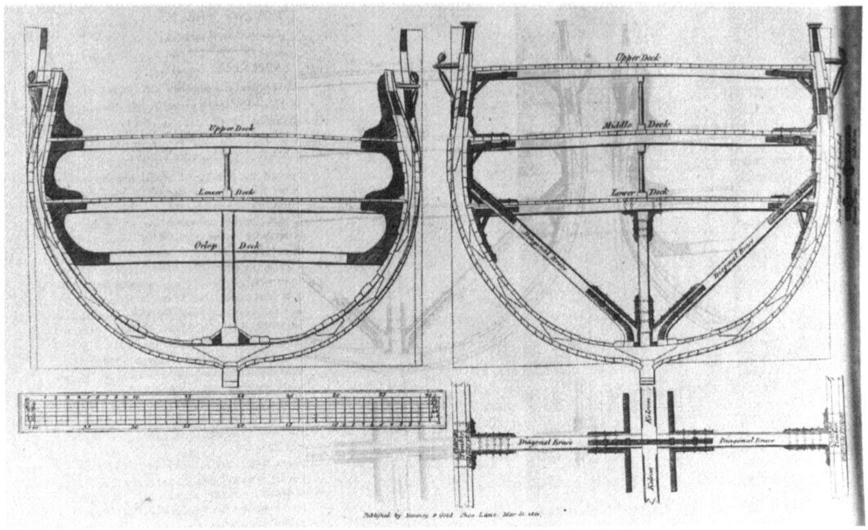

Figure 1: Comparison of midship sections of 38 gun frigates built in the ordinary British method (left), and one built to the method proposed by Mr. Gabriel Snodgrass before the House of Commons, March 1771. Line engraving, published in The Naval Chronicle, Vol. V From January to July. London: published by Bunney and Gold, 1801. Original in Navy Department Library. Catalog #: NH 66228, Naval History and Heritage Command, Washington, DC.

Nevertheless, as pointed out by Seppings, knees and rods worked only in the cross-section and did not confer any longitudinal stiffness. The longitudinal connections behaved like a free system if not suitably braced, as shown later by August Ferdinand Möbius (1790-1868) with his polyhedra in his work of 1837 [41]. Seppings' inspiration to avoid this condition came from the civil engineering. Indeed, as stated by Seppings himself, the developments of the wooden bridges structures, like the Schaffhausen bridge (1755-58) (Figure 2) designed by Ulrich Grubenmann (1709-1783), helped him to understand how a reticulated system with diagonal rods a solution could be. In that bridge, the truss is a composite of struts and tie rods, horizontal and vertical currents that form a tubular structure that discharges the most significant stresses on the piles. With this example in mind, Seppings designed a system consisting studded transverse continuous girders made of longitudinal and vertical transverse elements that formed a frame to reduce the hull deformation due to the weights of superstructures in the stern and the bow and of the whole sailing system that affected particularly the bow. Indeed, the stresses induced at the bow and stern generated a state of tensile deformation on the upper stream and compression on the lower one. Seppings' scheme recalled the intuition of Robert Hooke (1635-1703) on the deformation of elastic bodies and the presence of the so-called neutral line. It was a complex grid since the hull had a variable geometry of arches of circumferences of different radii, such as to configure the characteristic "tulip bulb" shape. Thus, the diagonal rods were not straight but curved. They formed lattices of rods with the function of tie rods or struts which curved in three-dimensional space. Seppings used the same bracing system to stiffen the panels between gunports, which were elements of great weakness, especially in combat since they were subject to the shot of guns that significantly deformed the geometry, creating disruptions in the sides of vessels. Again, Seppings proposed the insertion of the rod lattice similar to the following Möbius-style, anticipating the brilliant solution of the German scholar.

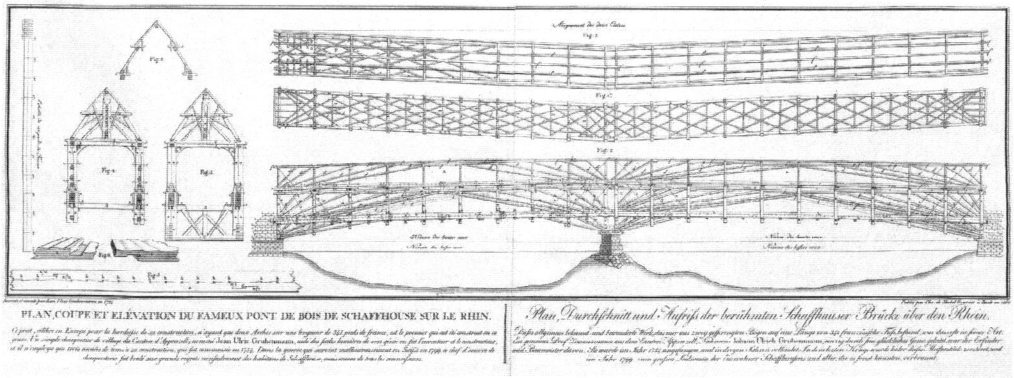

Figure 2: "Plan, Durchschnitt und Aufriss der berühmten Schaffhauser Brücke über den Rhein". Kupferstich nach den Originalplänen Grubenmanns von Christian von Mechel, veröffentlicht in Basel 1803 (Schweizerische Nationalbibliothek).

He also noted that the beams carried only the structure of decks since they were not connected to the internal planking, and consequently, there was no structural continuity between the longitudinal and transverse sections. The weaknesses of the system were evident when a vessel in the water was stressed by the pressures of the fluid on the hull and by waves and wind that generated rolling and pitching phenomena. If the rolling phenomena were sufficiently absorbed by the transverse stiffness of the hull, the pitch phenomenon was affected by the intrinsic weaknesses of the longitudinal structure since the resistance was delegated only to the planking and decks. The pressure on the hull was more amidships and less at the ends due to the arrangement of the loads. That caused bowing phenomena during launch and generated the hogging phenomenon during navigation, both so harmful to ships and added to the natural degradation of the materials and the loss of resistance of the joints. Seppings fully described these problems, but there was not enough knowledge of the mechanical behaviour of wooden structures and joints from a static and kinematic point of view, and therefore, of the possible loss of rigidity of ships and the consequent change of shape. Thus, to express his intuitions, Seppings used a simple example to show how a system formed by two vertical and three horizontal pieces of wood, joined with iron pins to form hinges, was a free system that, if subjected to vertical actions, behaved like a structure statically determinable while, if subjected to horizontal ones, it was completely deformable. The traditional system, configured as a frame with "rigid" rods hinged at the ends, showed many problems of deformability that Seppings solved by adding the "diagonal brace". The diagrams in Figure 3 represents the free system (A and B). He stated that by adding a diagonal rod, called by the naval carpenter brace, as in Figure 4, case Z, the system became sufficiently stiffened. However, the structure would effectively become rigid and symmetrical in behaviour only by adding two, not one, diagonal rods (Figure 3, on the right). It shows that although innovative, Seppings's solutions remain linked to a purely intuitive and constructive moment, unfortunately not supported by static analysis.

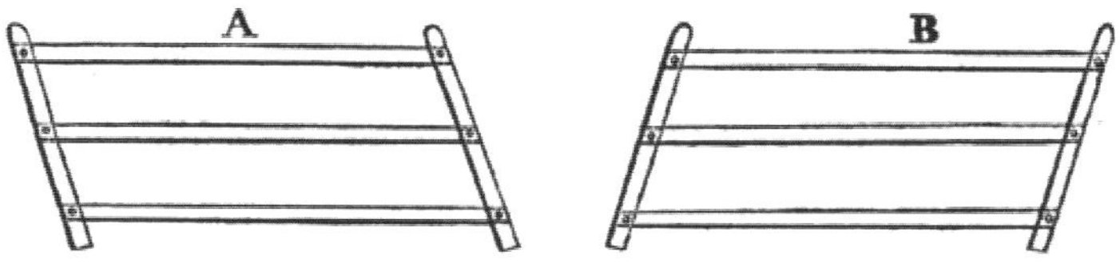

Figure 3: Free systems. In: Seppings, Robert. On a new principle of constructing His Majesty's ships of war. In: *Philosophical Transaction*, Vol. 104 (1814), pp. 285-302, cf. p. 290.

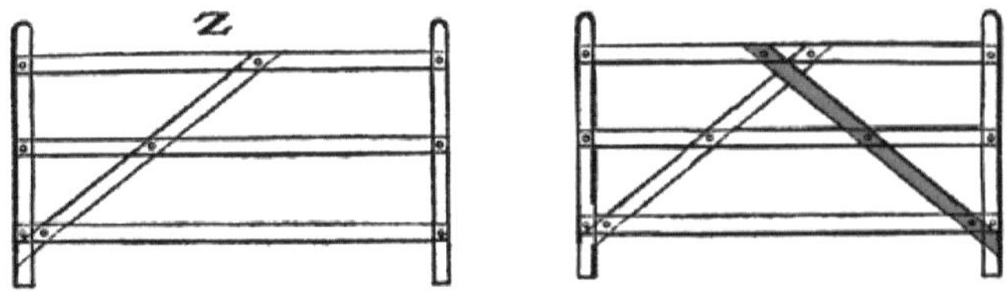

Figure 4: Free system on the left (in: Seppings, Robert. On a new principle of constructing His Majesty's ships of war. In: *Philosophical Transaction*, Vol. 104 (1814), pp. 285-302, cf. p. 290) and isostatic system on the right (revisited by the authors in red).

This is due to the complexity of the structural and geometric problem, the lack of analytical calculation tools and, above all, the not-yet-expressed graphic calculation that would have allowed him to reach the desired solution briefly. Indeed, the complex geometry of hulls required the knowledge of refined calculation algorithms, and the graphical static, in the absence of the analytical one, would not have allowed, even with appropriate geometric simplifications, to translate elements curved in linear structures and consequently arrive at a schematization of the system. However, Seppings's intuitions, expressed through schematic representations, demonstrated his attention to static problems and the goodness of his well-designed solutions is highlighted in the subsequent calculation schemes. Two examples are the schemes that he illustrated in his work of 1814 referred to a plane system of orthogonal rods (Figure 5).

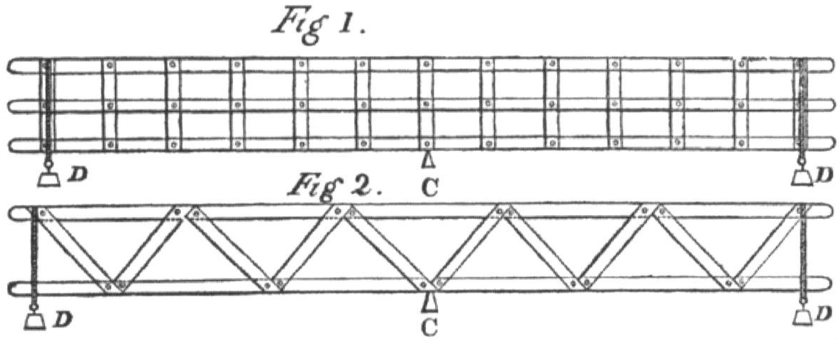

Figure 5: Plane systems of orthogonal rods. In: Seppings, Robert. On a new principle of constructing His Majesty's ships of war. In: Philosophical Transaction, Vol. 104 (1814), pp. 285-302, cf. p. 291.

The upper case (named by Seppings as "Fig. 1") represents the principle on which the traditional shipbuilding system was based, consisting of longitudinal and transverse girders. The lower scheme (named by Seppings as "Fig. 2") is the innovative system proposed by him and consists of two longitudinal girders and diagonal rods. These two schemes are suitable for analysis with numerical calculation, in the first case, and graphical in the second one. The first structural scheme is similar to the Vierendeel beam, with vertical elements "infinitely rigid" concerning the horizontal elements. In this specific case, it would be a double girder if the ties were rigid interlocking ones and not hinged ones, as the image might suggest; otherwise, the structure would be a free system to the horizontal stresses and statically determinable concerning the vertical ones. Considering the complexity of the stress acting on vessels, all joints must necessarily be rigid interlocking ties. In the second case, Seppings's proposal is similar to the Warren beam, a bridge beam with an isostatic structure widely used in the nineteenth century. In the first case, where the vertical elements are infinitely rigid with respect to the horizontal ones, the stresses are of the bending type because the structure behaves like a frame with fixed joints of the Vierendeel truss.

In the second case, the structure is a truss system that can be statically determined. The upper beam is stressed by an action equal to the value of the load acting on the extreme D (suppose D equals P, which substantially represents the ideal weight of stern and bow). The lower beam is equally stressed, but by compression, while the diagonal rods are inclined towards the central support C (fulcrum or point of support as the author says), diagonal rods are subject to the stress of compression equal to $P\sqrt{2}$, and traction stresses of the same magnitude. According to Seppings, the system represents a schematization of the hogging phenomenon. Seppings pointed out that the flexural stiffness of the second case, compared to that of the first, is in the ratio of 6 to 1, and the strength is in the ratio of 3 to 1 due to Warren type truss behavior. Seppings rightly understood that in the scheme that he proposed, the wooden structures were subject to stresses acting in the direction of fibres; differently, in the traditional case, the system was stressed transversely to fibres, creating shear stresses. In this regard, he stated that: «*To prevent any transverse action upon the fibre of the timber, is one of the benefits arising from the new system, and to impede a longitudinal extension of the structure, is another*» [42]. He imagined applying the same solution to stiffen the frames between gunports, in which he thought that the use of a frame stiffened by diagonal rods (arranged in a St. Andrew cross) significantly increased the resistance of the topside and that this gave it greater stiffness. In this way, the girders, the planking, and all the construction parts of the topside framework became a rigid system more resistant.

In the reinforcement system through diagonal rods (Figure 6), the stiffness of the ship increased significantly. In addition, wooden strips filled the space between the external and internal planking, almost at the height of the orlop. It significantly increased the mass of the hull structure, and in the case of impact, if the hull was well caulked, the damages were limited. Differently, using the traditional system, the thin planking could be seriously damaged and even lead to the sinking of the ship. Seppings suggested filling the small empty spaces of about three inches between the planking and the diagonal frame with wooden wedges, while in the larger areas, he proposed to use specially cut pieces of wood. These were obtained from the waste of shipyards and were in such quantity as to be sufficient, in his opinion, «*for the consumption of the whole navy*» [43]. Among the advantages of this idea, one of the most important was to increase the strength and durability of the wooden structures, less locally stressed by compound pressure-tension-flexion effort. By distributing the structural weights in larger quantities in the hull, Seppings' truss solution also made the ship heavier, requiring less ballast.

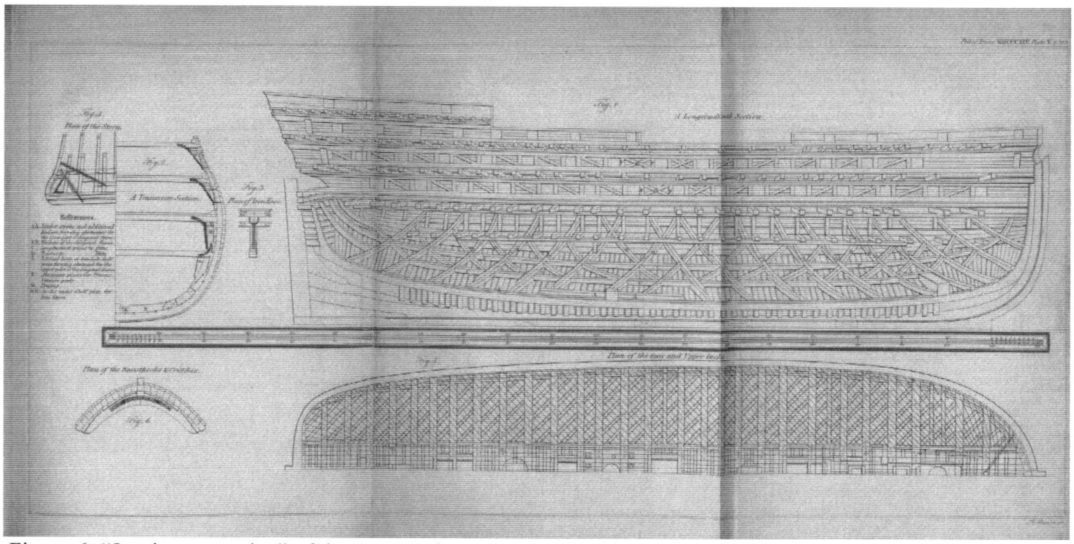

Figure 6: "Jesuit perspective" of the structural system proposed by Seppings (1814). In: Seppings, Robert. On a new principle of constructing His Majesty's ships of war. In: Philosophical Transaction, Vol. 104 (1814), plate X, p. 302.

The diagonal beams intersected the main frameworks at 45 degrees and were symmetrically disposed to the ship's longitudinal centre so that the diagonal elements were always taut. In Seppings' work, there is a detailed description of the complex arrangement system of the diagonal elements concerning the ship skeleton. This description highlights the careful attention to the construction elements to ensure that they behave structurally with the best static endeavour.

Interestingly, the modest-sized elements in the system of diagonal beams caused some parts to behave like wedges that strengthen the structures. That made the whole a complex mechanical system for which «*The truss pieces so introduced into the rhomboid, are to the diagonal frame what the keystone is to the arch*» [44]. Indeed, he imagined that the diagonal frame behaved just like an arched structure that opposes the stresses, especially the hydrostatic pressures on the hull. This lattice system required particular attention to the structural nodes through the use of coaks (such as tenons or dowels) and nails that fastened together the different parts to avoid damaging the structural continuity of the whole, as happened to the "heterogeneous and badly connected mass of materials" [45] of traditional ships. Already in the 17th century, a method had been developed to connect the individual beams that formed the transverse framework, and it became the norm throughout the 18th and until the second half of the 19th century. In his Royal Society article, he criticized the practice of bolting only alternating pairs of beams: «*In forming the frames or ribs, half the timbers only are united so as to constitute any part of an arch;*

every alternate couple only being connected together: the intermediate two timbers (termed fillings) being unconnected with each other, and merely resting on the outer planking, instead of giving support to it… This loose practice is, I believe, peculiar to the English merchant shipbuilder; and indeed was pursued till very lately even in His Majesty's Navy, while the preferable system of connecting the ribs was common to other maritime powers» [46].

Instead, Robert Seppings proposed in 1818 to connect futtocks with circular coaks. The system used shorter futtocks that reduced the required curvature in each piece and allowed them to be connected over the entire thickness [47]. Seppings was aware that «*when a body is constructed with such general unity and fixedness of all its parts, as that if one is moved the whole must move with it, then it may be said that all the parts of the structure bear their portion of the strain*» [48]. Indeed, a continuous perimeter beam shelf to support the deck beams appeared as a flat stiffening element that contoured the hull and increased its stiffness in the longitudinal direction. It also connected the diagonal elements and formed an integrated truss system. These shelves were also joined to the transverse skeleton by a system of wedges (placed under shelf beams) that connected the iron knees with ribs and beams. It effectively increased the lateral stiffness of the ship, making a better distribution of stress caused by the rolling phenomena or induced by the actions of the wind on the mast.

In Figure 6, the horizontal plane (named "Fig.5" of his plate X) shows that the deck beams were connected by short ledges. These were inclined at 45 degrees o the beams and, in the opposite direction, to the longitudinal ship axis. Also the deck planking was inclined at 45 degrees, in the opposite direction to the ledges, configuring a cross-mesh reticular system that guaranteed to the deck greater resistance to horizontal actions. In correspondence with the ship central axis, there was a longitudinal planking jointed to the diagonal planking on the sides. All parts were connected mainly through orthogonal crossing, combining "the whole in one homogeneous mass of strength" [49].

A further advantage of the solution proposed by Seppings was that it allowed the use of wood recovered from the dismantling of old ships. Indeed, that timber could be used profitably for the diagonal elements. According to Seppings, this could reduce the use of new oak wood required for building a ship by at least one-seventh. Furthermore, Seppings pointed out that all these stiffening parts could also be made of spruce lumber to replace oak with a significant cost saving.

In 1818, Seppings improved his system by proposing a Warren-type three-dimensional beam in two overlapping orders, integrating it with the Warren-type configuration and removing the vertical struts, as shown in Figure 7. In this way, he simplified the static scheme and improved the mechanical behaviour.

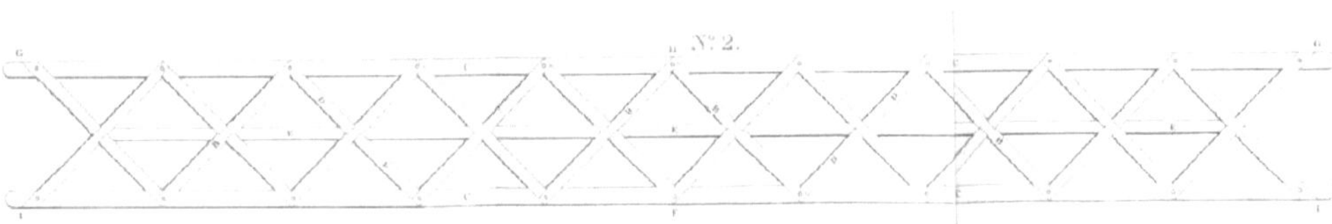

Figure 7: Seppings' truss similar to the 'double Warren girder'. In: Seppings, R. On a new principle of constructing ships in the mercantile. In Supplement to the Edinburgh Encyclopedia, Blumer and Nicol: London, 1820; Volume 18, pp. 133-143; Plate I, p. 8.

In the first quarter of the 19th century, in light of these considerations, in the building of three 74-gun vessels at Chatham some ideas proposed by Seppings were applied. Their construction received a favourable opinion from the Lords of the Admiralty, who hoped that other ships would be built following the same principles. In these vessels, the transverse skeleton was made up of more than eight hundred beams of different shapes and sizes arranged orthogonally on the keel. This "wooden skeleton" – to paraphrase Jacques Heyman (1925-Unk) with reference to architecture – was then covered with double external-internal planking, the latter stiffened by thicker elements, called *thick stuff at the futtock heads*, and *at the floor heads* and *limber stake*, which also had a longitudinal stiffening function. The transverse skeleton was also reinforced by beams connected to the ribs through wooden angle knees, later replaced by iron ones. Thus, the frame was formed by straight and, above all, curved beams connected with interlocking joints capable of giving rigidity to the whole ship. The transverse beams gave longitudinal stiffness to the hull structure and supported the weight of guns, which significantly affected the transverse stability, especially when used. The wooden and iron knees bolted to each beam gave greater strength to the single frame.

4. CONCLUSIONS

In the 1820s, the English shipbuilding industry was still in critical condition concerning the advancement and scientific research. In 1821, John Knowles (1781-1841), chief clerk of the Navy Office between 1806 and 1832, at the time Surveyor of the Navy, and a great friend of Robert Seppings, pointed out this situation. In his work *Inquiry into the Means which have been taken to Preserve the British Navy* he said: «*In the proportion that theoretic construction has been neglected, the practical art of building has been encouraged; and hence it is, that the workmanship in British ships has been equal, if not superior, to that of most of those built in any part of Europe. But, for the want of principles, little of alteration or improvement was for many years introduced in the disposition of the materials, or connection of the several parts, so as to increase the strength of the vessels*» [50]. He also stressed that Seppings was the one who advocated the introduction of a new scientific-mathematical method in ship design, no longer focusing attention on the merely practical aspects. Indeed, he wrote: «*Within the last ten years, a new system has been invented and introduced by Sir Robert Seppings, which has given to our ships of war, great additional strength to resist the effects of the waves, is likely to be conducive to their durability, and to ensure the safety of the mariners*» [51]. According to Knowles, although the innovative solutions proposed by Seppings proved to be of undoubted advantage in the construction of ships, they were not received with right attention by English shipwrights and the British Admiralty. Knowles blamed the fragility of human nature for this reluctance to apply new methods, arguing that: «*From the jealousy incident to human nature, in properly appreciating and applying the inventions of others, or the indolence of the mind in not bringing itself to examine new methods or combinations, these improvements, while they have been eagerly grasped by foreign nations, are but slowly introduced in the ships of our merchants, and, with an apathy hardly to be credited, are totally neglected by the first trading company in Europe* [Cf. East India Company (founded in 1600), authors' note]. *Although ignorance of their application, and bigotry to old customs, may for a time prevent their introduction, the evidence of their advantages must in the end prevail*» [52]. However, the British Navy's approach did not seem to change over the years.

Ten years later, the critical section in Knowles's text was reported by James Ballingall (active from 1832 to 1858) in the introduction to his essay *The Mercantile Navy Improved* [53] on the problems relating to British merchant vessels. Ballingall not only made a staunch reference to Knowles' comment, but like him, he looked again at the innovations that Seppings had introduced twenty years earlier. In his text, Ballingall did not just quote Seppings, but even chose to report an entire passage from his 1820 publication. From that text appears that one of the most serious and worrying problems of the construction of merchant ships concerned "the putting together of their ribs or frames, and the general arrangement of their materials" [54-55] and it is for this reason that Seppings, and therefore Ballingall, proposed to use the diagonal beam method, already introduced on warships thanks to the previous work of Seppings. The great value that Ballingall attributed to Seppings can be summarized in his words: «*The principles here laid down by Sir Robert Seppings, are of vast importance, and we earnestly hope will speedily and generally be adopted*» [56]. In the following pages, Ballingall referred several times to the experience of Seppings to legitimize his claims, and, in the case of the diagonal supports, he wrote that «*It would also unquestionably strengthen the vessel very much, to give her a few diagonal iron riders and braces, as recommended by Sir Robert Seppings, ...*» [57]. In 1833, twenty years after his first writing was published at the Royal Society, *On a new principle of constructing His Majesty's ships of war*, Robert Seppings was still recognized as the one who "introduced into the Royal Navy various improvements in ship building, which are universally allowed to have imparted great additional strength, safety and durability" on English warships [58].

Another twenty years passed, and Seppings was still a reference point in English shipbuilding for developing new theories with a scientific approach. Again, in 1854, Ballingall reported a series of Seppings testimonies in support of his ideas [59]. The problem, in this case, concerned the need to reinforce the bottom structure of ships to prevent serious damage resulting from accidents such as hitting the hull against rocks or shoals. Indeed, the hulls weakness of the bottom structures in traditional construction was the cause of many shipwrecks. Instead, adequate reinforcement allowed the damaged ships to resume navigation and safely reach the port of destination, ensuring the safety of the crew and the transport of goods even with the structures of the bottom severely damaged. In supporting this thesis, Ballingall reported some cases of accidents that would have been avoided thanks to the new construction method, recalling how Seppings, in a speech presented to the Committee of the House of Commons in September 1831, illustrated as an example the shipwreck of the steamboat *Rothsay Castle*. Ballingall added at the end of the quote: «*I may here state that I had it from Sir. R. Seppings's own mouth, that it was principally to protect the vessels against shipwreck that the improvement was introduced*» [60].

That of Seppings is a fascinating story in which the ingenuity of the shipwright, the vision of the structural engineer, the awareness of the builder, and the "practice of manufacturing" emerge, bringing the art of construction into the science of building. It happened in a century when science was abandoning empiricism, typical of the scientific Renaissance, to become an exact science. Mathematics will ultimately be the tool par excellence for developing knowledge in the structural field and, in particular, in the naval one. Graphical statics became fundamental for structural analysis, starting from the studies of Rankin, Maxwell, Culmann, Cremona, and others, at the end of the 19th century, but it only made a few appearances in shipyards. At that time, the treaties concerning shipbuilding, whether in wood or iron, were more influenced by analytical than graphic calculation. They were necessarily limited to structural considerations of an extremely general nature or single calculations, due to the complexity of the structure of a ship, which cannot be separated from a three-dimensional view. Therefore, it must be limited to local analyses that took into account the structural behaviour of individual elements. However, as aforementioned, the first elements of graphic calculation were sufficient to deal with the design and sizing of the ship's construction parts. For this reason, Seppings can be considered a forerunner, as he sensed the potential of static schemes when the effectiveness of static behaviour was not yet possible to determine in an analytical and complete way. Even if from an empirical point of view, he was aware of the goal he was looking for. Seppings' anticipations are certainly consistent with the architectural developments in wood and iron constructions that were taking place in those years, especially in the construction of bridges and, at the end of the century, with the universal exhibitions, in the structures of large spans [61].

In the naval field, Seppings proved to be a visionary structuralist. He understood the problem, identified the solution, and provided the operational tool for implementing his innovative precepts, which gave a breakthrough to wooden shipbuilding and had been widely used in the Royal Navy for over fifty years. Nonetheless, in 1833, one year after Sepping's retirement, a text was published [62] in which, even though the qualities of Seppings as a shipbuilder were recognized, within the Admiralty, the First Lord pointed out that as Surveyor of the Navy, it was better to choose a naval officer. Captain William Symonds (1782-1856) took the place of Seppings, and criticized his predecessor for his lack of knowledge from the point of view of calculation in the design of ships, remarking curiously how the projects of Seppings had caused damage to British shipbuilding. That was a curious controversy that certainly deserves an in-depth study. It recalls the dispute between builders and "number crunchers" that became increasingly evident in the nineteenth century thanks to the development of construction techniques and the entry of new materials, especially iron and steel, into shipbuilding. As is well known, the nineteenth century is also the century in which science faced a new chapter in its history and, as has been said, the mechanics of solids and the mechanics of structures found total fulfilment, entering, like a cyclone, in the project and construction of new ships.

5. ACKNOWLEDGEMENTS

This research was financed by the Department of Architecture and Design, with funds of the University of Genoa for scientific research, years 2019-2020.

6. REFERENCES

1. Original text: «An der auf die Pariser polytechnische Schule als Fachschule folgenden Artillerie- und Genieschule in Metz, hat Poncelet zuerst die graphischen Methoden mehr als sonst üblich zur Anwendung gebracht. Die Constructionen Poncelets waren jedoch immer nur Uebersetzungen vorher entwickelter analytischer Ausdrücke. Er hat zuerst, zur Ermittlung von Schwerpunkten das Seilpolygon, dieses fundamentale Hülfsmittel der graphischen Statik, angewendet, und zwar noch an der Artillerieschule in Metz, also vor 1837, in welchem Jahre er bereits schon in Paris wirkte.» CULMANN, K., 'Die graphische Statik', 2nd ed. (1875), *Zürich: Meyer & Zeller*, 1866; p. VIII.

2. VARIGNON, P., 'Projet d'une Nouvelle Mécanique', *Paris: Vve Edme Martin, Jran Boudot et Estienne Martin*, 1687.

3. VARIGNON, P., 'Nouvelle Mécanique ou Statique', *Paris: Chez Claude Jombert*, 1725.

4. GALILEI, G., 'Dialogues Concerning Two New Sciences', (Translated from the Italian and Latin into English by Henry Crew and Alfonso de Salvio) *New York: Macmillan, 1914, p. 1. Original version*: 'Discorsi e dimostrazioni matematiche intorno à due nuove scienza', *Leida: Appresso gli Elsevirii*, 1638.

5. POSPISIL, M., 'Les trios racines de la statique graphique, e-Phaïstos', *Revue d'historiques / Journal of the history technology*, 2021, IX-1, pp. 1-21.

6. BENVENUTO, E., 'The parallelogram of forces', *Meccanica*, 1985, 20-2, pp. 99-109.

7. MONGE, G., 'Mémoire sur la theorie des deblais et des remblais', *Histoire de l'Académie Royale des Sciences. Année M.DCC LXXXI. Avec les Mémoires de Mathématique et de Physique pour la même Année*, Paris: De l'Imprimerie Royale, 1784; pp. 666-704.

8. COUSINERY, B É., 'Le Calcul par le trait, ses elements et ses applications', *Paris: Carilian-Goeury et Darlmont*, 1893.

9. FINDEN, W G., 'Dynamics, Construction of Machinery, Equilibrium of Structures, and the Strength of Materials', *London: Robert Baldwin*, 1851.

10. BENARD, D., 'Gabriel Lamé et le polygone funiculaire', *Bulletin de la Sabix*, 2009, n. 44, pp.113-118.

11. MAXWELL, J C., 'On reciprocal figures and diagrams of forces', The *London, Edinburgh, and Dublin Philosophical Magazine and Journal of Science*, 1864, 27, pp. 250-261; MAXWELL, J.C., 'On the calculation of the equilibrium and stiffness of frames', *Philosophical Magazine*, 1864, 27, pp. 294-299.

12. MAXWELL, J C., 'On Reciprocal Figures, Frames, and Diagrams of Forces', *Transactions of the Royal Society of Edinburgh*, 1870, 26, pp. 1-40.

13. MAXWELL, J C., 'On Reciprocal Figures, Frames, and Diagrams of Forces', *Proceedings of the Royal Society of Edinburgh*, 1872, 7, N. 81, pp. 53-56.

14. RANKINE, W J M., 'Principle of the equilibrium of polyhedral frames', *The London, Edinburgh, and Dublin Philosophical Magazine and Journal of Science*, XVII, Series 4, N. 180, 1864, p. 92.

15. CAPECCHI, D., RUTA, G., 'La statica grafica in Italia', *La scienza delle costruzioni in Italia nell'Ottocento*, Springer, Milano, 2011, pp. 239-286.

16. BENVENUTO, E., 'La statica grafica', *La scienza delle costruzioni e il suo sviluppo storico*, Firenze: Sansoni, 1981, pp. 748-776.

17. For an in-depth study of the topics just mentioned Cf. CHATZIS, K., 'La réception de la statique graphique en France durant le dernier tiers du XIXe siècle', *Revue d'histoire des mathématiques*, 2004, n. 10, pp. 7-43.

18. LEE, S., Ed., *Dictionary of National Biography*, Vol. LI., London: Smith, Elder & Co., 1897, pp. 249-251.

19. Idem.

20. SEPPINGS, R., 'Method of obviating the necessity of lifting ships', *The Philosophical Magazine*, 1805, 22, n. 87, pp. 242-247.

21. CORRADI, M., 'Sir Robert Seppings (1767–1840): l'invenzione nella tecnica e l'arte della costruzione navale', *History of Engineering / Storia dell'Ingegneria*, D'Agostino, S., Romana, F., d'Ambrosio, A. (Eds), Napoli: Cuzzolin, 2020, vol. 1, pp. 203-218.

22 *Metropolitan Encyclopaedia*, Vol. VI; p. 378.

23. SMEDLEY, E., ROSE, H J., et alii, Eds. 'Encyclopaedia Metropolitana or Universal Dictionary Of Knowledge', Vol. 6*, London: B. Fellowes*, 1845, p. 378.

24. COTTERILL, J H., 'Obituary', *Nature*, 1922, 109, pp. 115-116.

25. ATWOOD, G., 'The Construction and Analysis of geometrical Propositions, determining the Positions assumed by homogeneal Bodies which float freely, and at rest, on a Fluid's Surface; also determining the Stability of Ships, and of other floating Bodies', *Philosophical Transactions, of the Royal Society of London*, 1796, Part I. London: Peter Elmsly, printer to the Royal Society, pp. 46-130.

26. ATWOOD, G., 'The A Disquisition on the Stability of Ships', *Philosophical Transactions, of the Royal Society of London*, 1798, Part II. London: Peter Elmsly, printer to the Royal Society, pp. 201-310.

27. SUTHERLAND, W., 'The Ship-builders assistant: or some asseys towards compleating the art of marine architecture', *London: R. Mount, A. Bell et R. Smith*, 1711.

28. On the developments in the history of naval science between the eighteenth and nineteenth centuries Cf.: CORRADI, M., 'Epitome della scienza navale in Atti del secondo convegno nazionale', *Cultura navale e marittima: transire mare*, Morozzo Della Rocca, M.C., Tiboni, F. (Eds.), Firenze: goWare, 2016, pp. 136-149.

29. LEMMERS, A., 'Shipworm, Hogbacks and Duck's Arses: The influence of William May on Sir Robert Seppings', *The Mariner's Mirror*, 2013, 99, n. 4, pp. 410-428.

30. SEPPINGS, R., 'On a new principle of constructing His Majesty's ships of war', *Philosophical Transaction*, 1814, 104, pp. 285-302; Cf. p. 287.

31. PACKARD, J J., 'Sir Robert Seppings and timber problem', *The Mariner's Mirror*, 1978, 64, n. 2, pp. 145-156.

32. The topic has been much debated and for an in-depth study the circular stern Cf.: HARVEY, G., 'On the circular stern of ships of war', *The Edinburgh Philosophical Journal*, 1824, 1, pp. 239-248; HARVEY, G., 'Observations on Sir Robert Seppings Plan for the Circular Stern of Ships of war', *The Edinburgh Philosophical Journal*, 1822, 7, pp. 19-28.

33. MÖBIUS, A F., 'Lehrbuch der Statik, Leipzig: Georg Joachim Göschen', 1837; pp. 286-287.

34. SEPPINGS, R., 'On a new principle of constructing His Majesty's ships of war', *Philosophical Transaction*, 1814, 104, pp. 285-302.

35. FALCONER, W., 'An Universal Dictionary of the Marine', *New Editions, Corrected. London: T. Cadell*, 1780.

36. GOODWIN, P., 'The influence of iron in ship construction: 1670-1830', *The Mariner's Mirror*, 1998, 84, n. 1, pp. 26-40.

37. STAMMER, M K., 'Iron knees in wooden vessels: an attempt at a typology', *The International of nautica archaeology*, 2001, 30 n. 1, pp. 115-121.

38. FINCHAM, J., 'A History of Naval Architecture', *London: Whittaker and Co.*, 1851; p. 114.

39. WRIGHT, T., 'Young and Robert Seppings: Science and ship Construction in the Early Nineteenth Century', *Transactions of the Newcomen Society*, 1981, 53, n.1, pp. 55-72.

40. WALTERS, J., 'Specification of the Patent granted to John Walter, of Fenchurch-buildings, in the City of London, Architect and Engineer; for certain Improvements in the Construction and Fastening of the Frame-timbers or Bends of Ships or Vessels, whether Building or under Repair', *Repertory of Arts, Manufactures and Agriculture*, 1816, XXVIII, 2nd series, pp. 336-341.

41. MÖBIUS, A.F., *Op. cit.*

42. SEPPINGS, R., 'On a new principle of constructing His Majesty's ships of war', *Philosophical Transaction*, 1814, 104, 285-302; Cf. p. 292.

43. Ibidem; p. 293

44. Ibidem; pp. 295-296

45. Ibidem; p. 296.

46. SEPPINGS, R., 'On a new principle of constructing ships in the mercantile', *Supplement to the Edinburgh Encyclopedia*, Vol. 18, London: Blumer and Nicol, 1820, pp. 133-143.

47. ADAMS, J., 'A Maritime Archaeology of Ships. Innovation and Social Change in Medieval and Early Modern Europe'., *Oxford and Oakville: Oxbow Books*, 2013.

48. SEPPINGS, R., 'On a new principle of constructing His Majesty's ships of war', *Philosophical Transaction*, 1814, 104, pp. 285-302.

49. Ibidem; p. 299.

50. KNOWLES, J., 'Inquiry into the Means which have been taken to Preserve the British Navy', *London: Winchester and Varnham*, 1821.

51. Idem.

52. Idem.

53. BALLINGALL, J., 'The Mercantile Navy Improved: Or a Plan for the Greater Safety of Lives and Property in Steam Vessels, Packets, Smacks, and Yachts', *London: W. Morrison*, 1832.

54. SEPPINGS, R., 'On a new principle of constructing ships in the mercantile', *Supplement to the Edinburgh Encyclopedia*, Vol. 18, London: Blumer and Nicol, 1820, pp. 133-143.

55. BALLINGALL, J., *Op. cit.*, pp. 133-134.

56. Ibidem; p. 9.

57. Ibidem; p. 52.

58. 'Improvements in Ship Building', *Journal of the Franklin Institute, of the State of Pennsylvania, for the Promotion of the Mechanic Arts; Devoted to Mechanical and Physical Science, Civil Engineering, the Arts and Manufactures, and the Recording of American and Other Patent Inventions* (1828-1851), Jones, T.P., Ed.; Philadelphia: The Franklin Institute, 1833, Vol. XII, Fasc. 1, pp. 61-65.

59. BALLINGALL, J., 'Unsafe ship-building, a national sin: a treatise in reply to a despatch of His Grace the Duke of Newcastle, upon the safe transmission of emigrants', *Melbourne: Argus Office*, 1854.

60. Ibidem; p. 11.

61. On the history of the History of theory of structure Cf.: CHARLTON, T.M., *A History of the Theory of Structures in the Nineteenth Century*, Cambridge: Cambridge University Press, 1982; TIMOSHENKO, S.P., *History of strength of materials*, New York: Dover Publications, 1983; KURRER, K.-E., *The History of the Theory of Structures: From Arch Analysis to Computational Mechanics*, Berlin: Ernst & Sohn, 2008.

62. 'An Apology for English Ship-builders; showing that it is not necessary the Country should look to the Navy for Naval Architects', *Descriptions of Various Plans for the Improvement of Naval Architecture*, Blake, R.F.S. Ed., London: J.W. Norie & Co., 1833, pp. 225-232

7. AUTHORS BIOGRAPHY

Massimo Corradi is Associate professor of History of Science at the University of Genoa. His research interests are developed about *History of science and techniques, History of engineering and structural mechanics, History of naval science and shipbuilding*. Visiting Professor at Faculty of Sciences at the Université catholique de Louvain (Belgique), Open University of London (Great Britain), Université de St. Etienne (France), Universidade de Brasilia (Brasil), School of Art and Architecture - Shiraz University (Iran), Art University of Isfahan (Iran), I.U.A.V. (Istituto Universitario di Architettura di Venezia. Lecturer: Universidad Nacional Autónoma de México / Patzcuaro (Mexico), PhD in 'Structural Engineering', Faculty of Engineering of the University of Pisa; College of Engineers of Tuscany, Florence; Iranian Architecture Center, Tehran (Iran). Past member of the Academic Board: PhD in 'Histoire des sciences' at the Faculty of Sciences of the Université catholique de Louvain (Belgique), PhD in 'Philosophy' at the Open University of London (Great Britain), HDR in Sciences Humaines et Sociales at the Université de St. Etienne (France, PhD at Escuela Tecnica Superior de Arquitecture, Escuela Politecnica de Madrid (España); member of PhD in Architecture and Design (University of Genoa). He is the author of over 140 scientific publications (See: IRIS UniGe). The latest book published is *Storia della nautica*. Genova: Edizioni di Storia, Scienza, Tecnica, &, 2023 (in collaborazione con Claudia Tacchella).

Claudia Tacchella is a Doctor of Philosophy in the field of History of Shipbuilding at the Polytechnic School of the University of Genoa (Italy), where she graduated with the thesis *Sailing Masterpieces. Ship design and aesthetics in the Naval Iconography of the 17th Century*. Her research interests concern the *History of science and techniques*, the *History of naval science and shipbuilding*, and she is the author of 15 papers on these subjects. She is co-editor of the volume *Scienza, Tecnica, Società. Dal Tardo medioevo all'età contemporanea* (Genova: città del silenzio, 2022) and co-author of *Storia della nautica* (Genova, 2023).

Historic Ships Conference
28th – 29th November 2023
London, United Kingdom

ON THE LOSS OF HMS HOOD

M Lawrence, NSWG Limited, Newcastle upon Tyne, UK.

1. INTRODUCTION

HMS Hood, the largest and most famous warship in the Royal Navy, blew up and sank while engaging the German Battleship Bismarck on the early morning of Saturday 24th May 1941 in what became known as the battle of the Denmark Strait. Casualties were extremely heavy: only 3 survivors were found from a ship's complement of 1,418 men. The tragedy represented the largest loss of life of any single Royal Navy vessel lost in the Second World War. It came as a great shock not only to the Admiralty but to the entire country, every outpost of empire and indeed the World.

The Battlecruiser was in its 21st year as a commissioned warship. She had become the darling of the fleet, and of the British population at large; she was known throughout the world. "The Mighty Hood" was the biggest warship in the Royal Navy [1].

In the battle with Bismarck, Hood suffered a huge explosion and sank just eight minutes after opening fire. It is known that she sustained a hit from a high explosive shell fired by the heavy cruiser Prinz Eugen very early on in the exchange, which started a blaze amongst some light ammunition on the boat deck that was seen and widely recorded, but this was the only German shell fired in the exchange that appeared for certain to have detonated correctly [2]. Apart from this hit enemy shells landing close by were observed even from the German squadron's first salvoes and Hood was repeatedly near missed but eventually a shell from Bismarck's fifth salvo was believed to have hit Hood near the rear of the ship, somewhere forward of the mainmast [3] and this is thought to have been the trigger for the magazine explosion. Hood, having fired the first shots in the engagement was regularly straddled by both German vessels while not landing any blows herself. The action has been pored over ever since [4].

Following the sinking, two boards of inquiry were convened [5] the second considerably more comprehensive than the first. This (second board) took evidence from the three survivors, Signalman Ted Briggs, Midshipman William Dundas and Able Seaman Robert Tilburn. However, both boards reached essentially the same conclusion, that a shell from Bismarck found its way into the magazine of "X" turret (either directly or by setting off a neighbouring smaller calibre magazine) causing a fire which instantly became hugely powerful and this caused the explosion that destroyed most of the after part of the ship. Since then enormous effort has been expended in trying to find out, in forensic detail, the sequence of events in the last minutes of the great vessel. Given that combat had been joined, the explosion of a heavy shell that had penetrated deep into the bowels of the ship was automatically suspected as the cause; Bismarck was believed to have already hit Hood once (the spotting top) [6] and was bracketing her. Bismarck had the range, and her shooting was supremely accurate, especially given the sea state and weather conditions prevailing, recorded variously as winds ranging from strong breezes of 15 – 20 mph up to gale force of over 30 mph and a heavy, irregular swell of fifteen to twenty foot waves (corroborated by photographs below) [7,8.9].

The fire on the boat deck while widely seen was not considered as being powerful enough or near enough to the magazines to have caused the more major fire that led to the subsequent magazine explosion. Two other possibilities that could have caused the loss were considered in detail by the boards of enquiry, but ultimately rejected; namely the explosion of the ships' own torpedoes on the upper deck, or a serious mishap in "X" turret. The torpedo theory was initially put forward by Captain [Alfred] Phillips of HMS Norfolk who was an eye witness, but also, and more importantly perhaps, by no less a person than Stanley Goodall, the Director of Naval Construction and without doubt at the time the world's foremost expert on warship design [10]. However, after much debate, it was felt that the torpedoes themselves, firstly being in the open air where the explosive force, and the fragments, would spread out widely and secondly (and obviously connected to the first) such fragments would simply lack sufficient kinetic capacity to cause the fatal explosion that took the vessel down. To trigger a main magazine to blow up, metal fragments would need to travel a considerable physical distance, and taking into account the presence of various parts of the ship's structure, including armour, between the torpedo area on the upper deck and the principal cordite magazines right down near the keel this seemed unlikely. The second part of the theory, that the torpedoes exploding would lead to extensive hull damage and possibly to the vessel breaking her back, had a considerable level of mechanical credibility but still did not support the observation of a magazine explosion.

The torpedo theory was also understandable in that a fire was clearly raging amongst some light ammunition on the boat deck in the vicinity of the launching tubes but additionally to the doubts over the outcome of a torpedo explosion it was noted there would have been some time interval (short, but certainly observable) between a visible explosion of the warheads on the deck and the far larger one that took down the ship and no sighting of such was made by any of the large number of witnesses who reported to the boards of enquiries. There was also expressed some doubt that a German 15" shell even exploding right beside a torpedo in its housing tube would actually have triggered a sympathetic warhead detonation and the conclusion as to whether the explosion of the ship's own torpedoes could have caused the loss of Hood reached a consensus that this possibility did not match up with observations of the event.

The theory around a serious fire or misfire occurring in "X" turret had perhaps greater weight as it was clear the main explosion was towards the stern of the ship but amongst the eye witnesses no especial sight was made of a prior indication by any of flames, flash or smoke coming from or near the after turrets and given the number of interlocks and anti flash doors between turret and magazine (probably the main lesson from the losses at Jutland) [11] it was again felt that this was unlikely. A misfire causing an explosion in a turret (HMS Devonshire in 1929) [12] and a fire in a turret (HMS Barham, off Crete on 27th May 1941, when one of the rear turrets was penetrated by a 250kg bomb) [13] had not caused subsequent magazine explosions while earlier in the war – April, 1940, the year before the loss of Hood - a bomb hitting and going off aboard HMS Suffolk [14] had similarly not caused a major explosion despite setting off a cordite charge in the handling room (between the magazine and gun turrets), all of which led considerable weight to the argument that the gun loading and ammunition handling equipment as well as the formal operating procedures in use at the time provided substantial resistance to a "flash" – an intense thermal shock wave - from an explosion in or around the turret which travelled down into the magazines causing the cordite there to catch fire and explode itself. It was therefore felt unlikely that even a serious cordite fire developing within the turret (a misfire would have been seen immediately) could have led to the magazine explosion that destroyed the ship (while the incident on HMS Barham happened three days after the sinking of HMS Hood the information would have been available, certainly at least to the second board of inquiry which sat between late August and early September 1941). In terms of "flash" and cordite's combustability, Admiralty experiments with burning the material [15] had led to a belief it would not explode if ignited (i.e. if the material was set on fire it would only burn until it burnt itself out) (and more specifically these studies had shown the propellant would only explode under 'explosive' conditions (i.e. through a detonator)) but this work had only been with half a ton of cordite and Hood's rearmost magazines contained 112 tons in confined spaces. There was an existing counter to this experimental result however as it was already believed the explosion of magazines at Jutland was based on flash (but with an earlier form of cordite) and a strong reinforcement of the view that cordite could explode without being 'formally' detonated came six months and a day after the sinking of Hood when the rear magazines on HMS Barham were seen to blow up more than a full four minutes after she was torpedoed [16] (and this being the newer and more stable form of the propellant). The Admiralty view on this is surprising in the light of the Royal Navy having over 100 years of experience in handling gunpowder; if ignited in the open air gunpowder burns, fiercely (as does cordite); if confined in a tight space it explodes. Gunpowder and cordite are course different propellants but the general principal does not seem to have been investigated further [17]. The witnessed events (Hood, then Barham) clearly support explosions occurring when a fire reached a high enough temperature in confined spaces where large amounts of burning cordite was present (many tens of tons or more) without a primer activating the detonation, in other words a real operational situation in conditions far beyond those in the tests the Admiralty performed on cordite.

From these deliberations, which in the case of the second board of enquiry were based on a very extensive gathering of facts from a wide range of witnesses, the survivors themselves, and experts, the conclusion was drawn that a shell from Bismarck penetrated Hood somewhere forward of the mainmast, travelled over seventy feet into the vessel and exploded in or very close to either the 4" or 5.5" cordite magazines which were set off themselves and this explosion in turn caused the huge and intense fire seen venting from HMS Hood that led almost immediately to the detonation of at least one of the rearmost 15" magazines. This combined explosive force was of immense power, completely destroying this part of the ship, and, it was concluded, would have been the direct cause of the sinking. Admiral [Tom] Phillips (at the time Tovey's deputy (following the death of Vice-Admiral Holland on HMS Hood)) and particularly Stanley Goodall had been the prime movers behind the setting up of the much more comprehensive second enquiry [18] but essentially it reached the same conclusion. Goodall remained dissatisfied and pointed out that this did not explain the widely observed emission of a huge tongue of flame from a long way forward of the magazine spaces and particularly the German shell would have needed a very long fuse delay; by the time of the second enquiry he had some knowledge of these weapons' performance because a 15" armour piercing type from Bismarck had struck and entered Prince of Wales during the action without exploding [19]. The shell was recovered while the ship was being repaired at Rosyth in June and examined by bomb disposal from which some data on the fusing mechanism and fuse delay was ascertained [20]. (Even if the shell did not explode it must not be overlooked that it still possessed massive kinetic energy (a piece of machined steel weighing two thirds of a ton and travelling at over 2,000 feet per second) and could have smashed its way into a magazine but it is difficult to see how this would cause such a huge fire). Goodall clearly had some doubts, but of the choices on the table the theory that a hit from one of Bismarck's shells was responsible for the loss of Hood was accepted as being the most likely, and so it has remained. Goodall himself referred to this only as "the probable cause" in subsequent correspondence [21].

There was a fourth possibility however, one that never seems to have been considered at all. This possibility was HMS Hood suffering catastrophic mechanical failure as action was joined.

Given the known poor state of firstly the vessel's overall structural and mechanical condition [22, 23] and secondly her marginal seaworthiness (HMS Hood was widely recorded as having her quarter deck awash while at speed in anything but a flat calm sea, a situation that would bring into question the seaworthiness of any vessel) [24] (within the Navy in general Hood was affectionately referred to as "the largest submarine in the fleet") it is surprising these quite serious, and widely known issues, were not at least aired. Additional to dockside visits (considered in slightly more detail below) a longer term matter was Hood continuously shipping some water aft. While this was generally dealt with by her powerful pumps the continuous presence of salt water in numerous places around the ship (and especially on the lower decks as the seawater would inevitably work its way downwards) led to much corrosion and is likely to have been an additional factor in reducing the strength of already worn components in the propulsion system, particularly items such as shaft supports (and the point where these supports and bearings (called plummer blocks) are fixed into the ship's hull) leading to the potential for a stress corrosion crack leading to a fracture and very quickly, in a matter of just a split-second, to the failure of the whole plummer block and the supported propeller shaft. Additionally of course there was also the possibility of a tensile stress induced fracture (basically, ageing; carrying stresses, day in and day out, until the metal fails). Critically, in Hood's case, these parts of the ship's propulsion system, carrying a quite immense amount of power, operated right alongside (and right underneath) the cordite magazines [25, 26, 27], literally only five feet away and if anything went wrong these magazine spaces became highly susceptible to 'internal' damage from flying metal fragments, parts of a damaged shaft, plummer block or mount (see figures 1a and b and figure 2), which would have smashed through the magazine walls in a shower of red hot sparks. Metal fatigue is insidious and can be extremely hard to detect until it is too late; the crack that caused the Silver Bridge disaster in the United States in 1967 was only 3mm long [28] and the tiny casting impurity that caused disintegration of a titanium fan blade leading to the Sioux City crash of United Airlines flight 232 in 1989 had not been correctly identified at maintenance inspections despite a tracer dye indication and despite the airline industry having a fanatical devotion to safety [29]. It is of course pure coincidence, but HMS Hood was commissioned into the fleet in 1920, the year before the first comprehensive study of brittle fracture was published [30], in other words, when metal fatigue was "born" as an academic subject showing patterns of cause and effect with at least reasonably repeatable results. At the time the naval design philosophy was to make all parts overstrength; a certain amount of "ad hoc" knowledge of stress fracturing had led to the adoption of techniques such as cold rolling and surface polishing but what was known about metal fatigue at the time Hood was built was almost wholly empirical and even the most modern of today's techniques such as ultrasound and X-ray are not foolproof. While certainly much more is known about metal fatigue now than Hood's designers or dockyard maintenance staff would have, it is still a highly complex subject [31] and whether the underlying cause is corrosion or stress (or a combination) this sort of structural decay can still go undetected and cause parts to fail. More recently (although assessment work is still underway), the collapse of the 51 year old Morandi Bridge in Genoa, Italy, in 2018, already has salt-air corrosion being a major suspect as the cause; as early as 1979 the designer (Riccardo Morandi) expressed concern that the bridge was ageing faster than anticipated and said remedial intervention was needed to protect the structure from polluting elements and sea air [32].

In her last hours Hood was being driven to the limit both in reaching Bismarck and in manoeuvring to engage, so much so that the new Prince of Wales could scarcely keep pace [33]. Hood's twenty year old machinery was being worked flat out, had been for some time, and no matter how much "overstrength" is built into components the recent history of the vessel showed many of her engine room fitments had reached the end (or very near the end) of their working lives. Given all of this, quite how the possibility of mechanical failure being a cause of Hood's demise was apparently not even thought about in the first place does seem remiss of the Admiralty although in their defence it would have been almost impossible for observers to see any physical indication of such ahead of the magazine explosion given the very short time interval that would have elapsed between the physical failure of a major part, the penetration of the magazine (or magazines) by red hot fragments and subsequent fire, the enlargement of this fire into a conflagration then the magazine explosion; perhaps sixty to one hundred seconds. There would have been a loss of speed, but not much in such a short time; in calm seas Hood was capable of 25 knots on just two engines [34]. She would have dropped back slightly towards Prince of Wales but her own momentum would have carried her for a while so the distance would be no more than 50 yards and most probably less; an un-noticeable 10 or 20. Given the sea state and the high mental stress levels caused by the ongoing action it is unlikely this would have been noticed by Prince of Wales, for while keeping an eagle eye on the flagship for station-keeping and signals there was also a need to observe the enemy and fall of shot, all the time with enormous sheets of spray coming over the flat bow and limiting visibility and of course there were no survivors from this part of Hood to provide witness accounts.

This leaves the only clues to work on being the intense vibration felt in the whole vessel after Bismarck's second salvo landed close by (evidence of AB Tilburn) [35] and the huge column of flame that emerged from the ship in her last seconds (almost certainly through one of the engine room vents) near to and as high as the mainmast; some witnesses said higher [36]. This sequence was strongly reminiscent of Jutland, as referred to above, as well as the observed catastrophes that overtook HMS Natal and HMS Vanguard while both were at anchor in safe waters.

An evaluation is presented of the possibility of HMS Hood suffering a major failure in the propulsion mechanism which tore a large hole in the engine room bulkhead that soon after became the route of the observed venting and this mechanical failure simultaneously causing the penetration of the ship's rear magazines by heavy fragments (or indeed a dislocated shaft itself)) which smashed through the magazine floor and walls in a shower of red hot sparks. These sparks caused a fire in the stored bags of cordite that expanded rapidly and quickly – a minute, two minutes - leading to a huge blaze that was responsible for the massive flame that was seen venting, this followed seconds later by the explosion and subsequently the vessel's loss. This sequence is the subject of this paper. It is not meant to decry the theory of a hit from Bismarck, but to present an alternative possibility based on this outlined mechanical failure that also fits recorded and observed facts about the loss of HMS Hood, overcoming Goodall's dissatisfaction over the venting being so far away from the magazines and presenting a reason for AB Tilburn's observation of "Hood being shaken with a great vibration" early on in the action.

2. DEVELOPMENT OF THE MECHANICAL FAILURE THEORY

It was widely known that Hood was in need of a full scale refit on the level of those already undertaken on some other capital ships of 1914 -1918 vintage. The vessel was overweight to the point of making her seaworthiness marginal and the same description could be made about the operating condition of much of her machinery. She leaked, sometimes quite badly [37] and sea water entered the ship in almost any sea state other than flat calm. There was much rust and signs of decay. Just as bad as the hull was the state of her machinery; her condensers had required urgent major repair following a very near miss (or maybe hit) from a German bomb close to her principal armour strake on 24th September 1939 (which possibly caused weakening to the hull in the locality as well)(the work needed not being completed until a refit that lasted from early April until the end of May 1940), her underwater hull and deck plates had required patching and re-riveting at the same time and after the operation against the French fleet at Mers-el—Kebir one of Hood's turbines had partially failed (it was the starboard inner turbine and this is very relevant because the two major causes of drive train failure are (a) imperfectly balanced components within the drive train, including the turbines and the propellers or (b) shaft misalignment) [38, 39]. (This is not to discard other causes such as bearing mount failure). The turbine was only restored at Rosyth early in 1941[40] just some 4 months before the vessel was lost. While officially classified as 'refits' examining this list closely shows these dockyard visits could more accurately be described as "patching up" as all were of a heavy and serious nature, happening one after the other. Indeed, Hood was in dockyard hands for other varying lengths of time as well because apart from Rosyth as above and the April – May 1940 repairs at Devonport (completed at Liverpool), she was also worked on just before the war, in January 1939 – August 1939, a "special large repair" where the concentration was on the engines and re-riveting the outer hull. At this time Hood had actually only just emerged from a spell in dock at Malta in the period November-December 1938, in which "Y" turret received attention, and once war was underway she was also having work undertaken between

13th – 25th November 1939 (at Plymouth) and 15th – 25th August 1940 at Rosyth (for a gun change) [41]. Six dockyard visits for heavy repairs in just over two years (November 1938 to January 1941) spoke tellingly of the state of the ship.

The operational state of HMS Hood now forms the basis for development of the mechanical failure theory.

As introduced earlier, the most important fact here is the shafts from the inner turbines pass very close to the magazine walls, in fact the floor of "X" magazine is only 1½ metres (about 5 feet) above and about 2m (about 6 feet six inches) diagonally removed from either the port or starboard inner shaft (which were themselves some 28" in diameter meaning the distance between the edge of the propeller shaft was less than four feet from the magazine floor overhead). Also at least three of the plummer blocks – shaft support bearings – are close by as can be seen from the figures 2 a, b and 3.

Failure by metal fatigue of the drive shaft couplings or any of their support bearings, themselves of very substantial construction, has the potential to cause a large amount of damage. This would involve big, heavy pieces of solid metal, probably hot from being riven apart (almost exploding)(20 tons and possibly more if this was a propeller shaft section) coming adrift from their fixings at some speed and carrying enough force to easily penetrate the local structure and in doing so generate sheets of red hot metal fragments. At the same time it is probable that the engine room bulkhead would itself suffer serious damage because if this was a propeller shaft shorn of its bearings the remaining shaft connected to the engine would instantly sag under its own weight and the change in momentum would lead it to 'windmill' until it breaks. In doing so it would cause the large vibration reported by A.B. Tilburn with the'windmilling' reaming open the bulkhead between engine room and the compartments to the rear of the ship (principally the shaft tunnels and the magazines). The destruction of this part of the engine room bulkhead opens up the large aperture that soon becomes the route for the combustion products from the magazine fire (or fires) through to the engine room and out through the vents which were situated near the mainmast. As well as the magazines it is possible the flying fragments could have penetrated the shell side (hull) the keel or possibly the wing fuel tanks which would explain their 'blown outwards' appearance within the wreckage on the seabed (see below) [42]. Each propeller shaft was effectively driving 12,000 tons (a quarter of Hood's weight) at 28 knots in heavy seas; this represents a quite colossal amount of power.

Another cause of drive train failure however, and different to metal fatigue, could have been a propeller shaft being bent out of true (physically bent, as opposed to misaligned) by the close explosion of a heavy shell. Such an occurrence had happened to the 3 year old German Dreadnought Markgraf at Jutland [43] and was to happen later in 1941 to the still almost new HMS Prince of Wales after she was hit by a torpedo dropped by Japanese aircraft off Malaya [44]. It could also have been damage to a propeller, for example a blade being bent (by the force of a close explosion), heavily chipped or even completely sheered off either by the force of a close explosion or simply by fatigue. The external shaft supports were themselves another area of structural weakness. Such a widespread variety of potential machinery failures is a long list and not yet complete, yet each mentioned is serious, capable of causing major damage or even the complete loss of a vessel in a worse-case [45], and this would be without setting off a large store of combustible and potentially explosive material.

It is not known, nor never will be, how many of Hood's secondary systems were fully operational but the refits (repairs) listed above are themselves major and while it could be assumed other minor faults might be corrected in these dockside periods this has to be seen against the overwhelming need to get HMS Hood back "on the front line" as quickly as possible. Given the urgency of the war situation it is highly likely that the 'defect list,' handed to the chief shipwright at docking, would probably have run to many, many pages and with pressure on the repairers to get vessels fixed and back to sea as quickly as possible, once the major problem had been dealt with and armament upgrades carried out anything else would be treated as "very second order" and the ship would sail immediately these primary tasks were completed, with a high chance that many of the more minor defects would remain on the 'repairs' list with a note "can wait for next time."

Another serious cause of stress, introduced above, was that Hood's quarterdeck was often awash, actually submerged sometimes for a minute or more, meaning the on-off nature of this additional load (some hundreds of tons) would have to be borne by the hull, with the bending moment being somewhere between the rear of "Y" turret and the rear of "X" turret. At some point during the 1930's there had been a recommendation for a full structural survey of the vessel's entire hull as it was known to be highly stressed [46]. It never happened. The strain of this additional weight would have caused extra flexing, above that caused by the movement of the sea, leading to minute cracks in the hull and its supporting ribs in the region of "X" turret, reaching right down to the keel causing a cumulative structural weakness in this area.

It is just above the keel where the propulsion machinery lies. HMS Hood needed a tremendous amount of energy to drive her 48,000 tons through even calm seas and the action in the Denmark straits was fought in seas anything but calm. The last known photographs of Hood, taken from HMS Prince of Wales shortly before the action commenced (see below) shows the vessels steaming in what appears to be sea state 6 with large and irregular waves, symptomatic of seas that have been driven for many hours by gale-force winds arising from a deep depression tracking rapidly eastwards, with further confirmation of this coming from the same photograph where "A" turret of HMS Prince of Wales can be seen to be trained astern of the beam.

Figure 1: Taken from HMS Prince of Wales, believed to be shortly before action was joined although may have been the evening before. Hood's funnel smoke is blown almost flat in the gale force winds and the roughness of the sea can be observed. Note the sea-spray in photograph 2 and the roll to starboard on Hood. (IWM).

It was standard operating practice when driving through heavy seas to have the ship's main forward guns train well away from the "fore-and-aft" position to stop the barrels and the turrets themselves filling with water from the bow spray. Further evidence on the sea state comes from the fact that HMS Norfolk and Suffolk, shadowing Bismarck, could only just keep pace with her, despite having a nominal 3 or 4 knot superiority in speed, a paper advantage that had helped them shadow the ship in the calmer waters to the north west of Iceland but was reduced in rougher weather. One further, final, indication of the sea state was that Hood had had to detach her escorting destroyers during the night because the high waves made it almost impossible for these smaller vessels to keep up.

Hood's badly worn hull and machinery had carried it through these increasingly heavy seas since soon after leaving Scapa Flow some 40 hours previously, as she had done on many other occasions since the outbreak of war.

The battlecruiser only had one major refit, and that had been ten years earlier. In the intervening time the ship had regularly been at sea, whatever the weather, no matter how big the waves. She had undertaken practice gunnery shoots as well as firing her weapons in anger, particularly at Mers –el -Kebir. This introduces another major source of stress, the requirement of absorbing the huge blast force of her own 15" guns into the frame of the ship. The effect of this could be judged against the damage reports on Rodney and King George V after their own action against Bismarck. For Rodney, when on arriving at Boston on 12th June 1941 following the engagement (after having made a necessary return to Greenock for refuelling and some immediate repairs) the American shipyard engineers found distorted framework, buckled plates and popped riveting all over the forward section of the vessel (including some hull plates), purely inflicted by the massive percussive effects of the repetitive firing of her own guns shaking the whole ship [47].These forces amount to very considerable levels of strain being placed on the fundamental structure of the vessel and Rodney was a full seven years younger than Hood and had steamed a lot less miles in the years between joining the fleet and war breaking out. To all intents and purposes H.M.S. Rodney had to be retired (placed on the 'reserve list') a little over two years after the action with Bismarck despite the repairs at the Boston Naval Yard and a series of refits and further repairs in the UK. HMS King George V was still under repair in July although 'ready for sea' and while her battle damage was less severe than Rodney she suffered from numerous fractured pipes, broken wiring, twisted brackets and some splits in her main deck. The Department for Naval Construction arranged for a party of 60 specialist workmen to be drafted to Scapa Flow to work on the vessel so she could remain in full active service [48]. After the battle with Hood the German vessels reported widespread internal damage from their own firing although this was largely cosmetic [49].

In Hood's case one final, additional, serious physical factor must be brought into consideration: the combat situation. As indicated above, there were a number of close near misses by Bismarck. Water is incompressible, so the shock wave from these near misses would have a hammering impact effect against the hull and its frames, and it was onto these frames the various parts of the propulsion machinery were fixed. Should these shock waves have caused damage to the propellers or shafts then automatic action would be taken to close down the driving turbine ('designed in' as this state of affairs can happen in a rough sea where the propeller temporarily comes out of the water), then go on, if felt necessary, to lock the shaft, but this would not happen instantaneously and if the shaft is 'out of true' or 'highly imbalanced' it could have been seconds only before a gland was damaged, a bearing displaced and very soon after, again a period of only seconds, that major damage could have been wrought with bearings breaking and parts of the propeller shaft lifting out of its mounts in the shaft tunnel and potentially doing very heavy impact damage to anything in the vicinity as despite the turbine shut down (and unless the propeller had been completely sheered off) the damaged shaft would continue to rotate either due to the ships' speed through the water or the shaft becoming discontinuous (so – called "windmilling") and does further damage, although clamping could limit this if it could have been applied quickly enough but then such an action could also sheer off the whole propeller shaft, as happened with Prince of Wales in her own final action some months later.

All of these mechanical issues were well known to the Admiralty. A prime symptom of any of these matters actually occurring would be a massive amount of vibration, duly observed and reported by AB Tilburn [33]. Hood, like all vessels, was prone to various vibrations but this was clearly heavy enough and recently onset to have been particularly remembered.

A final observation is that Hood was part way into a turn to port – the famous 20^0 turn – (the second turn) when she was lost; the ship's rudder is of course also at the stern, the force needed to turn a 48,000 ton ship at speed is large. If at this time serious structural or mechanical damage had occurred (or was occurring) in the rear of the vessel the forces due to this turn could hardly not make matters worse.

In the last perhaps one or two minutes of Hood's fight against Bismarck it is therefore quite possible that a series of connected events could have occurred that led to the loss of the ship with the fundamental issue now being mechanical failure, a failure brought on (either, or both) by the battle situation, ('external' to the ship), internally by a major mechanical breakage caused by a combination of the long term accumulation of heavy and continuous wear outlined above, an old, highly stressed, metal fatigued and structurally imbalanced hull carrying this aged and possibly corroded machinery and particularly its associated footings; machinery, which, despite its age, was generating and handling extremely high power loads; the possible – even minute – misalignment or incorrect balancing of parts of the starboard inner turbine after the January repairs at Rosyth, the recent prolonged flexing of the ship's structure by steaming towards Bismarck in increasingly heavy seas at high speed with the stern awash and the accumulated stresses, over many years, of having to absorb the recoil of her own heavy guns into the ship's frame. That there are many potential causes of metal fatigue damage or battle damage is clear; the result would have been the same but the two main causes are worth very brief further consideration.

3. BATTLE DAMAGE AND MECHANICAL WEAR

In trying to evaluate whether either of these is the most likely culprit (if indeed HMS Hood did succumb to mechanical failure and not a 15" shell from Bismarck) then the observed facts that hint at mechanical failure consist firstly of AB Tilburn's statement that after a salvo landed from Bismarck HMS Hood was "shaking with a great vibration" so accepting this as fact then the only cause of such vibration (such vibration, of sufficient strength to make the whole ship shake) could only have been caused by a major fault developing in the ships' propulsion machinery. There is just nothing else powerful enough to cause a 48,000 ton ship to vibrate so noticeably.

Secondly was the huge tongue of smoke and flame that escaped from the ship just seconds before the explosion, widely seen and reported by eye witnesses several of whom specifically described it as being of yellowish smoke with red fire streaks and proposing it came from a cordite fire. This extremely powerful emission must have been routing through the engine room vent which was close to the mainmast, and while this leads to questions about how this fire became so intense so quickly (see below), it also indicates that there must have been some major damage to the watertight bulkhead at the rear of the engine rooms to allow the flames to escape from what was reported as a position close to the mainmast. This location is some sixty five feet away from the nearest magazine.

Metal fatigue or battle damage to the propulsion system: either, or quite possibly both, could have been responsible for the vessel's loss. We will never know. Both explain the heavy vibration but then must go on to explain the fire being started amongst the cordite, a fire which grew rapidly. The start has to be a fire, not an immediate explosion, to explain the huge venting. As was observed, the explosion follows the fire. A fire of the magnitude observed would have needed time to grow.

Mechanical failure of one of the plummer blocks (there are three near the magazines) would lead to the load previously taken by this bearing having to be borne by the next in line but the lack of support on the shaft would lead to it bending and breaking soon afterwards and this could happen more quickly than the turbine could be shut down. Quite apart from metal fatigue, it should also be pointed out that the failure of the bearing mount could have been brought on by faulty lubrication [50, 51], but the result is the same; right next to the cordite magazines – within the span of a man's arms - there is a failed and almost certainly fractured plummer block, a large and heavy construction of metal in its own right, and a heavy and now unsupported propeller shaft still rotating 4 times a second. At the point where the propeller shaft passes into the engine room through the watertight bulkhead, the few seconds of non-supported rotation before failing would ream a large hole here, but additionally and most critically, the fragments from the disintegration also penetrate one or more of the cordite magazines (only feet away) and set some of the cordite on fire. Probably at this stage this involves only a limited number of bags but importantly (as mentioned above) the penetration of the engine room bulkhead opens an outlet for the combustion products as they are drawn through the hole in the magazine walls and floor and into the shaft tunnel, then the engine room by the powerful engine room ventilation fans. Once this fire is underway it quickly spreads; Hood was not believed to have been fitted with sprinkler systems.

Another possible source of mechanical failure concerns the percussive effects of one of Bismarck's shells landing close by and damaging a propeller, a propeller shaft or a shaft support bracket external to the hull. This does have support from the direct observations of HMS Hood being close-missed by Bismarck from the second salvo onwards. Some circumstantial evidence (although based on explosions that were closer) that occurred later (still in 1941) was Bismarck herself basically being lost by a torpedo hitting very near to her steering gear and HMS Prince of Wales lost in large part due to serious damage to one of the propeller shafts which kept rotating out of true and tore a hole large in the stern gland, enough to allow water to pass through the shaft tunnel and into the engine room through the equally damaged bulkhead gland, with the propeller shaft itself eventually splitting in two. The breach of the shaft gland was the specific, key damage that caused the large amounts of flooding that was the main reason Prince of Wales was lost. In any vessel the propellers, rudder and assorted glands, shafts and shaft supports are known areas of vulnerability, and while Bismarck and Prince of Wales suffered direct hits from torpedoes and Hood was only close-missed by Bismarck's 15" shells (at least until the end), Hood's machinery and hull were in badly worn condition. Prince of Wales and Bismarck were brand-new.

If the close-miss caused damage similar to that suffered by Prince of Wales then firstly a propeller having blades bent or distorted could possibly cause a turbine shut-down as the force into the ship is measured (as introduced above, propellers can come out of the water in rough seas; unlikely in Hoods' case however as the ship's quarter deck was only level with the water 'at best,' but in 'more normal' cases there are governors on the engines to prevent over-speeding when this happens). However as long as the propeller remains attached after shut down (even if badly damaged) the speed of the ship will cause it to turn as before but it is then due to the speed of the ship so it is revolving "against the engine." While it might be possible to lock it, it is more likely that parts of the moving shaft, certainly those nearest the stern glands, will vibrate massively due the imbalance caused by the damaged propeller and in turn the pressure caused by the uneven vibration and intense off-axis forces will almost certainly either make the shaft damage its bearings so badly these will fail and the forces on the shaft will make it tear out of these mounts, or the shaft will split at one of its coupling points and cause additional local damage.

If the damage is to a shaft however (purely the shaft and not the propeller, again this is considered to be closer to the stern than the engine) the most likely initial result will be major deformation to the stern gland which in turn will cause flooding. However the shaft itself, being bent or rotating out of true will not only damage the stern gland but also its local bearings and the bearing mounts which will fail in the same way as described above. If the cordite had already been ignited by metal debris it is unlikely the water flooding through the stern gland would be enough to extinguish it. While the inflow would be intense it would still only cause flooding on a timescale far longer than the intervals between Bismarck's second and fifth salvoes, and the cordite magazines were on the "second floor," located above the shell rooms, themselves above the keel. The magazine floor being holed would also prevent the magazine itself from being deliberately flooded, the water would simply pour out through this hole and into the shaft tunnel.

Both of these 'battle damage' situations will cause breakage most likely closer to the very rear of the ship but the thrust blocks (the parts of the drive train that take the shaft power and transmit it into the ship's frame) are located in the engine rooms and so if the disturbance of a seriously imbalanced drive shaft got through to the engine room the shaft breakage could have occurred there, physically closer to the smaller calibre magazines (the 4 inch and the 5.5 inch weapons) and also the engine room watertight bulkhead. All of these possibilities would cause a very large amount of vibration as well as serious damage to this bulkhead. The only counter to this possibility is the distance involved means that collateral damage (i.e. igniting cordite via shattered parts penetrating into the magazines) is slightly less likely.

The general metal fatigue proposal (i.e. wear or corrosion) would make the range of possible failure broader however; to the possibility of a shaft or its supports being bent or damaged by a near miss from Bismarck should be added the further possibility that while this might not directly damage either of these, the heavy shock wave was nonetheless sufficient to cause a brittle fracture of one of the worn bearing mounts at the position where it is fixed into the frame of the ship; the straw that broke the camel's back. The source of damage being inside the ship adds physical proximity to the key areas of magazine walls and engine room bulkhead. It will of course never be known, but as long as it can be accepted that shortly before HMS Hood was lost a heavy vibration began which is taken to be major mechanical failure, then the consequences of this are as already outlined; penetration of the magazine walls by hot, shattered, metal parts that also perforate the brass storage bins and ignite some bags of cordite propellant, and the tearing open of the engine room bulkhead. A failure of the starboard inner turbine itself would cause similar damage in that the drive to the propeller would be lost but the 'windmilling' effect would cause the shaft to continue to rotate against the drive and break with similar consequences.

Turning now to the cause of intense, "blowlamp-like" flame observed by many witnesses: if the start of the cordite fire has been caused by red hot sparks from broken metal parts breaking through the magazine wall and the brass cordite containers holding the actual bags of propellant within then taking this as matching in time the arrival of Bismarck's second salvo (according to AB Tilburn's testimony) it gives a time period for the evolution of the fire to a point where flashover occurred and the magazines exploded.

Fire goes through stages of development and while these can be quite rapid in the right circumstances (principally the presence of plentiful low flashpoint materials such as petrol or acetone to 'get the fire going', or some substance that once ignited is difficult to extinguish, such as cordite, some free air, unimpeded heat transport by any of conduction, convection or radiation, and a venting route) (the venting route is usually important as there are also circumstances where poor venting can limit or even smother a fire but this is probably less important in this case). Here however there is a specific low flashpoint material, acetone, which was a constituent in the manufacture of cordite and this would certainly provide a small amount of acceleration, but cordite itself, once ignited, burns extremely fiercely and is hard to extinguish.

The appearance of such a powerful burst of flame from what was almost certainly the starboard inner engine room vent shortly before the magazine explosion indicates that (a) the fire was issuing from a very high pressure situation, indeed it must have been inferno-like and the exit route had became established through the huge extraction fans used to maintain airflow in the engine room itself, and (b) such a fire must have been evolving from an earlier phase as it is difficult to see it becoming so powerful in just a few seconds.

Previous evidence for massive flame venting shortly before a cordite magazine explosion (other than Jutland) was already available in the Royal Navy; a series of eye – witness accounts of the explosion of HMS Vanguard in Scapa Flow in July 1917 had described the last moments of the ship as 'a large flame issued immediately before the vessel blew up.'[52]. Loss of life in this incident was also very heavy; there were only three survivors from Vanguard, one of whom died shortly after being rescued. An explosion on the cruiser HMS Natal, at anchor in the Cromarty Firth on the second last day of 1915 [53] saw reports of flames rising to up to the height of the mainmast (similar to HMS Hood) just seconds before the vessel blew up. The source of the explosion was clearly one of the rear magazines and at the inquiry suspicion fell on some of the cordite stored there self – igniting and this fire spreading rapidly to other cordite until a sufficient temperature was reached and flashover occurred, triggering the explosion; again, there was heavy loss of life. As a result of these mishaps the formula for cordite manufacture was changed and procedures tightened; there were no instances of cordite self – igniting post World War 1. The temperatures inside the magazines were regularly taken every few hours but from the start of a cordite fire to an explosion is a time scale of minutes, possibly less.

In the case of HMS Hood, for the fire (any fire) to have progressed in this way it would have needed flammable material; here it is cordite, available in abundance, then oxygen, to burn (cordite can however supply its own), a mechanism for heat transport, here it is conduction (everything on Hood was made of metal, even the cases for cordite bags) or radiation (there was only minimal air flow so convection was limited until later) and as stated a venting route. The need for a venting route is necessary for fire evolution and presents strong circumstantial evidence for a major hole having been formed in the engine room bulkhead. The cordite bags were dispersed in the largely enclosed magazines and surroundings and this could initially slow the fire (note that this is all happening on a timescale of seconds). At this stage the fire is spreading with flame but also with intense smouldering combining to cause a strong local temperature rise and initially promoting the very rapid production of a certain amount of carbon monoxide that would almost immediately disable the magazine crew [54]. It could also be considered that the magazine fire alarm might not be working; many of Hood's systems were known to be faulty. At this stage, with Bismarck's third salvo having probably arrived, Hood now has a growing and dangerous fire in one of her cordite magazines.

With the venting route established via the engine room the heavy fumes, dense smoke and noxious gases pass through the damaged bulkhead and thence out of the ship. This layout is shown in figure 4. Given Hood's speed of 28 knots and her heading into a strong wind of at least 15 knots and probably more the air movement over the deck would have been in excess of 40 knots (gale force 9 on the Beaufort scale) and the heavy yellowish brown cordite smoke, at least at this stage, would have been flattened and dispersed instantaneously. The only ship near enough to have possibly seen it (Prince of Wales) would have had to view it against a backdrop of Hood's boat deck fire, itself producing smoke, and the brightness of which in the dim and stormy early morning conditions would have been enough to have rendered nearly invisible any additional smoke in the foreground that in any case the wind was dispersing as soon as it appeared (and even if it had been seen it would, for almost certain, have been ascribed to the visible and very obvious blaze in the line of sight because the wind direction had the smoke and flames being blown towards the stern of Hood which of course is where Prince of Wales was stationed). Down below however the limited airflow and time needed to spread heat across the metal shelving temporarily limits the expansion of the fire but it is still enough to cause other bags of cordite to ignite (this will shortly accelerate and become a chain reaction) and consequently pressure and temperature increases in the enclosed compartment consistently, with the fire ready to expand [55]; the next sixty seconds are critical.

One of Hood's stern turrets fired; it is unclear from the evidence presented which stern turret fired first or not at all; at least one of them did, but almost certainly not both; this is a third clue, and a strong one, (although second in chronological sequence: heavy vibration / stern turret failure / large jet of flame) to there being something seriously wrong at the stern of the ship (it was however known that "Y" turret's roller bearings were very badly corroded and this might have prevented full rotation, additionally there had been problems with the hydraulics on board) [23, 41]. From now on it will be assumed that only "X" turret fired.

The massive percussion of the guns discharging will probably cause further damage below. This could cause more cordite to catch fire, at the same time further damaging the bulkhead between the shaft tunnels and the engine room (only some 23 feet (7 metres) away (figures 2 and 3) and possibly causing additional damage (for example further opening the wing fuel tanks or oil handling compartments; heavy fuel oil has quite a high flashpoint so if such damage did happen the oil would be unlikely to catch fire, at least immediately; but the flashover before the magazine explosion would have sent out a massive thermal shock wave capable of igniting almost anything). However the magazine fire is the critical factor here; at this time it is still accelerating but being held in check to some degree by the limited heat transfer deep in the ship.

"X" turret firing could indeed have been the final cause of the loss, but not by a misfire in the turret. The shock of the firing would almost certainly have caused extension to existing damage, but the opening of the anti-flash doors to facilitate reloading of the guns has the effect of introducing a large free air channel in the otherwise battened down ship, feeding the fire in the cordite magazine with more air, but far more critically giving rise almost instantaneously to a 'fire tunnel' or chimney effect. While the anti-flash doors and screens were not specifically air-tight they were nonetheless heavy and close fitting and would have seriously impeded air flow; once open however the fire could draw in whatever air it wanted, through the gun ports then down the ammunition trunking, through the magazine and exiting through the engine room vent via the large hole torn by the damaged shaft, this hugely powerful 'through draft' causing massive acceleration of the existing blaze.

4. THE LOSS OF HOOD

If the beginning of this sequence was indeed around the time of Bismarck's second salvo landing this gives a time period of somewhere between ninety to around one hundred and twenty seconds from the damage occurring and initiating the cordite fire, to the vessel exploding. By the time Bismarck's third salvo lands and the fourth is in the air, Hood is probably already lost.

After "X" turret fires the air blast operates to clear the gun barrels of debris such as bits of the cordite bags, then the anti – flash doors open to allow reloading of the guns, allowing a free supply of air to flow down from the turret and into the cordite magazine. This gives the fire a large-scale air current and the burn quickly becomes a blaze causing even more of the yellow cordite smoke to pour through the propeller shaft tunnels into the rear engine room and out through the ventilation ducting. In the next perhaps ten seconds the burn accelerates massively and becomes a fire of raging intensity that continues venting, for a short while, through to the open air via the engine room, the fierce, now blowlamp-like flames emerging in the huge column of reddish fire seen shooting up near the mainmast. Within just a few seconds more the fire grows even further and the pressure release through the engine room ventilators is nowhere near enough to prevent the temperature in the magazine continuing to rise and rise until very shortly afterwards, two seconds, maybe three or four, the heat causes the burning cordite to spawn its own thermal shock wave – "flash" [56] –and the whole mass of propellant explodes, taking the entire ship down. From the initial mechanical failure to the fatal magazine explosion, from what was initially a fire, then a blaze, then a conflagration, an inferno, and finally an explosion caused by a thermal flashover event, would have taken at most approximately one hundred to one hundred and twenty seconds.

5. CONCLUSION

It is postulated here that the starting point of the loss of HMS Hood was a major mechanical failure involving a propeller shaft and a main bearing mount, this being initiated by any of several possibilities all of which had roots in metal fatigue and the existence on the vessel of components in the propulsion system that had become seriously worn and overworked. This failure led to catastrophic damage involving the disintegration of support bearings and the severing of a propeller shaft. As a result of this, a heavy vibration is caused by the remaining part of propeller shaft flailing about, this vibration being powerful to shake the whole ship and additionally the hot metallic fragments of the broken bearing mount, each of substantial weight, perhaps several tons, penetrate one or both of the nearby cordite magazines serving H.M.S. Hood's main stern guns and the red hot sparks ignite some of the propellant. The flailing propeller shaft also causes the bulkhead at the rear of the engine room to be reamed wide open which provides the required large venting route for the developing fire, the fumes and smoke, and later flames, being drawn through the engine room by the powerful extraction fans.

The magazine fire is initially small scale but expands rapidly amongst the highly flammable material, the burning cordite producing carbon monoxide which disables the magazine crew almost instantaneously and it is also possible that some crew members were killed outright by flying debris. Alarm systems, if working at all, could have been cut (possibly by the boat deck fire or the local damage), or if they were active and the control positions alerted to a magazine fire then remedial action – flooding – would not have happened quickly enough to prevent the magazine explosion and in any case would have been negated if a large hole had been opened in the magazine floor. It could also have been a situation where alarms from the boat deck fire were concentrating the minds of those elsewhere in the ship in control positions, it could also be that alarms from the magazines were mistakenly attributed to the boat deck in the heat of the situation.

In the rearmost turbine room the engineers would have understood quickly that something had gone wrong and it is probable that the engine would have been shut down (or procedure started) however the same problem would also have affected the crew here; the cordite smoke entering the engine room would also have contained carbon monoxide; additionally between Bismarck's second salvo landing, the damage to whichever part of the propulsion suffered it, and the turbine being shut down, would have taken at least a minute and the third salvo would have landed and the fourth on its way. At about this time "X" turret fires and the reload procedure allows a huge downdraught of fresh air, establishing the chimney effect as described above. The engine room crew may have been able to report their problems to the bridge but whether or not this happened is something that will never be known. In the consideration presented here, the time period from the fire starting and expanding into a conflagration which bursts out near the mainmast, followed by the violent and catastrophic explosion that sank Hood, is a time period not dissimilar from that seen at Jutland, when hits were observed but magazine explosions did not happen immediately [57], the 'train of events' being the shell explosion on the turret causing some cordite to catch fire and expand, the thermal shock wave from which travelled down into the magazines, insufficient to cause the whole stock of cordite to explode but enough to ignite the nearest bags of it, this burn spreading quickly, mostly through conduction, until flashover occurred causing all the cordite present to explode. The Jutland delay was of the order of somewhere over sixty seconds up to about to one hundred and twenty seconds.

There is no other real explanation for the huge tongue of flame erupting from "the vicinity of the mainmast" by many of the eye witnesses other than a large scale cordite fire that must have been building up for a time, a short time, probably that between Bismarck's second to fifth salvos, a fire that developed huge pressure behind it and had an already established route to the outside via the hole torn in the engine room bulkhead by the shaft failure and the large ventilation trunk in the starboard inner turbine engine room. Flame is a gas phase phenomenon [58] and the sheer intensity observed, the "column of fire," reported to be greater than the height of the mainmast, physically rising this high straight into the air against a wind of gale force 9 over the deck indicates the fire was absolutely rampant for in order for flame to reach this height it almost certainly had to contain a large quantity of unburned, hot, combustible volatiles which only ignited after ejection at great force into the free air outside the ship [59]. The huge heat build up in the confined space of the magazine then initiates its own "flash" and an explosion follows, the shock from which sets off the cordite in any (probably all) of the neighbouring magazines (that of "Y" turret sternward and the 5.5" and 4" forward) a split second later, a huge, unstoppable force blasting upwards, downwards and outwards, taking off the hull plating, blowing the turret into the air and almost vaporising this part of the ship. If the wing tanks (or any oil handling equipment) had earlier been penetrated by flying fragments the flashover would easily have been of sufficient power to cause instant ignition, probably explosion, of the fuel on this side of the ship.

The real reason for the loss of Hood will never be know for sure and the "lucky hit" hypothesis is very plausible given firstly how accurate German shooting was and secondly through the extremely detailed forensic analysis [4] that shows a clear track – a small aperture to be sure, but a track – whereby a 15" shell from Bismarck could evade (or penetrate) Hood's principal armour sections, and given a long enough fuse delay, travel to the vicinity of the aft magazines before exploding. It was widely recorded that Hood was being bracketed and Bismarck had the range and had probably already hit her. It is perhaps worth noting that if Hood's magazine explosion was indeed caused by such a detonation it is the only heavy calibre (15") armour piercing shell fired by the German Navy to have exploded correctly in the whole war, although the sample is obviously small.

However, the possibility of a bearing mount breaking apart, serious damage or disintegration of the recently repaired starboard inner turbine or a propeller shaft bending or failing close to one of Hood's rearmost magazines never seems to have been considered. A train of events has been detailed here, initiated by such a major and critical mechanical failure followed by specific occurrences happening consequentially that match the key witness observations and critically align almost exactly with recorded sequences of earlier magazine explosions.

The last resting place of H.M.S. Hood lies 3km down on the seabed of the North Atlantic and other than knowing the vessel lies in 3 parts with a large midships-to-stern section almost completely missing there are – unsurprisingly, given the violence of the sinking - no clues in the wreck as to how the magazine explosion was initiated. It is proposed here however that mechanical failure cannot be ruled out, despite it apparently never having been considered before; slightly the more surprising really given that the First Sea

Lord is recorded as stating in 1939 [22]: "...the likelihood of her being able to steam at high speed for any length of time was considered remote."

6. ACKNOWLEDGEMENTS

I would like to thank the officers of the HMS Hood association for their invaluable website and for permission to publish this paper. I would similarly like to thank the staff of the HMS Barham association for their own website presenting information on the life and loss of that vessel, which along with Hood was the only other capital ship of the Royal Navy to suffer a magazine explosion in World War 2.

I would also like to acknowledge with grateful thanks the help from staff of the Literary and Philosophical society of Newcastle upon Tyne where most of this manuscript (and the as-yet-unpublished book that begat this idea) was written, the staff of the National Archive, Kew, London, staff of the John Brown archive at the Thurso Road site of the University of Glasgow and a number of staff members of the Marine Engineering department of Newcastle University.

In particular I also need to thank Mr Alex Grover, Assistant Curator of Historic Photographs and Ship's Plans at Royal Museums, Greenwich, London, for his help in arranging my appointment and then the viewing and measuring of the generously--scaled plans of H.M.S. Hood that allowed me to find the direct distances between Hood's propeller shafts and associated plummer blocks and the cordite magazines nearby.

Finally I would also like to gratefully acknowledge the assistance of my sons Chris and James Lawrence, who helped enormously with proof reading, various discussions, reference checking, rewriting of sections where the original text was clumsy and at times drew my attention to relevant reports and papers that helped further my own understanding both of metal fatigue and especially the flammability and burning characteristics of some materials in confined spaces and the consequent speed at which fires develop to the point where flashover conditions are reached.

7. REFERENCES

1. Bradford, E., "The Mighty Hood", *Hodder and Stoughton, London*, 1959

2. ADM267/111 and ADM229/14, also A. Raven and J. Roberts, British Battleships of World War 2, Arms and Armour press, London, 1976 P351. Prince of Wales took three hits from Bismarck and four from Prinz Eugen. One each of these was close to the aircraft catapult which caused local damage but there was also some damage in this area caused by the blast from Prince of Wales' rearmost turret firing forward of the beam. The passage of a 15" shell through the compass platform is known and it was assumed this shell did not explode because it did not hit anything robust enough to trigger the fuse but the same rule could hold for the 15" shell near the aircraft catapult which (if it did explode) must have done so on contact, i.e. performing as a high explosive shell, not an armour – piercing type as the former, with a much greater explosive charge, would have done far more damage. This particular hit is mentioned in the "diary" of HMS Prince of Wales ('One year of Life: the story of HMS Prince of Wales' Franklin A.G.C and Franklin G, William Blackwood and sons, London and Edinburgh 1944 page 31) but it is stated that the hit happened *before* the loss of HMS Hood which would indicate (if this statement is correct) that the hit on the aircraft launcher came from Prinz Eugen which mainly fired high explosive shells during the engagement (there was a switch made to armour piercing later); given the damage noted on Hood from a single 8" HE shell it is unlikely that Prinz Eugen's shell detonated properly. Bismarck fired at Hood until the explosion took down the latter, only thereafter switching her fire to Prince of Wales. There was additionally an underwater hit from a 15" shell that also failed to explode and was later recovered and gave very good intelligence on German shell design especially fusing. Other shells failed to detonate, or it is reported, detonated only partially. Even when not exploding, the damage that a large and solid mass of steel (the shell) travelling at supersonic speed is always considerable, for example two further hits from Prinz Eugen near the stern of Prince of Wales were responsible for partial flooding of the Battleship's stern by the ingress of some 2,000 tons of seawater. One of these was recovered intact and thrown overboard.

3. Naval-History.net: https://www.naval-history.net/xGM-Chrono-01BB-HMS_Prince_of_Wales.htm. Accounts of where the shells from Bismarck's 5th salvo landed all

place them in a line level with somewhere between Hood's mainmast and rearmost funnel (from observations of the shell splashes from rounds that missed), see also Captain Leach's statement in ADM 116/4351

4. JURENS, W J., "The Loss of HMS Hood: A re-examination", *Warship International vol 2 1987* p122 – 180 and references therein

5. ADM 116 / 4351 and ADM 116 / 4352. (Reports of the boards of inquiry into the sinking of HMS Hood). ADM 116/4351 in particular is very extensive and includes not only the conclusions of the boards but a great deal of the constituent as well as widespread background information

6. TAYLOR, B., "The Battlecruiser HMS Hood", Published by *Naval Institute Press / Seaforth* 2004 P 221.
   'Body parts and debris' that 'showered the lower deck, indicating a hit to the spotting top' (attributed to AB Tilburn).

   Additionally in this text there are numerous references to Hood's mechanical condition throughout the later chapters but perhaps of particular significance to the subject matter of this paper is just a sentence in chapter 8 P203: ".......one of the ventilation fans which began to disintegrate in clouds of rust once the Hood's main armament opened fire" (the primary reference for this is cited as a tape held by the Imperial War Museum (IWM/SA 12422 reel 3).

7. TAYLOR, B., "The Battlecruiser HMS Hood", Published by *Naval Institute Press / Seaforth* 2004 P 222.

8. KENNEDY, L., "Pursuit", *Collins, London* 1974 P. 67

9. ADM 234/509. The sea state is also discussed in the official summary of the action. In Prince of Wales the foremost turret had its floor awash with sea water pouring in through the gun ports and turret roof, the ammunition handling room below was also partially flooded with sea water coming in around the barbette of "A" turret.

10. BUXTON, I., (ed) "Diary of a Wartime Naval Constructor", *Sir Stanley Goodall, Seaforth Publishing / Pen and Sword books, Barnsley, UK* 2022.

    There are references throughout to Goodall's outstanding mastery of Naval matters from the very fundamentals of warship design to issues of building such as steel supply and quality through to the availability of dry docks, manufacture of ammunition and to regular updating of weapons and radar.

11. LAMBERT, N A., "Our bloody ships or our bloody system: Jutland and the loss of the Battlecruisers 1916", *Journal of Military History* 62 P29 – 56, January 1998.

    The loss in rapid time of Beatty's three battlecruisers at Jutland, all to magazine explosions, stirred much controversy and is widely covered in many texts, see for example "The Jutland Scandal" R. Bacon, Frontline books (Imprint of Pen and Sword publishing) Barnsley UK. Originally publications of this book date from 1916. It is regularly reprinted, and the destruction of the battlecruisers has rumbled on elsewhere through the years: see New York Times May 2nd 1989: https://www.nytimes.com/1989/05/02/opinion/l-disaster-of-1916-saved-lives-on-the-iowa-600189.html. It should also be pointed out that cordite fires occurred on other Royal Navy ships during this battle, most notably on some of the 6" gunned cruisers which came in for severe punishment but these fires were of smaller extent (ADM 229/25). A good summary of the timings of hits on, explosions of and the wreck positions of Beattie's Battlecruisers can be found in I. McCartney: The Opening and Closing Sequences of the Battle of Jutland 1916 Reexamined: archaeological investigations of the wrecks of HMS Indefatigable and SMS V4, downloadable from https://core.ac.uk/download/pdf/83943391.pdf

12. HOLMAN, G., "The King's Cruisers", Chapter 7 P102, *Hodder and Stoughton, London*, 1947.

The formal investigation into the explosion in Devonshire's turret is carried in Admiralty records ADM 1/8738/102

13. CUNNINGHAM, A B., "A Sailor's Odyssey", chapter 29 P377, *Hutchinson, London and World-wide*, 195, also JONES, G., "Battleship Barham", chapter 11 P219, *William Kimber, London* 1979

14. MACINTYRE, D., "Narvik" P119-12, *Evans Brothers, London*, 1959, also DERRY, T K., "History of the Seond World War: The campaign in Norway" P74-75, *HMSO London* 1952, The UK National Archives online document /pdfs/small/cab-66-7-wp-40-131-11.pdf (page 6) carries the official recording of the damage to HMS Suffolk as presented to the War Cabinet.

15. ADM 116 / 4351 carries a reference to these experiments. Quite why comparisons to Gunpowder handling and storage were never made is surprising.

16. JONES, G., "Battleship Barham", *William Kimber London* 1979, p255-263.

17. In stark contrast to the extensive boards of enquiry that followed the loss of HMS Hood, the inquiry into the loss of HMS Barham was carried out on the depot ship Woolwich moored in Alexandria harbour. The board reported on 18th December 1941. Very substantially different to Hood's case however was not only a considerable set of eyewitness accounts of the torpedo hits but also the physical appearance of the U boat on the surface nearby after it temporarily lost buoyancy control, which made the source of the torpedoes extremely obvious. Similar to Hood however was the magazine explosion; here it tore through HMS Barham well after the torpedoes themselves exploded giving more circumstantial evidence to the possibility of cordite exploding while burning under pressure and without a primer.

    See U.S. Environmental Protection agency: https://epa.gov/default/documents for specific issues regarding the burning and explosion of black powder (https://www.epa.gov/sites/default/files/2015-05/documents/9530608.pdf)

    A document on the properties and handling of black powder is provided here by the GOEX Powder Corporation (2009). The change in the burning properties of gunpowder from "open air" to "confined space" is explained in some detail.

18. CHESNAU, R., "Life and death of a Battlecruiser", *Cassell, London* 2002 P165.

    Phillips and Goodall's drive for a second (and larger scale) inquiry into the loss of HMS Hood was correct and highly laudable, but the conclusions reached were the same.

19. Widely recorded. See ADM 229/14 (DNC's correspondence) also in Brooke, G., "Alarm Starboard" *Patrick Stevens, Cambridge, U.K.* 1982 P74.

20. Details of German 15" Armour Piercing shell properties are given on the KBismarck website: http://www.kbismarck.com/armament.html.

    The important figures (as far as penetration is concerned) are an initial muzzle velocity of (close to) 2700 feet per second and a fuse delay of 0.0035 sec. This would give a distance of penetration into a ship, before the shell detonated, of only 9.45 feet which seems too short. The figure Goodall used (Bomb disposal's estimate of 0.01 second) would give a penetration before detonation of some 27 feet. Given the variables of obliquity and aspect ratio (particularly the latter) then 27 feet is closer to what would be expected although this is neglecting the shell slowing down while en route. It is possible of course to make an educated guess as to what the fuse designers would have thought optimal. Once 'through' the armour the desire would be for the shell to explode somewhere near the middle of the ship. Working on a typical "maximum beam" of British Capital Ships of the era (e.g. Queen Elizabeth and Royal Oak classes) of about 90 feet then an 'optimal fuse delay' of 45 feet would see the shell explode in the centre of the vessel. However the chance of hitting the very broadest part of the ship was small, so taking an 'average' beam over (say) the central half of the vessel would be more sensible. This looks to be of the order of 60 - 65 feet so a fuse delay sufficient to cause the explosion of the shell after penetrating around 30 - 35 feet would allow some margin for the shell slowing down while travelling through the air (then smashing through the armour plate) and also for penetration at angles other than 90^0. Goodall's figure for the fuse delay in Bismarck's 15" armour piercing shells being 'around 0.01 seconds' seems accurate enough if probably slightly low (but only slightly). It is not really possible to go into extended detail here but there exists an excellent online resource at http://www.navweaps.com/index_nathan/index_nathan.php and there is a good and well written summary of the complexities of shell aiming, including deceleration in the air, and acceleration as the shell "falls" towards its target, gravity effects and 'yaw' due to the shell spinning, in this article: https://nigelef.tripod.com/fc_ballistics.htm

21. ADM 229/25.

This document contains mainly correspondence to and from the DNC's office where "a probable hit" or "a probable cause" are phrases that indicate Goodall's overall dissatisfaction with the conclusion of the enquiry.

22. RAVEN, A., and ROBERTS, J., "British Battleships of World War 2", *Arms and Armour Press, London* 1976 P197.

Probably the most damning assessment of Hood's overall condition is a statement recorded here from the First Sea Lord in 1939: "...the likelihood of her being able to steam at high speed for any length of time was considered remote." There are additional remarks on Hood requiring a complete and total change of machinery as hers "was 20 years old" and that "her displacement is too high, her stability is too low."

23. TAYLOR, B., "The Battlecruiser HMS Hood". Published by *Naval Institute Press / Seaforth 2004* P182.

Of note here is a series of remarks about the condition of Hood's seaworthiness, battleworthiness and overall mechanical state, specific issues including, for example, the need to remove deck planking on the topside of the hull, to re-edge and caulk seams to stop sea water pouring in through breaks in them. This is on top of the overall tone in which the dockings for repairs (some of which were heavy) and maintenance, especially to her engines but also to her steam plant (the condensers as well as the boilers), her electrics, and shortly before sailing to fight Bismarck failures in her hydraulics were recorded. This list cannot lead to anything other than a conclusion that the vessel was in a poor and badly worn state.

24. Almost all texts on Hood refer to her having water on, and sometimes sweeping over her quarter deck in any sea state other than calm; while the sheer of the stern afforded some additional buoyancy at the rear of the ship the general view from various professional Marine bodies (e.g. the UK's Institute of Marine Engineering, Science and Technology) is that a vessel having water regularly not clearing away from the deck in anything other than severe storm conditions would "seriously bring into doubt" the vessel's seaworthiness. See also PARKES, O., "British Battleships", *Seeley, Service and Co. Limited, London*, P647 (1966): "Floating some 3 feet deeper than originally intended, the reduced freeboard was partially offset by a marked sheer at the bow and stern, but insufficient to keep the quarter deck from being flooded in a seaway."

25. Drawing NPB 3893 (HMS Hood, Builder's plans). National Maritime Museum collection, Greenwich, U.K.

26. Drawing UCS 110/460/12 John Brown Archive, University of Glasgow.

27. ROBERTS, J., "The Battlecruiser Hood" *Conway Maritime Press, Greenwich, London*, 1982, contains an excellent set of diagrams and drawings of the vessel.

28. Silver bridge disaster: National Transportation Safety Board (NTSB) "Highway Accident Report: Collapse of U.S. 35 Highway Bridge" (NTSB 1970). See also LICHTENSTEIN, A.G. "The Silver Bridge Collapse recounted" *American Society of Civil Engineers* 1993. Additionally there is an excellent documentary by the Open University with very understandable coverage of brittle fracture, and posted to YouTube: https://www.youtube.com/watch?v=8w5Fjouvma8&t=1042s

29. Sioux City disaster: NTSB/AAR-90/06 (1st November 1990)

30. GRIFFITH, A A., "The phenomenon of rupture and flow in solids", *Philosophical Transactions of the Royal Society A: (Mathematical Physical and Engineering Sciences)* 221 P582-593 (1921). Griffith was at the time a member of staff at the Royal Aircraft Establishment (he later moved to Rolls Royce). While his paper is generally accepted as the first major treatise on brittle fracture, work had been going on elsewhere as it had become widely known that metal sometimes failed at levels far below those predicted by the understanding of the time (as exposed by the references in Griffith's paper). The subject itself is complex but perhaps of significant interest here is the known fracturing of steel due to its exposure to low temperatures. During World War 2 as welding became more and more the method of choice for constructing ships some severe cracking - bad enough to cause a number of liberty ships to actually break apart and sink - was traced to the steel used becoming embrittled by cold water. In Hood's case, while her construction relied in riveting rather than welding (so cracking of this type would most likely be localised to one particular plate), the fitting of engine beds and shaft bearing mounts directly into the ships frame, itself in direct contact with the very cold Atlantic water could be yet another factor that caused fracturing, leading to the ship's loss. See Wikipedia article: https://en.wikipedia.org/wiki/Constance_Tipper. On perhaps a less academic note, the snapping of the steel hinges on a heavy hatch cover because of freezing cold conditions is mentioned in Alastair MacLean's outstanding novel about the Russian convoys, H.M.S. Ulysses.

31. SIMONS, E N., "Metal Fatigue", *David and Charles, Newton Abbot*, 1972. This is an informative and well written general introduction to this subject.

32. The Guardian February 26th 2019: "What caused the Genoa Morandi bridge Collapse, and the end of an Italian national myth".

33. TAYLOR, B., "The Battlecruiser HMS Hood". Published by *Naval Institute Press / Seaforth 2004* P 218.

34. Janes Fighting Ships, pub Sampson Low, Marston and Co. Limited, London 1937 edition p28-29.

35. ADM 116 / 4351. A.B. Tilburn evidence as reported to the second board of enquiry: "As regards the
explosion AB states that tremendous vibration resulted from another hit..." this was presumed to be the near miss from Bismarck's second or third salvos. The words in inverted commas are taken verbatim from ADM 116/4351

36. ADM 116 / 4351. Specific evidence given by Captain J. Leach of the Prince of Wales which includes a drawing showing the venting of a major fire seconds before HMS Hood blew up. Variations on this were reported by many other eye witnesses. Captain Leach's diagram, while simple, is very clear in all ways of the height, general extent and ferocity, and critically the position of the fire, close to the mainmast and thus some 65 feet ahead of the nearest magazine.

37. CHESNAU, R., "Life and Death of a Battlecruiser" *Cassell, London* 2002 P 109.

38. CARLTON, J S., "Propeller, Ship and Rudder interaction", *Elsevier Press, Amsterdam and New York* 2018. (This work deals with ship dynamics and causes of drive train failure)

39. ZHANG, X., and GU, X., "Effect of misaligned bearing support performance on Natural Frequencies of Marine Propulsion shafting" in *Journal of Vibro engineering, Vol 19 issue 3*, 2017 pages 1854 – 1866.

40. ROBERTSON, R G., "Ships Monthly" vol 16 number 4 page 28 (1981)

41. HMS Hood website at www.hmshood.org.uk

42. See https://warfarehistorynetwork.com/article/the-hood-has-blown-up/.

The observations of the wreckage of HMS Hood on the seabed with the starboard wing tanks blown open came from the expeditions to the wreck site after it was discovered by the explorer David Mearns in 2001 (and visited again in 2015 when the ship's bell was recovered). There are numerous articles online about these missions and photographs on the HMS Hood Association website at www.hmshood.org.uk .

43. CAMPBELL, J., "Jutland", *Conway Maritime Press, London* 1998 P384

44. BARNETT, C., "Engage the Enemy More Closely", *Hodder and Stoughton*, 1991 P416, also GARZKE, W H., DULIN, R O., and DENLAY, K V., "Death of a Battleship: a marine forensic analysis of the sinking of H.M.S", *Prince of Wales* available online as a PDF.

45. VIZENTIN, G., VUKELIC, G, MURAWSKI, L., RECHO, G., and OROVIC, J., "Journal of Marine Science and Engineering", Vol 8, 662, 2020. This work contains extensive coverage of ships engine room failures especially corrosion cracking.

46. ADM 116 / 4351 carries a letter from Admiral Sir John Tovey specifically regarding this but the matter is widely publicised and well known.

47. BALLANTYNE, I., "HMS Rodney", *Pen and Sword Books, London*, 2008 P144

48. ADM229/14 carries correspondence from the Office of Naval Construction regarding this work. Amplified in Commanding Officer / HMS King George V papers (Paper 4026/192 13th June 1941).

49. KENNEDY, L., "Pursuit", *Collins, London* 1974 P97

50. POOLMAN, K., "Ark Royal", *Tempus Publishing, Stroud, Gloucestershire*, 2000, P118. The partial bearing failure on Ark Royal was controlled by pouring (and then having to recover and re-use) cold olive oil over the unit for hour after hour to prevent it catching fire (or seizing completely).

51. Failure in ship's propulsion systems (which can include propeller shafts as well as associated equipment and fittings such as bearing mounts, support brackets, pressurised lubrication injectors, engine governors, and others) are not necessarily always associated just with old vessels. HMS Prince of Wales of the Royal Navy, in this case a nearly new Aircraft carrier commissioned into the fleet only in 2019, has suffered a widely publicised engine room breakdown due to shaft and propeller problems at present being blamed on faulty lubrication (see reference 50, above) and consequent failure of a bearing mount that was serious enough to lead to the propeller becoming so misaligned it hit the rudder, almost certainly causing some deformation in the ship's hull in the area of the main stern gland. See (for example) https://www.itv.com/news/meridian/2022-09-03/warship-returns-to-portsmouth-after-suffering-significant-damage-to-propeller or https://www.telegraph.co.uk/news/2022/08/30/hms-prince-wales-warship-grinds-halt-failure-grease-propeller/

While the new Prince of Wales is a larger ship than Hood, she was proceeding very slowly down the Solent, a busy and shallow sea lane, making 3 or 4 knots and in calm seas. Hood was travelling flat out at 28 knots in a storm. It takes little imagination to 'scale up' the damage a broken shaft mount could have caused on HMS Hood given the serious damage Prince of Wales has suffered.

52. SCHLIEHAUF, W., "Canadian Nautical Research Society / SCRN.org/Northern Mariner" vol 10 No 3 (July 2000) pages 57 – 89.

53. HMS Natal explosion. In Technical History and Index, Volume 2 part 24 (1919). (An Admiralty publication). A summary can be found here: http://www.gwpda.org/naval/thist24.htm

54. Fire Research Abstracts and Reviews, National Academies press, 1960. The production of Carbon Monoxide from incomplete combustion of carbon- containing products is well known in chemistry

55. OHLEMILLER, T J., "Smouldering Combustion in SPFE Fire Protection Engineering", ed. DINENNO P J., (a National Fire Protection Association publication)(U.S.A.) 1995.

56. QUINTIERE, J G., "Principles of fire behaviour", *Delmar Publishing, part of ITP, Albany, New York and London* 2016, P 38-42

57. BACKER, S., "Grand Fleet Battlecruisers", *Seaforth Publishing* 2011, Chapter 2 P31. At Jutland it had been variously noted that a delay occurred between shell hits and magazine explosions most notably on HMS Indefatigable but also HMS Queen Mary and HMS Invincible.

58. DRYSDALE, D., "Introduction to fire dynamics", *John Wiley and Sons, Wiley Interscience, New York*, 1985 Chapter 1, P2 – 10. Part of the definitions of 'fire science' associates "live flame" with the burning of vapour (gas phase) and specifically not with solids or liquids themselves; thermal decomposition must occur first.

59. DRYSDALE, D., "Introduction to fire dynamics". *John Wiley and Sons, Wiley Interscience, New York*, 1985. Chapter 10 pages 312-313. While beyond the scope of this paper, in earlier chapters this book also considers fire ignition and fire development in confined spaces and outlines in some detail the conditions needed for flashover to occur, especially the role of convective heat transfer. There can be no doubt that these conditions would have been met on Hood and convective heat transfer was almost certainly accelerated by the 'through draft' ('chimney effect') caused by the opening of the ammunition hoist doors serving "X" turret.

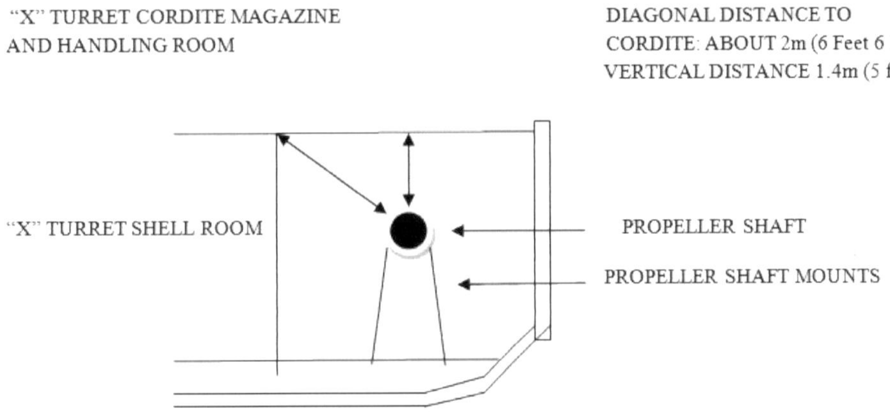

(2a)

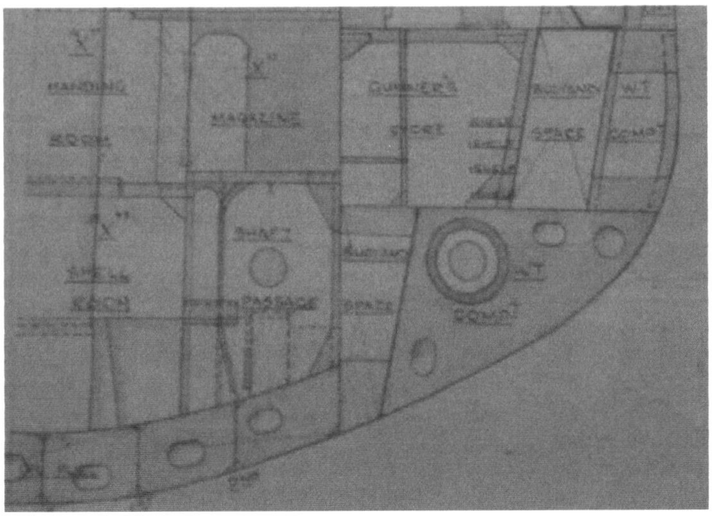

(2b)

Figure 2a and 2b: Proximity of Hood's magazines to the propeller shafts and shaft mounts. (a): Simplified drawing (see figure 2b, taken from the original builder's drawing). Starboard side of the ship, looking forwards. The cordite propellant at its closest point is only 1.4m (5feet) away from the propeller shaft. Note also the propeller shaft mounts (plummer blocks) are solid structures. Weight of shaft section between 20 and 28tons, the plummer blocks between 12 and 15tons. (b): From original builder's plans (reproduced courtesy of the University of Glasgow John Brown Archive). HMS Hood, cross section at frame 320, looking forwards. The floor of the magazine is only five feet above the shaft. (Same dimensions are seen at "Y" magazine).

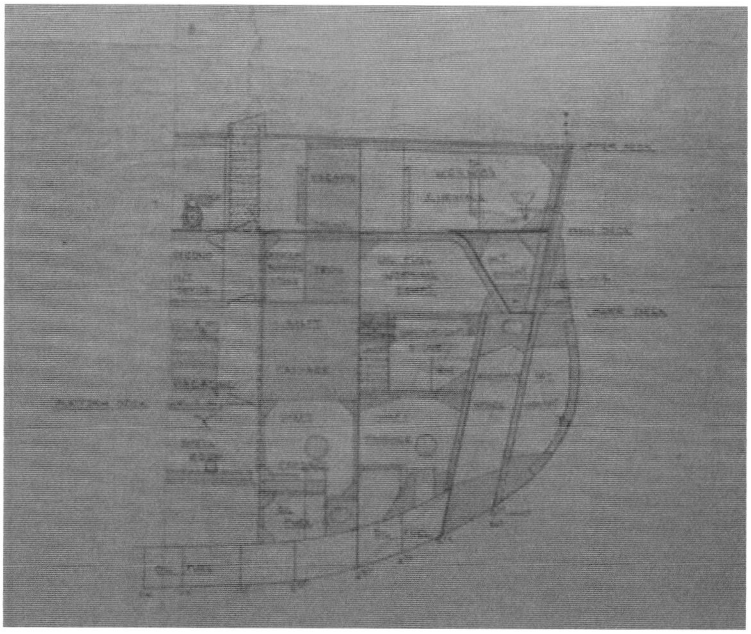

Figure 2b: As above, view from frame 310 looking forwards

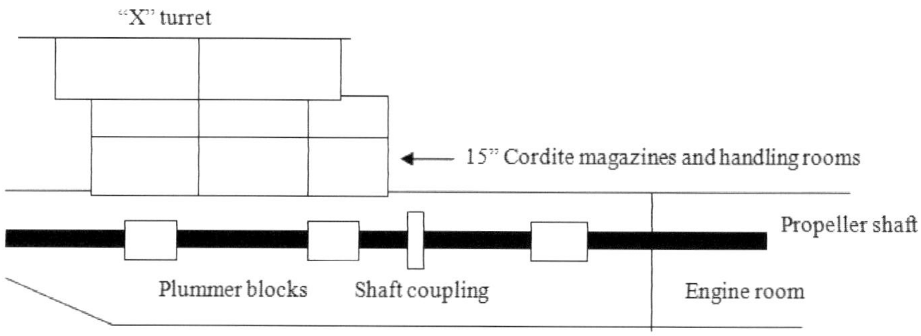

Figure 3: General (much simplified) arrangement of the rear of HMS Hood showing key compartments.

HMS Hood, elevation, stem section, considerable simplification of a part of drawing NMM NPB 3893 to show the proximity of the starboard inner drive shaft, the plummer blocks and shaft couplings, to the 15" magazines. The 15" shell room is behind the drive shaft and the 5.5" and 4" cordite magazines are also alongside and over the shaft immediately forward of the 15" magazine.

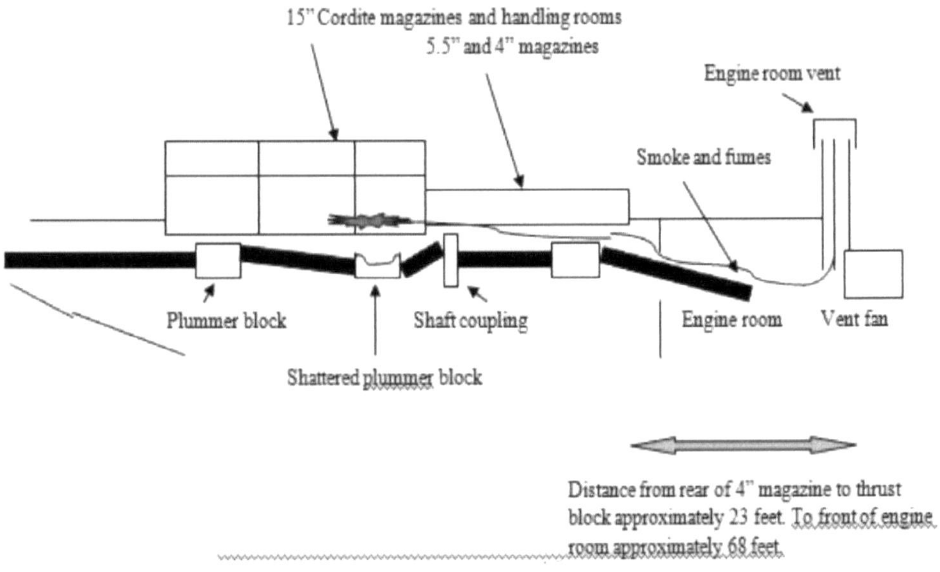

Figure 4: Fire in cordite (assumed here to be in "X" magazine) with fire venting through the engine room

HMS Hood, elevation, stem section, considerable simplification of a part of drawing NMM NPB 3893. Impression of damaged plummer block, early magazine fire, collapsed propeller shaft. The powerful ventilation fan in the engine room draws the smoke and flames through holes in the engine room bulkhead and outwards through the engine room vent. The main-mast, not shown in this diagram, is just to the right of the engine room vent. Also not shown, the distance from the rear of "X" turret to the engine room bulkhead is approximately 60 feet.

THE CUTTER IN THE FIELD: RESCUING A 1911 BOMBAY NAVAL CUTTER

N Ball, K S Hennum and **J Morgan,** Chatham Historic Dockyard Trust, UK

SUMMARY

The paper tells the story of how Chatham Historic Dockyard rescued a 1911 teak-built Royal Navy cutter from a field in Suffolk. It outlines the history of the vessel, how the Trust moved the cutter to the Dockyard, how they plan to conserve and display her to the public.

1. INTRODUCTION

In autumn 2021, Chatham Historic Dockyard Trust received a mysterious phone call from a Mr John Wilson about a boat languishing in a field in Suffolk. Mr Wilson had stumbled across the boat while out walking with his wife. Described by Mr Wilson as 'a motor/sailing RN cutter… in reasonable condition, made of teak and currently abandoned in a wood near Ipswich'. The Dockyard team were intrigued.

Never one to turn down an adventure, in February 2022, James Morgan the Trust's Historic Ships Manager, with his reliable sidekick Mark Smith, went to investigate. Mr Wilson had made contact with the owner of the field, Roger Bearcroft. He had bought the boat for firewood, luckily he didn't burn it, instead deciding to keep it. Clearing away years of debris, they uncovered something special – it was a 27 foot teak cutter. Mr Bearcroft kindly agreed to offer to donate the boat to the Trust. Now they just needed to work out how to get it out of the field and to the Dockyard.

Figure 1: The Cutter on a rusting trailer in a field near Stowmarket, Suffolk. February 2022.

2. CUTTERS

The name 'cutter' can sometimes cause confusion. In different periods it meant different things, and at some points several things. There are commonly held to be two definitions. The first is a broad single-masted vessel employed chiefly in the English Channel in the 18th and 19th centuries. This paper is not concerned with this definition, but instead the second definition, described in Falconer's *Marine Dictionary* of 1769 as a subsection in the entry for 'boat':

> *Cutters of a ship,* bateaux, *are broader, deeper, and shorter than the barges and pinnaces; they are fitter for sailing, and are commonly employed in carrying stores, provisions, passengers, &c. to and from the ship. In the structure of this sort of boats, the lower-edge of every plank in the side over-lays the upper-edge of the plank below, which is called by shipwrights clinch-work.*

The origin of these vessels is quite clear. They originated as the vernacular boats of the seafaring inhabitants of Deal in Kent.

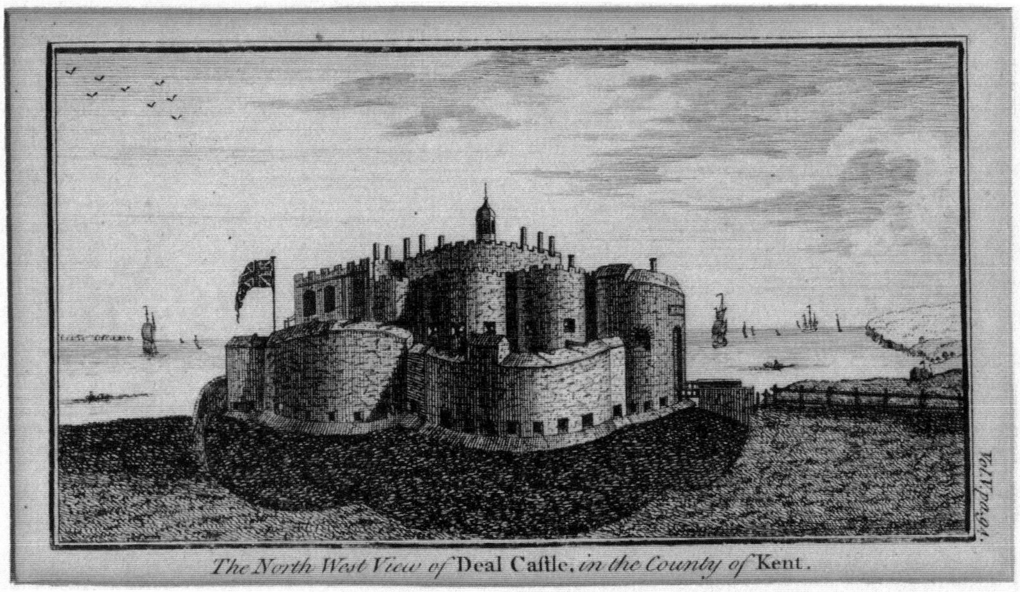

Figure 2: North West View of Deal Castle in the county of Kent by Samuel and Nathaniel Buck, 1735. CHDT Collections.

Cutters were clinker-built open boats, with a transom stern and a curved stempost and pointed bow. Relatively flat across the bottom, they could be pulled up the steep shingle beaches along the east coast of Kent. They could carry one or two masts, as well as be pulled by up to twelve oars. An early reference to a 'cutter' being used upon a Royal Navy vessel dates to 1712, when the Navy Board expressed disapproval of the captain of HMS *Rochester* for buying a cutter to replace the pinnace. [1] However, by the middle of the 18th century cutters were regularly carried aboard RN ships. In 1740 the Admiralty approved a Navy Board plan to equip ach First, Second and Third Rates with a cutter. These boats were ordered from boatbuilders in Deal. The overall design changed relatively little from the late 18th to early 20th century.

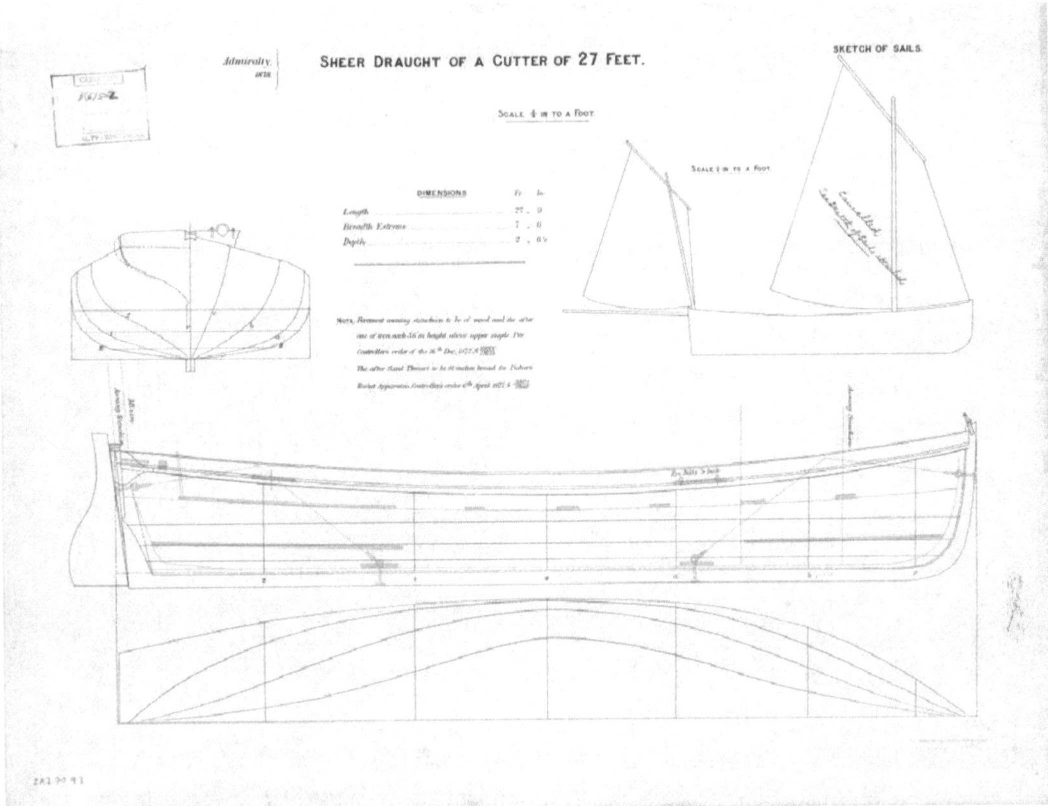

Figure 3: J0652, Object: ZAZ7093, 27 ft Cutter, Caption: Scale: 1:16, and 1:48. Lithograph showing the body plan, sheer lines with inboard detail, longitudinal half-breadth, and sail plan [cancelled] for a 27 ft Cutter. ©National Maritime Museum, Greenwich, London

3. THE RECOVERY

After an initial reconnaissance visit in February 2022, the team returned to the field in July 2022. To prepare the boat for moving, the hull was cleaned inside to remove mud leaves and detritus. Some 6x8 timbers were added to strengthen hull. A specialist boat moving company, was hired to move it. The boat was stropped under the hull and lifted off the rusting trailer onto a flatbed lorry. It was unloaded at the Dockyard, alongside HMS Gannet, after several hours on the road.

Figure 4: The cutter on the back of the lorry prior to departure to Chatham.

4. THE CUTTER ITSELF

An initial assessment of the cutter told us two things, that it was built of teak and was 27ft long. But a closer inspection revealed quite a few details. The stempost has markings.

Figure 5 and 6: The date 1911 and BOM and Admiralty Borad Arrow can clearly be seen on the starboard side of the stempost, indicating that the boat was built at Bombay Royal Dockyard in 1911. (The carved figures have been highlighted blue on the photograph for clarity.)

Figure 7: The prop shaft can be seen in this image, just above a short skeg.

Known as a 'three in one' open cutter, it was powered by oar, sail or engine (using an early paraffin engine). The engine, propellor or engine did not survive in the boat. The rudder was present inside the boat. Most of the thwarts survive. The are five rowlocks built into the wash strake, this added strake giving a bit more freeboard when under sail. More research is needed into the origin of the cutter.

Figure 8: The cutter prior to conservation treatment.

5. BOMBAY ROYAL DOCKYARD

Founded by the East India Company in the late 17th century, Bombay Dockyard became Bombay Royal Dockyard in 1811 when the Admiralty formally took charge. It was managed, just like the British Dockyards, by a Resident Commissioner sent by the Navy Board. Upon Indian independence, the Dockyard was taken over by the Indian Navy, and of course the city is called Mumbai. The Dockyard benefited from two things, firstly access to Malabar teak. Teak was strong, oily and flexible and had been known in shipbuilding in the region since at least the 5th century BC.[2] Secondly, and the skill of the local shipwrights who knew how to work the tropical wood. Famously the Wadia family of shipwrights rose to form a shipbuilding dynasty that contributed no less than nine Master Shipwrights to Bombay Dockyard between 1736 and 1884.[3]

6. CONSERVATION AND DISPLAY

This project is a combination of restoration and conservation. The cutter currently represents two separate histories: its archaeological burial environment and its historical purpose as a cutter used by the Royal Navy. To be able to showcase both histories, the naval cutter will be divided into two halves, whereas one will represent its archaeological context (conservation) while the other will bring it back to what its former glory might have been (restoration).

The cutter has, since arriving Chatham Dockyard, been stored in a water tank to ensure it does not dry out rapidly, which would eventually lead to the catastrophic event where the wooden structure starts collapsing. Further investigation of the cutter has shown that some areas are dry and stable while others are not, which means that the more unstable areas will require extra treatment. As the cutter has been stored in wet conditions for many years, water has replaced the degraded wood cells that once would have held the wood's structure. The water will therefore be replaced with a consolidation material to prevent the wooden structure from collapsing upon drying. The consolidation material will dry together with the wood and will provide a synthetic structure which will maintain the shape of the wood.

Figure 9: The cutter in its water tank. A spraying system was also set up to keep the boat moist.

After drying, there will most likely be some areas that will need further consolidation. In these areas, a similar type of consolidation material will be inserted into the wood using a syringe with a thin needle. To complete the conservation treatment, the cutter will undergo a thorough surface clean. However, the sturdy nature of the teak, has actually enabled us to conserve the boat with minimal intervention, and it is not thought necessary to do a full immersion in polyethylene glycol as originally considered.

Upon completion of the conservation treatment, the cutter will go into its next phase, the restoration project. Half of the cutter will be covered up with a breathable water and paint proof cover and will become the half that will represent the archaeological history of the cutter. The uncovered half will then undergo restoration to bring one half back to what the cutter might have looked like back in 1911. Once in its final position on display, this will quite literally allow us to tell the two sides of the story for the cutter in the field.

7. ACKNOWLEDGEMENTS

The authors would like to acknowledge: John Wilson for giving us the tip off about the boat. Roger Bearscroft for kindly donating the boat to the Chatham Historic Dockyard Trust. Simon Stephens for assisting with the research. Alfie Windsor of HMS Conway.

8. REFERENCES

1. MAY, W E., *Boats of Men The Boats of Men-of-war*, (1999)

2. TRIPATI, S., SHUKLA, S R., SHASHIKALA, S., and SARDAR, A., ‘Role of teak and other hardwoods in shipbuilding as evidenced from literature and shipwrecks’ *Current Science* Vol. 111, No. 7, pp. 1262-1268 (7 pages) Published By: Current Science Association (10 October 2016)

3. WADIA, R A., *The Bombay Dockyard and the Wadia Master Builders*, (1983)

9. AUTHORS BIOGRAPHY

Nick Ball holds the current position of Collections, Galleries and Interpretation Manager for Chatham Historic Dockyard Trust (CHDT). Ball is responsible for managing CHDT's collection of nearly 80,000 items, curating the permanent galleries, and temporary exhibitions across the 80-acre Historic Dockyard. A master's graduate in Maritime Archaeology from the University of Southern Denmark, he was formerly Archaeological Data Manager for HMS *Victory* at the National Museum of the Royal Navy, Portsmouth and Assistant Curator of Ship Models at the National Maritime Museum. He co-authored *The Navy Board Ship Models* with Simon Stephens, published by Seaforth in 2018.

Karoline Sofie Hennum holds the currently position of Conservator for Chatham Historic Dockyard Trust. A master's graduate in Conservation from the University of Oslo, she holds accreditation from the Nordic Association of Conservators and is a member of the UK's Institute of Conservation. Before taking charge of CHDT's Conservation in January 2022, Hennum has undertaken conservation work at the Mary Rose Trust and The National Trust.

James Morgan holds the current position of Heritage Engineering and Historic Ships Manager for Chatham Historic Dockyard Trust. Morgan is responsible for the maintenance programme for the three historic warships, HMS *Cavalier*, HM Sub *Ocelot*, and HMS *Gannet*, as well as CHDT's boats, machinery and historic working Ropery. Morgan completed two Advanced Apprenticeships, one Engineering and one in traditional boatbuilding, and previously worked for Heritage Marine and Anglian Water.

Historic Ships Conference
28th – 29th November 2023
London, United Kingdom

SS GREAT BRITAIN: VOYAGE TO CARBON NEUTRAL

N Grahamslaw, SS Great Britain Trust, UK

SUMMARY

Brunel's SS Great Britain, the first iron ship, is conserved and displayed in the historic dry dock where she was built. The most vulnerable parts of the original wrought iron hull are preserved by keeping the surrounding air at 20% relative humidity to prevent corrosion. A glass "sea" at waterline level creates a seal, and bespoke desiccant dehumidifiers desiccate the environment inside and underneath the ship. This conservation method relies on natural gas for drying the air, and electricity for circulating it.

A specialist engineer was appointed to develop system-level understanding and ownership of the desiccant plant and develop a strategy to decarbonise the conservation system. Improvements to controls and instrumentation provide new insight into the system's behaviour. Using this information, optimisation and adaptation of the system have both directly reduced emissions and altered the heat demand, introducing the possibility of future decarbonisation via electrification overcoming reliance on natural gas.

1. INTRODUCTION

1.1 THE SS GREAT BRITAIN

Launched in 1843, Isambard Kingdom Brunel's ss Great Britain was the first ocean-going vessel to be driven with a screw propeller and the first to be made from wrought iron, allowing her to be significantly larger than had previously been possible and providing the blueprint for modern-day hull design. She survived being run aground early in her career and went on to enjoy an extraordinarily long and impactful working life, thanks to the strength and longevity of this exciting new shipbuilding material and technique [1]. Ending her working life as a hulk in the Falkland Islands, the ship was salvaged in 1970 and returned to the original Bristol dry dock in which she was constructed, the combination of ship and dock creating a site of international cultural significance which has since been almost continuously accessible to the public.

1.2 THE CONSERVATION CHALLENGE

On her return to Bristol the SS Great Britain was cared for by traditional engineers, but analysis of the hull revealed that corrosion, accelerated by chlorides in the wrought iron structure, meant that without intervention within 25 years the ship would become structurally unsafe and original fabric irretrievably lost [2].

A report commissioned to examine the importance of the ship's historical, technological, and cultural contexts presented an evidence-based argument that preservation of the ship and her original dock was of international importance, with survival of the original iron hull central to any conservation plan [3].

Condition surveys revealed a marked contrast between the topsides and the lower hull and interior. The hull below the waterline and the interior surfaces of the ship were found to be severely corroded and the material infused with chloride, and the exterior of the upper hull less corroded. A conservation strategy was developed adopting two different regimes for the two distinct parts of the structure.

2. CONSERVING THE SHIP

2.1 THE LOWER HULL AND INTERIOR

Previous attempts at protecting the lower hull and interior had proven ineffective, infused chloride causing corrosion to continue beneath any coating which would become compromised, re-exposing the material to ambient air. No practically attainable or sufficiently proven method of removing embedded chloride on the scale of the ship and grade II*-listed historic dry dock structure was available.

Conservation efforts therefore focus on creating a desiccated microclimate surrounding the most vulnerable parts of the iron, to achieve a negligible rate of corrosion while allowing visitor access [4]. A horizontal glass roof creates a seal between the hull waterline and the side of the dry dock, allowing desiccation of the hull exterior below the waterline (Figure 1). Airlock-style doors create an equivalent space within the ship's interior (Figure 2).

Figure 1: The glass plate creates a seal around the hull below the waterline. Ducts and vents are positioned to create a layer of dried air adjacent to the historic iron.

Figure 2: Interior of lower ship's hull, showing iron structure and desiccated air delivery ducts.

Two bespoke desiccant dehumidifiers condition the air inside and underneath the ship, with vents and ducts positioned to form a protective desiccated air curtain adjacent to the hull. This design allows parts of the space furthest from the most vulnerable material to deviate from the target conditions [5], reducing the impact of visitor presence and the energy requirement of the desiccation plant compared with uniform desiccation of the entire envelope.

To derive target conditions for the desiccated microclimate, research was commissioned to identify both 'no corrosion' points and corrosion rates for chloride-infested iron as a function of relative humidity (RH). Laboratory modelling of corrosion mechanisms within chloride-infested iron identified that at a temperature of 20°C, corrosion ceased below 15% RH, was negligible at 20% RH and did not produce a significant impact until 25% RH [6, 7]. A target condition of 20% RH was selected.

The ship's galley, dining saloon and other catering areas in the centre of the interior are sealed in a separate environment which does not include any of the vulnerable original iron. This space is vented via the ship's funnel to prevent moisture from catering activities being incorporated into the climate-controlled envelope.

2.2 THE UPPER EXTERIOR

The upper exterior of the hull has been stripped of corrosion products and is protected from the external environment using a multi-layered coating, formulated to combine the chemical resistance of epoxy resins with the UV resistance of polyurethanes.

The separation of the two regions with differing conservation protocols allows the energy-intensive desiccation treatment to be employed only where the condition of the material necessitates it. This approach minimises the size of the required climate-controlled envelope, offering energy savings compared to encasement of the whole ship in a building, and allows the upper portion of the ship to be displayed as part of the skyline of Bristol's historic harbour.

The glass roof is flooded with a 50mm layer of water, creating the appearance from above that the ship is ready to set sail (Figure 3), enhancing the visitor experience as they descend "underwater" to enter the desiccated dry dock.

Figure 3: A seal at waterline level with a 50mm layer of water on the glass plate provides insulation while creating the appearance that the ship is ready to set sail.

3. SUSTAINABILITY STRATEGY

3.1 DESICCATION SYSTEM, AS DESIGNED

The energy requirement of such a conservation system was carefully considered and justification sought prior to installation [8]. Nevertheless, the solution was designed to be non-interventive and reversible, with no chemical or physical interference with the material. If at some future time the argument for conserving the ship is called into question and the energy expenditure found to no longer be justifiable, the equipment can simply be switched off.

The climate-controlled envelope and desiccation plant were also designed to be energy-efficient. The flooded glass plate acts as insulation and reduces solar gain, providing an estimated 20% energy saving for the desiccation plant, which also utilises water from the floating harbour for cooling. Only the most vulnerable parts of the iron are enclosed in the climate-controlled envelope, and the desiccated air is carefully directed to form a curtain over the iron hull, allowing the fabric of the ship to be kept at 20% RH, while air closer to the dry dock walls is allowed to deviate from the target.

3.2 FUTURE SUSTAINABILITY PLANNING

Following a 15-year period of successful operation, a dedicated "Ship's Conservation Engineer" was appointed by the SS Great Britain Trust to assess the effectiveness of the conservation system, with a view to ensuring it remains fit-for-purpose in the very long term. A particular objective was identified to reduce the energy consumption and environmental footprint associated with the system. Creating an in-house role was intended to allow the Trust to develop understanding of the conservation at a whole system level, in contrast to the previous approach of supplementing the work of maintenance contractors with ad-hoc consultant-led support. This new role would facilitate a programme of research, adaptation, and optimisation to ensure the long-term stability and sustainability of the ship's conservation.

4. 'VOYAGE TO CARBON NEUTRAL' PROJECT

4.1 OPERATION OF THE DESICCATION PLANT

Each of the two desiccant dehumidifiers comprises two streams of air:

- The "supply" stream dries air by forcing it through an adsorbent desiccant, circulates the dried air around the ship or the dry dock, and then recycles it for re-drying and recirculating, mixing in fresh air only if the CO_2 concentration increases beyond a given threshold.
- The "reactivation" stream takes in external air, heats the air using a natural gas burner then uses it to remove moisture from the desiccant, exhausting the now damp air from the desiccant to the external environment. Heat exchangers recover heat from the exhaust air, pre-heating the incoming air.

There is no air exchange between the two streams, but the desiccant material is mounted on a rotor continually moving each segment through the supply stream to extract moisture from the supply air, and then through the gas-heated reactivation stream where moisture is extracted from the desiccant material and exhausted to the atmosphere (Figure 4).

To control the system, relative humidity sensors are positioned in a variety of locations adjacent to the surface of the iron, and the difference between the average of the sensor readings and the target RH is used to control the gas supply to the reactivation heaters via a simple Proportional-Integral (PI) control algorithm. When the RH increases above the target, the supply of natural gas to the burners is increased and the reactivation stream air becomes hotter, removing more moisture from the desiccant material which in turn removes more moisture from the supply air.

The key carbon footprint contributors of this system are the gas supply providing heat to the reactivation stream, and the four fans used to circulate air through the supply and reactivation streams and around the ship and dry dock.

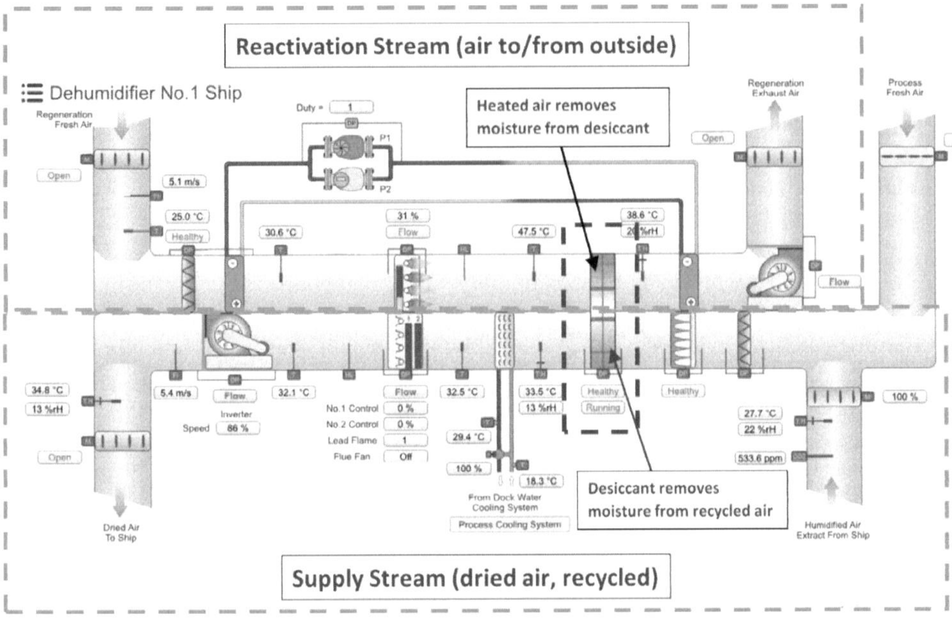

Figure 4: Schematic showing one of the two bespoke desiccant dehumidifiers that create the controlled environment.

4.2 REVIEW OF PLANT PERFORMANCE

An initial review of the system data showed that the average RH was being held broadly stable, but further assessment of the detailed data led to the following key observations:

4.2 (a) Relative Humidity Sensor Performance

Over time, approximately half of the wireless relative humidity sensors had failed. The frequency of data transmission by the remaining sensors was low, resulting in apparent sudden step changes in the average relative humidity. The control algorithm tuning parameters were not appropriately set for the reduced frequency of data transmission. This combination resulted in the system controls exhibiting 'cycling' behaviour, repeatedly over- and under-shooting the target RH. During periods when the gas demand was too low, moisture was being actively introduced into the system via cool and damp air in the reactivation stream, and during periods when the gas demand was too high, unnecessary gas was being burned to dry the desiccant, and therefore the space, more than required (Figure 5).

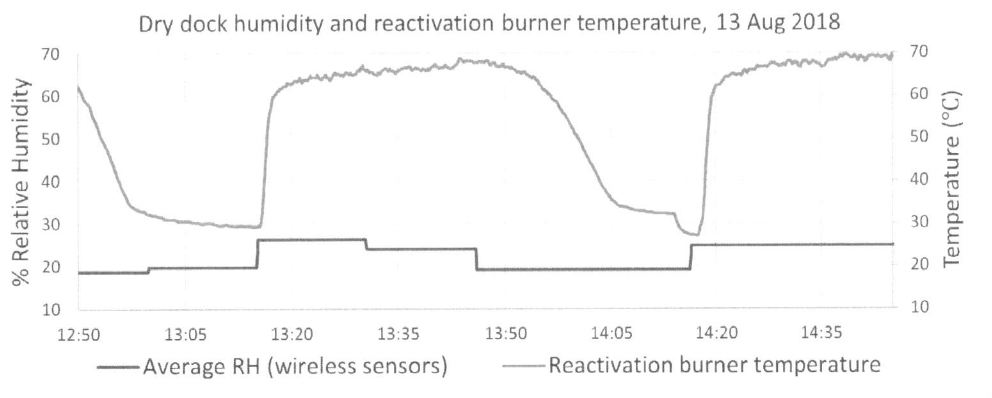

Figure 5: Dry dock dehumidifier reactivation air temperature and average RH cycling.

4.2 (b) Plant Instrumentation

Each sub-component of the air handling systems was found to be instrumented with temperature and relative humidity sensors, with the data from all sensors available via building management system (BMS) software. Prior to appointment of the Ship's Conservation Engineer, the Trust did not have any staff members with the capacity or the experience to utilise the full functionality of the BMS software, which itself was shortly due to become obsolete. Information from gas and electricity meters was also available but had not been studied in detail.

4.3 CONTROL AND INSTRUMENTATION UPGRADES

To address these two challenges, the 'Help the Hull' project was initiated to improve the ship and dry dock RH sensors, and to facilitate easier data collection and analysis from the instrumentation elsewhere in the system. The obsolete BMS was migrated to a newer and more user-friendly software package. The layout and functionality of the migrated system was collaboratively designed by BMS Engineer contractors, the Ship's Conservation Engineer, and the Trust's team of Site Technicians, resulting in a system which is more intuitively navigated and key information more easily accessed by both technically experienced and non-specialist users.

The wireless relative humidity sensors in more accessible ship and dry dock locations were replaced with wired sensors, overcoming the previously identified communication issues, and introducing the additional benefit of not requiring regular battery changes. An additional set of portable wireless sensors were also installed, offering better signal transmission than their predecessors to monitor less easily accessible parts of the ship and dry dock.

4.4 RETRO-FITTING ENERGY EFFICIENCY TECHNOLOGY

The modular nature of the desiccation plant and BMS control system offers the flexibility for individual components to be modified independently of the rest of the system, minimising the expense and embodied energy requirement of plant modernisation. Aided by easy access to data from the new BMS, the system-level understanding developed by the Trust allowed for two relatively simple design modifications to be planned. Energy payback and carbon footprint reduction calculations provided compelling evidence to stakeholders allowing grant funding to be secured.

4.4 (a) Desiccated Air Supply Fan Replacement

On approaching the end of their originally anticipated service life, the belt-driven scroll fans circulating desiccated air around the ship and dry dock were replaced with a more reliable and energy-efficient new design. A grid of smaller direct-drive fans provides an overall electricity saving of more than 20% without the need to modify any other part of the air handling plant or controls, resulting in an energy payback time of less than 3 years.

Figure 6: Belt-driven scroll fan (left, 2005-2021) replaced by a grid of nine direct-drive fans (right, 2022-present).

4.4 (b) Energy Recovery Purge Systems

An energy recovery purge system, an additional heat recovery device in the reactivation air stream, was also installed on each dehumidifier (Figure 7). This device fits onto the desiccant rotors, recovering heat from the desiccant material as it moves from the reactivation stream into the supply stream. The recovered heat reduces demand on the gas burners, and cooling of the desiccant material as it moves into the supply air stream also improves its adsorption performance, as well as reducing the load on the subsequent supply air cooling system.

For the energy recovery purge system to deliver its full potential benefit, it was also necessary to eliminate the "cycling" behaviour in the control algorithms. The combination of higher quality RH sensor data and improved understanding of the system controls which had now become available was used to inform adjustments to the dehumidifier gas supply and the PI controller tuning parameters, improving the stability of the system (Figure 8).

Figure 7: Energy Recovery Purge system fitted to the desiccant rotor on the dry dock dehumidifier.

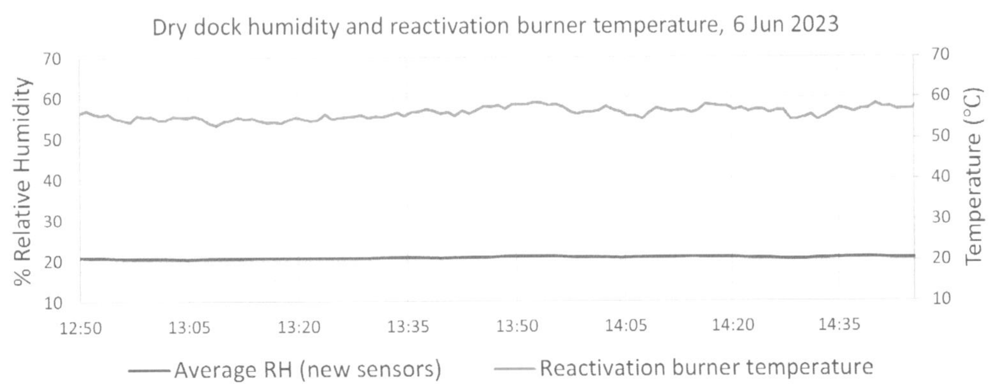

Figure 8: Dry dock dehumidifier reactivation air temperature and average RH, showing improved stability.

5. RESULTS

With the PI control algorithms appropriately tuned to avoid overshoots and the energy recovery purge system installed, the natural gas consumption of the dehumidifiers has been reduced by an average of 28% over a period of 1 year. In addition to delivering energy savings, the combined improvements to the system also result in the desiccant reconditioning taking place at lower temperatures compared with previous operation (Figure 9). This is significant because lower operating temperatures can more easily be achieved with lower-carbon heating methods, such as heat pumps, instead of natural gas. Work is currently underway to determine whether it would be feasible to supply some or all of this heat via heat pumps, perhaps even utilising waste heat from other parts of the system, which would allow for further energy savings and, potentially, complete decarbonisation.

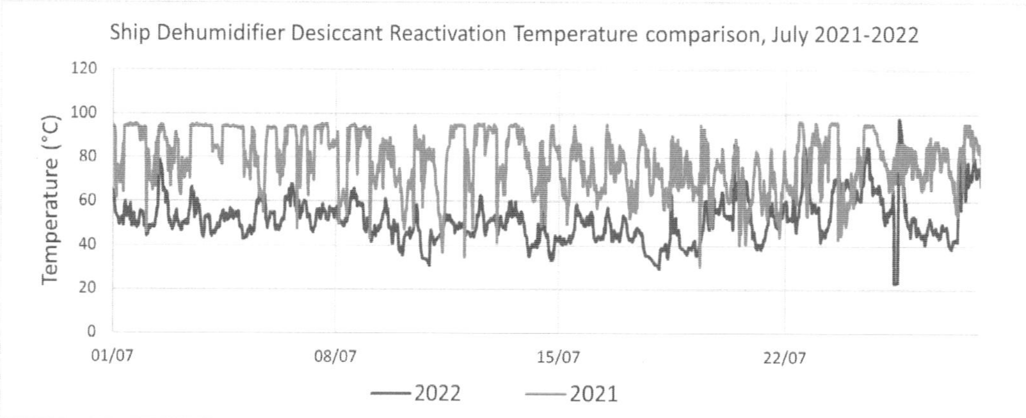

Figure 9: Installation of an energy recovery purge system, combined with control algorithm optimisation, has reduced the temperature at which heat is required for the drying action of the desiccation plant.

6. CONCLUSIONS AND FUTURE WORK

The only feasible method identified to conserve Brunel's SS Great Britain, the first iron ship, is to create a desiccated microclimate around vulnerable parts of the original wrought iron hull to prevent corrosion. A desiccation system was designed to achieve this objective and the environment around the hull has been held broadly stable for more than a decade, albeit with appreciable energy costs and emissions.

The Ship's Conservation Engineer is a new specialist role introduced to review and develop the long-term strategy to sustainably conserve the ship. Following a review of the desiccation plant's performance, improved instrumentation and a new building management system have been designed and implemented. The dehumidifiers have been retro-fitted with new energy-efficient technology and, using new data, the control algorithms have been optimised. This has directly reduced energy consumption and associated emissions, and reduced the temperature at which heat is required for the drying action of the dehumidifiers, introducing the possibility of further decarbonisation, and overcoming reliance on natural gas.

Continuation of this work will develop a deeper understanding of the relationship between factors such as external weather conditions or visitor numbers and dehumidifier behaviour to allow further fine-tuning of the controls, for example using weather forecasting, visitor opening hours and events schedules to pre-empt changes in conditions and further optimise the performance of the system. It is anticipated that this optimisation will work alongside a further design development to recover more waste heat from the system, utilising heat pump technology, to further reduce gas consumption and associated emissions.

7. ACKNOWLEDGEMENTS

The author wishes to acknowledge the DCMS Wolfson Museums and Galleries Improvement Fund, the West of England Combined Authority Green Business Grants scheme, and the SS Great Britain Trust's members and supporters for providing the funding to make this work possible.

8. REFERENCES

1. CORLETT, E., 'The Iron Ship: The Story of Brunel's SS Great Britain', *Park Lane Press Ltd*, 2012.

2. TURNER, R., TANNER, M., and CASEY, S., 'Conservation plan for the Great Western Steamship Company Dockyard and the ss Great Britain. Volume 2 – Condition report and recommendations for the SS Great Britain', *SS Great Britain Trust*, 1999. (unpublished) https://www.ssgreatbritain.org/collections-and-research/research-articles-papers/ (accessed September 2023)

3. COX, J. and TANNER, M., 'Conservation plan for the Great Western Steamship Company Dockyard and the SS Great Britain. Volume 1', *SS Great Britain Trust*, 1999. (unpublished) https://www.ssgreatbritain.org/collections-and-research/research-articles-papers/ (accessed September 2023)

4. WATKINSON, D., TANNER, M., TURNER, R. and LEWIS, M., 'SS Great Britain: Teamwork as a platform for innovative conservation', *The Conservator. 29. 10.1080/01410096.2005.9995214*, 2005.

5. GOODARZI ARDAKANI, V., GAMBARUTO, A., GRAHAMSLAW, N., GARSED-BRAND, G., AULT, R., VAQUERO, E. AND PREGNOLATO, M., 'Heritage conservation with a surface air curtain: Use of CFD and sensor data with a near-wall region flow analysis' *Building and Environment, Volume 243*, 2023.

6. WATKINSON, D. and LEWIS, M., 'SS Great Britain iron hull: modelling corrosion to define storage relative humidity', *Metal 04 Proceedings of the International Conference on Metals Conservation, ed. J. Ashton and D. Hallam, National Museum of Australia, Canberra*, 2004.

7. WATKINSON, D. and LEWIS, M., 'Desiccated storage of chloride contaminated archaeological iron objects', *Studies in Conservation, 50 1-12*, 2005.

8. WATKINSON, D. and TANNER, M., 'SS GREAT BRITAIN: CONSERVATION AND ACCESS – SYNERGY AND COST', *Studies in Conservation, 53:sup1, 109-114, DOI: 10.1179/sic.2008.53.Supplement-1.109*, 2008.

9. AUTHOR BIOGRAPHY

Nicola Grahamslaw graduated with an MEng from St Catharine's College, Cambridge, before working as an analyst and project engineer in aerospace and nuclear energy. In 2018 she was appointed to the newly created "Ship's Conservation Engineer" position at the SS Great Britain Trust, developing a sustainable strategy for the future care of this unique object. She is a Chartered Engineer and holds an MBA. She sits on the International Congress of Maritime Museums Ships Committee and the Institute of Conservation Heritage Science Group Committee. Nicola was awarded Institution of Mechanical Engineers "Young Member of the Year" in 2019 for her work with the ship and as a volunteer STEM ambassador, speaker, and mentor.

PRESERVING OUR SHIPS FOR POSTERITY – A HOLISTIC APPROACH TO MANAGING HISTORIC SHIP PROJECTS

O Millar, M Thomas and **J Monasta**, Beckett Rankine, UK

SUMMARY

Historic ships are an essential part of the United Kingdom's maritime heritage, but their ongoing conservation faces many challenges. These include an ongoing requirement for maintenance that extends beyond the vessel's working life, unsustainable financial cycles, and the loss of skills necessary for repairs and restoration. To support the owners of historic ships through conservation projects, Beckett Rankine has developed the "Engineered Conservation" method. This applies a holistic perspective, a focus on the requirements of the vessel and owner, consideration of all key stakeholders, and commitment to proactive and continuous maintenance to achieve sustainable conservation. The principles and application of this method are presented through four case studies, using vessels at different stages in their conservation projects. Through these, the successes and limitations of the engineered conservation method are identified and discussed, allowing conclusions to be drawn on its effectiveness.

1. INTRODUCTION

1.1 BACKGROUND

The United Kingdom has a unique relationship with its maritime heritage. The nation's history is memorialised by the many artefacts and structures of great significance still present across the country. The UK's historic fleet is a critical component of this maritime heritage, providing communities with gateways into their past, sharing stories, skills and ways of life that are rapidly disappearing. The National Register of Historic Vessels (NRHV), managed by National Historic Ships UK (NHS-UK), contains over 1,200 ships, diverse in age, background, and origin. Many ships on the NRHV are no longer used for their designed purpose, having been superseded by technological advancements [1]. Instead, they are retained for their historical significance, such as HMS *Victory,* which was secured in 1921 as a memorial to British Naval Power, long after the end of its military career [2].

1.2 UNSUSTAINABLE CYCLES OF CONSERVATION

Historic ships are powerful symbols that are emotive and important in transferring historical knowledge [3]. However, historic ships face various challenges to their conservation. The nature of their design and construction means that historic ships have a continuing requirement for maintenance, extending beyond their design lives. Ongoing maintenance places a financial burden on the vessels' owners, who often rely upon grant schemes to secure the necessary funding [4]. Accordingly, many historic vessels are trapped in financially and environmentally unsustainable cycles of large capital investment, followed by periods of insufficient maintenance. This cyclic nature often leads to loss of historic fabric and reputational damage, with no clear path to greater stability.

1.3 NATIONAL HISTORIC SHIPS AND THE CONSERVATION PATHWAY

National Historic Ships UK is a government-funded, independent organisation providing a vital source of support to the historic ships sector. To assist vessel owners, NHS-UK has developed the Historical Ships Conservation Pathway, consisting of a series of guidelines and suggested practices for the conservation process [5]. NHS-UK's Conservation Pathway is an essential resource for the owners of historic ships, providing flexibility in selecting the appropriate strategy for specific vessels and projects. Flexibility is necessary due to the inherent difficulties in identifying a vessel's correct conservation path, as owners of historic ships often face highly complex situations. The various conflicting requirements include the needs of the vessel's structure, available finances and funding, future financial profile, geographical considerations, and external factors such as community support and availability of skilled labour.

1.4 AN ENGINEERING APPROACH TO HISTORIC VESSEL MANAGEMENT

NHS-UK's Conservation Pathway and wider guidance provide a strong foundation for managing historic vessels. However, the practical delivery of conservation projects remains an extremely complicated process. A novel engineering-based conservation approach has been developed to assist owners with these challenges. This method aims to provide the owners and curators with the necessary support at any stage in a historic vessel's life, from the early steps of the Conservation Pathway to ongoing maintenance.

The "Engineered Conservation" method is the subject of this paper. It has been developed by Beckett Rankine (BR), a marine engineering consultancy, over multiple conservation projects on historic ships. The underlying principles are presented, and their practical application is demonstrated through several case studies. The results are discussed and evaluated, allowing conclusions to be drawn on its applicability and advantages, alongside its limitations. Through the practical application of engineering principles, the method provides historic ship owners with a roadmap to sustainable conservation.

2. ENGINEERED CONSERVATION

2.1 OBJECTIVES

The aim of engineered conservation is to support owners in conserving their vessels. This requires developing and implementing a tailored approach specific to each ship. The demands of all key stakeholders should be taken into account, including curators, owners, funders, public users and those carrying out the conservation works. A realistic set of options for the vessel can then be developed, including the work paths necessary for each. A proactive maintenance regime should be established, which functions within the financial and geographical constraints. Through these means, a sustainable long-term vision can be established for the vessel, securing its conservation for the future.

2.2 PRINCIPLES

The engineered conservation approach is founded on five key principles, reflected throughout its implementation. These are:

- **A Holistic Perspective**
  The consequences of all actions are examined throughout their lifecycle, allowing effective evaluation and decision-making. The priorities of all key stakeholders are incorporated, ensuring that their concerns are addressed during delivery.

- **Vessel and Owner-Focused**
  The core needs of the vessel and owner are identified, allowing for the creation of a tailored approach focused on their requirements. This approach relies upon a technical understanding of the structural demands of the ship and a close and collaborative relationship with those responsible for the vessel.

- **Balancing Conflicting Demands**
  Conservation is a central concern for any historic vessel, and historic fabric preservation is often vital. However, this must be achieved within the present financial limits, and must balance potential revenue streams and public interest. The need for ongoing maintenance, including its costs and method, must be considered.

- **Proactive and Continuous Maintenance**
  Proactive maintenance is essential, as it minimises loss of historic fabric and structural decay, reducing major intervention requirements. A suitable maintenance programme must be implemented within the existing financial and geographical constraints.

- **Sustainability-Driven**
  Historic vessels must be conserved through environmentally and financially sustainable techniques to ensure their longevity. The long-term consequences of actions must be considered to ensure successful intervention and guarantee the vessels' survival for posterity.

2.3 PRACTICAL IMPLEMENTATION

The application of the engineered conservation method depends on the vessel's situation. However, it can be divided into general stages, which reflect a full project lifecycle, from the assessment of the vessel through to ongoing maintenance.

2.3 (a) Information Gathering

A strong understanding of the situation of a historic ship is the foundation of engineered conservation. In line with NHS-UK guidance, through the Statement of Significance, the Conservation Management Plan and discussions with the owner, the conservation priorities are determined, serving as a decision hierarchy for the conservation process. The ship's condition can be ascertained through reference to recent survey reports and consultation with those responsible for maintenance, or combined with direct inspection of the assets. Additionally, any constraints on the vessel must be established. These are often vital in selecting conservation paths, as they determine the viability of the options.

When all relevant information has been collated and reviewed, a gap analysis is conducted to identify areas of uncertainty. Further investigations are then recommended to address any significant gaps. These can include:

- Archive research
- Hydrographic and LIDAR scans
- Additional surveys into the condition
- Material testing
- Living testimonies
- Consultation with experts

The information gathering phase aims to minimise risk to the vessel and maximise the efficiency and effectiveness of work. A full understanding of the vessel and its' situation is essential to adopting a holistic perspective and is at the heart of the decision-making process.

2.3 (b) Optioneering

With a comprehensive understanding of the vessel, all viable options can be identified and examined. The optioneering process is based on a holistic approach to problem-solving, with the perspectives of all key stakeholders being recognised and integrated, and the full spectrum of final outcomes and the work path required to reach each one clearly set out. These are visualised through the conservation map, a vital tool for decision-making, identifying key points at which different paths diverge. It should be noted that sustainable maintenance is an ongoing element that must continue throughout the project lifecycle.

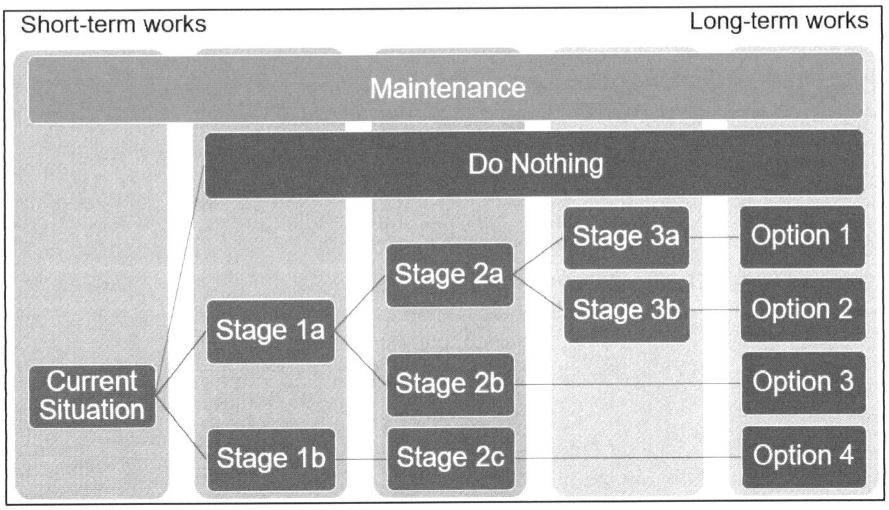

Figure 1: Example Conservation Map.

With the full paths of works understood, their requirements, risks and results can be established. This can include qualitative analysis, risk assessments and cost estimates. The options can be compared in a thorough and objective manner, and the demands of stakeholders examined and balanced accordingly. This allows for fully informed decision-making for sustainable conservation.

The decision-making process can be achieved through various means, depending on the preference of the vessel owner. This includes direct choice by the owner or curator, an impartial scoring procedure, or consultation with key stakeholders.

Optioneering is at the centre of the engineered conservation method. It seeks to break down and simplify complex situations, presenting all viable options in an objective and comparable form. All decisions in the engineered conservation process are made through optioneering, either through formal options studies or a process of value engineering. This results in a thorough decision-making process, providing a basis for future planning, funding applications and promotional campaigns.

2.3 (c) Project Management

Once a preferred conservation path is selected, the engineered conservation method can be implemented throughout its delivery. Through constructive collaboration with all parties, including contractors, curators, and operators, challenges are overcome such that their demands are balanced. Preservation of historic fabric and quality of the works is maximised while the project is kept financially viable.

In this phase, the deep knowledge of the ship, developed throughout the information-gathering stage, is vital. The conservation map allows key decision points to be understood and negotiated. By retaining a holistic perspective, the wider trajectory of conservation is appreciated, allowing balanced consideration of different measures. Furthermore, the conservation map allows for flexibility in managing project change, both internal and external to the project. It also provides the tools to flexibly alter the project trajectory while ensuring the principles and priorities identified during the optioneering phase are preserved. Crucially, by developing solutions that facilitate and support ongoing maintenance, the long-term sustainability of the vessel is prioritised.

2.3 (d) Ongoing Maintenance

After completion of the works, long-term management and maintenance plans are developed for the vessel. These involve a combination of monitoring regimes and specific interventions required at various intervals. Such measures are designed to minimise degradation while operating within the present funding constraints. If implemented correctly, they substantially increase the interval at which major restoration works are required.

Volunteer groups are often vital in managing and operating historic ships within financial constraints. However, a recurring challenge associated with long-term maintenance is the lack of suitably skilled companies and individuals. The restoration and management of historic ships require technical specialists, but due to the long-term decline of the industries that created the vessels, the necessary abilities are now scarce in the UK [6]. This causes maintenance costs to rise and increases the likelihood of poor management and degradation. As such, skills are a key concern in the ongoing conservation of historic ships. This issue is discussed further in the case studies below.

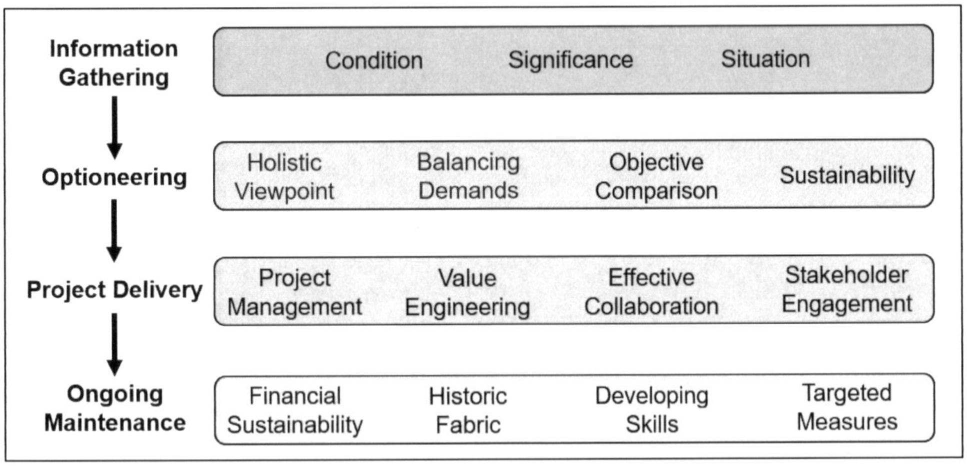

Figure 2: The stages of the engineered conservation method.

3. CASE STUDIES

In this section, four case studies are used to demonstrate the implementation of the engineered conservation method. This includes assessment of its successes and limitations.

3.1 HMS *UNICORN*

HMS *Unicorn* is a Napoleonic-era frigate constructed in Chatham in the 1820s. It spent the first decades of its life laid up in ordinary before being relocated to Dundee to serve as a training ship for naval reserves. It remains in Dundee at present, under the stewardship of the Unicorn Preservation Society (UPS) [6]. Due to a long period of limited maintenance, many of *Unicorn*'s timbers are heavily decayed, and its' structural stability is threatened by hogging. Major intervention is required to ensure the vessel's ongoing conservation and preserve the exceptional amounts of original material remaining.

The engineered conservation method was applied to assist HMS *Unicorn* through the Conservation Gateway, through a formal options study into the long-term conservation of the vessel. The unique geographical circumstances of HMS *Unicorn* dictated the viable options. Due to a faulty dock gate to the River Tay, *Unicorn* is confined to its current dock system, with only one dry dock available for carrying out work. However, this dock has been abandoned since the 1990s, with minimal information available on its construction or condition. The dock is a vital component in considering options for the vessel, and as such, significant effort was invested in gathering further information on it. A dock survey was carried out, involving scans above and below the waterline. Archive research was carried out at Dundee City Hall, supported by volunteers from UPS. Together, these concluded that the dock can be suitably restored in a reasonable timeframe.

Figure 3: HMS *Unicorn* undergoing repairs to the historic timber roof, January 2023.

The options study considered six long-term conservation options. Each option and work path was subject to holistic cost estimates, risk assessments and qualitative analysis, allowing thorough and objective comparison. UPS selected the preferred long-term conservation option through consultation with stakeholders. With a managed session involving scoring and open discussion, consensus was gained around a specific choice.

The engineered conservation method was successfully applied to help HMS *Unicorn* through the Conservation Gateway. The options study provides a strong and comprehensive basis for the sustainable conservation of HMS *Unicorn*. The process was tailored to reflect UPS' priorities. A specific requirement was to deliver it in an open and transparent manner. Accordingly, two stages of stakeholder engagement were carried out, allowing feedback and demands to be incorporated into the study. UPS now has a long-term vision for the vessel and a structured plan for achieving it. This has enabled the charity to receive an initial National Lottery Heritage Fund (NLHF) support for the first "pre-development" stage of the project.

However, the engineered conservation method cannot eliminate uncertainty from the conservation process. As the project progresses, more information will be obtained, new funding opportunities may arise, and the financial profile and priorities of the charity may evolve. To allow for this, key decision points have been identified for re-evaluating the initial goal. In this way, engineered conservation provides flexibility for changes in situation over the project's lifetime.

3.2 RRS *DISCOVERY*

RRS *Discovery* is a polar exploration vessel constructed specifically for the 1901 British National Antarctic Expedition. Since 1986, the vessel has been in Dundee, where it is now at the heart of the Discovery Point centre. Repairs and restoration works have been carried out to RRS *Discovery* regularly, with the most recent campaign around 2018. However, significant areas of decay have formed in the stern and port hull, which are being exacerbated by leaks through the deck.

The engineered conservation method has been applied to several stages of this conservation project. The first component was a detailed inspection in 2021, identifying the regions of concern listed above. An options study on long-term conservation solutions was then delivered, setting a target of returning the vessel to a floating condition. This work path requires immediate remedial works to the stern, port hull and deck. To address these, a detailed specification of works was assembled, and a tender package was issued in 2022.

Commencing the works has proved challenging and has highlighted numerous issues. The initial tender package received no submissions. A second tender package was prepared and issued in December 2022, receiving three bids. However, all parties were in the south of England, with the associated mobilisation costs causing large increases to the budget. Additional funding was secured, but negotiations have highlighted fundamental differences around typical contractual arrangements and approaches to safety legislation.

Figure 4: RRS *Discovery* in the Discovery Point Centre, Dundee.

This situation reflects the need for highly specialist skills in maintaining and restoring historic vessels, and their long-term decline. To deliver the complex works necessary for RRS *Discovery* to the desired quality requires timber experts with extensive experience and strong technical capabilities. Very few companies can or are willing to do this work. Local shipbuilders who previously assisted with maintenance and repairs no longer feel that they have the competence. As a result, contractors must travel large distances to carry out the work. This has significant financial implications, making it more challenging for historic ship owners to raise the necessary funds. Ongoing maintenance is more likely to be neglected or carried out to lower standards, causing a reduction in durability and loss of historic fabric.

When applying the engineered conservation method to RRS *Discovery,* the severity of this skills shortage was not initially understood. It was decided with the owner, Dundee Heritage Trust, to adopt a rigorous approach to the required works, with a flexible scope to account for uncertainties in the vessel's condition and an emphasis on health and safety through CDM Regulations. This holistic method was laid out to encompass the complexities of the project and deliver the work effectively and in a financially sustainable manner. However, these aspects represent a change from the standard operating procedure in the shipbuilding industry and have contributed to the difficult resourcing of suitable tenderers.

To address these issues, alternative options are being investigated in Dundee. Due to the presence of two significant timber ships in the city (RRS *Discovery* and HMS *Unicorn)*, there will be a continued requirement for shipwrights and timber specialists for the foreseeable future. Therefore, the local development of these abilities would be of great benefit to both vessels and assist their sustainable conservation. Opportunities to achieve this were explored through the HMS *Unicorn* options study, and the requirement to train staff in shipwright skills was incorporated into the specification of works for RRS *Discovery*. The perilous situation both vessels now find themselves in has been caused by a lack of long-term maintenance, exacerbated by the disappearance of local skills. Sustainable conservation of both vessels cannot be fully achieved until the necessary skills have been fostered in Dundee.

3.3 PS *WINGFIELD CASTLE*

The Paddle Steamer *Wingfield Castle* was built in 1934 for the London and North Eastern Railway as a Humber ferry. Although its sister, the *Tattershall Castle* still exists, it has undergone significant modification and is now a floating bar and restaurant on the River Thames. This leaves *Wingfield Castle* as the only surviving vessel completely built in Hartlepool [8].

The ship was returned to Hartlepool in 1986 and is currently located in the Hartlepool Marina. A project in 2005 aimed to increase the functionality of the ship through the creation of two multi-purpose activity rooms in the lower deck, a fully accessible elevator, and a functioning cafeteria in the first-class saloon. However, the works used non-marine materials for the plumbing, including underfloor heating for the classrooms. Leakage from these systems degraded the hull steelwork and led to the weakening of the lower decks. The degradation has been accelerated by progressive cuts to the maintenance budget, caused by financial pressures on the local authority, who own the ship, and shortages of the specialist technical skills required for maintenance. Finally, the marina entrance has been narrowed since the arrival of PS *Wingfield Castle*. This has effectively locked the ship in the marina basin, with the only available drydock being the historic one currently used as the home for HMS *Trincomalee*.

Figure 5: PS *Wingfield Castle* and HMS *Trincomalee* in the Hartlepool Marina

Wingfield Castle is now in an advanced state of disrepair and requires major work to preserve the original fabric, which includes the original steam engine and boiler. The council has a continued interest in a capital conservation project but has struggled to secure the finances and wider approval necessary.

The example of the PS *Wingfield Castle* demonstrates a number of important issues with conservation projects at large. From analysis of the previous projects, it can be observed that:

- The narrowing of the marina entrance was progressed without fully appreciating the long-term consequences of preserving the historic ships in the basin. This has restricted the options for these vessels and increased costs for maintaining and conserving the ships in the marina. This has been caused by a lack of a holistic perspective, focusing on the marina entrance rather than the stakeholders' needs.
- The owner secured sufficient capital for a large conservation project in 2005, including innovative solutions and expensive systems. However, cost savings resulted in the use of non-specialist labour, which has had long-term consequences for the ship's maintenance.
- Councils, as owners of historic vessels, can often secure and mobilise significant amounts of capital funding to enable large conservation projects but may lack the ability to sustain long-term maintenance budgets.

The above demonstrates that the long-term sustainability of maintaining PS *Wingfield Castle* was not considered a priority during previous projects, with significant consequences.

This case study is considered particularly informative as it provides a historic conservation project from which lessons can be gained. It is also a particularly complex case, where the restricted finances of the owner, the geographical constraints of the site and the advanced state of degradation of the vessel combined to significantly limit the options available for conservation.

When assessing the project, it can be seen that applying the principles described in Section 2.1 of this paper at the appropriate stage of the Conservation Gateway would have allowed a better understanding of the critical requirements of the conservation project. A fundamental principle in this is the "Vessel and Owner-Focused" approach and how investing at the capital stage in solutions that would minimise maintenance and maximise sustainability should have been a key objective of previous conservation projects.

The current project, where the engineered conservation method has been applied to support the council, explored a large number of options to conserve and maintain the ship through a formal study. While solutions such as lifting the ship or floating her on a pontoon were discussed, the only viable option identified was to drydock PS *Wingfield Castle* in the Trincomalee dock. By narrowing the focus of the project to this single option, it was possible to concentrate the resources of Hartlepool Borough Council on preparing the drydock for the conservation works and stabilising the ship in the interim.

An unintended consequence of the study has been to create a large long-term capital project, which has moved the focus of the owner further away from ongoing maintenance, leading to a degrading ship that requires immediate attention. The need for a package of emergency works is now vital to stabilise the ship and avoid significant damage or failure of critical structures. This is a key takeaway as it demonstrates the need to prioritise and maintain ongoing maintenance of the ships, even when planning large-scale conservation projects, as this may result in the short-term maintenance being perceived as abortive work. Managing and prioritising key maintenance activities to focus on critical concerns would ensure that sustainable maintenance is preserved and delivered effectively.

3.4 HULL MARITIME

The Hull Maritime project is a major council-led project aiming to conserve the story of Hull's seafaring past. A key element of the project is the conservation of two ships, the *Arctic Corsair*, a historic deep-water trawler, and the *Spurn Light Vessel*. Hull City Council (HCC), who own both vessels, have successfully operated them as museums for several years, over which time they have experienced long-term degradation. A group of active volunteers maintains both ships, however, these groups do not have the specific skills necessary to deliver the major repairs required for the long-term conservation of the vessels. A significant capital investment was necessary to support the restoration and ensure the vessels' longevity, raised by HCC and the NLHF [9]. This case study discusses the support offered to the vessel owners through the engineered conservation method across the delivery of conservation projects.

Figure 6: Restoration works to the *Arctic Corsair.*

HCC is experienced in managing large capital projects but lacks specialist knowledge of historic ships. As such, they enlisted the assistance of BR early in the project. To ensure the effectiveness of the works, investigations were commissioned, including internal scans and dive surveys. These enabled a detailed specification to be produced. As with all historic ships, the required repairs could not be precisely quantified. To secure funding from both HCC and NLHF, a financial case was presented to demonstrate how the ships could be maintained for the future while delivering a net economic benefit to the city. Long-term maintenance requirements were considered at every stage of the restoration.

In the Autumn of 2021, a local shipyard was contracted to carry out the works, with BR acting as the owner's on-site representative. The adoption of a holistic perspective was essential to the management of this work, as it required balancing the demands of different parties. In this case, the key stakeholders were:

- The curators, seeking to ensure that the ships were restored in a sympathetic manner, preserving the vessel's historic fabric and cultural value.
- The engineers, working to preserve each ship's longevity and its structures and ensure the safety of the vessels for workers and future users.
- The shipyard, interpreting the requirements of the specifications and contract to carry out the works on the ship using industry standards and experience.
- The owner, delivering the conservation project on time and within budget while balancing the requirements of the other stakeholders.

Although these parties often had conflicting interests, each played an equally significant role in the restoration works. Their concerns were balanced through close and effective collaboration, enabling advantageous outcomes for the project and playing a crucial role in its successful delivery. This required continuous discussions between parties, and a holistic approach which balanced the preferred outcomes of all parties to achieve a suitable solution that met the project's requirements.

However, the differing perspectives also presented challenges, with differing views on achieving the best outcome, particularly while balancing capital spend and long-term maintenance. In these instances, constructive problem-solving resulted in many deviations from the original specification of works.

Typically, shipyard contracts are not structured to handle significant changes and variations to the scope. This was a key source of tension, with contractual deviations often resulting in direct increases in the delivery cost. To mitigate against this, a value engineering process was used. This demonstrated the need for appropriate contractual controls to facilitate these changes. Moreover, it became apparent that contractual methods of conflict resolution were absent when disagreements on scope developed, with these being a key source of learning for the team.

Ultimately, the balance between the three main stakeholders was recognised as a key element of the successful delivery of the project. The tension between the varying parties was identified as a source of collaborative problem-solving and creative resolutions, but also one that required constant management and resources.

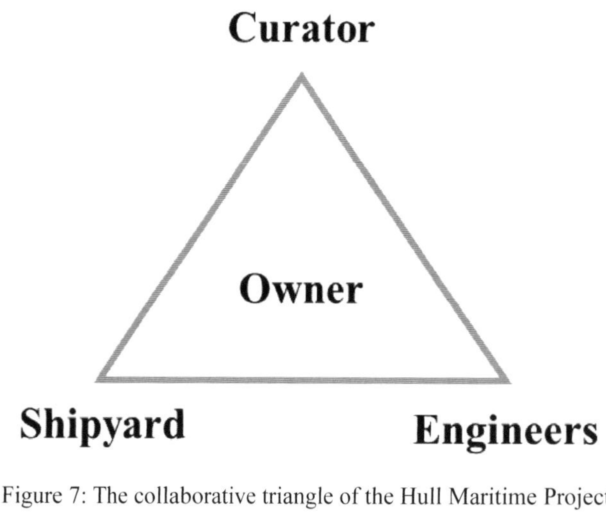

Figure 7: The collaborative triangle of the Hull Maritime Project.

4. CONCLUSIONS

4.1 DISCUSSION

Section 3 demonstrates the application of the engineered conservation method across four historic ships. These unique vessels are each subject to their own particular constraints and are at different points on the Conservation Pathway. This diverse range allows analysis of the engineering conservation method and wider issues in the conservation of historic ships.

The case studies demonstrate that the engineered conservation method provides effective support to historic vessels. In each case study, BR's work has assisted the owners in making decisions and progressing their vessels towards sustainable conservation. With HMS *Unicorn,* this has been achieved by setting a long-term goal for the conservation process and providing a structured but flexible plan to reach it. In the Hull Maritime Project, the practical delivery of works has been supervised and managed, helping to complete them in a manner that satisfies all parties while remaining financially sustainable.

In addition, the case studies show that the overarching principles at the centre of the engineered conservation method are applicable at any stage of a project lifecycle. PS *Wingfield Castle* has been supported in the early stages of its conservation project, while RRS *Discovery* has progressed from an initial inspection to the verge of the construction phase. Vessels at all stages of their conservation are approached using these principles, allowing a cohesive but tailored approach.

Finally, the engineered conservation method is successful at achieving sustainable conservation. On PS *Wingfield Castle* and RRS *Discovery,* works specifications have been developed to restore the vessel in a manner which meets the funding constraints of the owner. Major conservation works have been delivered effectively in the Hull Maritime Project, with the works kept within budget. Alongside financial sustainability, the engineered conservation method also contributes towards reducing the environmental impact of work. By targeting works to the areas requiring intervention and holistically identifying the most appropriate long-term solution, material quantities and the need for further works in the future are minimised. The selection of materials and processes balances the needs for preservation, quality, longevity and affordability.

However, the engineered conservation method also faces challenges. Many of these are general to the historic ships sector and not unique to this approach. For example:

- The practical delivery of conservation works has been complicated by the lack of public funding in PS *Wingfield Castle*, with this being a constant issue for owners.
- On RRS *Discovery,* delays have been caused by the shortage of appropriate skills, which is an industry wide issue, that is progressively worsening as these skills disappear.
- The difficulties in using traditional shipyards to deliver complex projects, with changing scopes, have highlighted the absence of suitable contracts that are balanced but easily used by all parties. This has been highlighted in both the RRS *Discovery* and the Hull Maritime projects, where the use of various existing contracts have not provided the required terms to effectively deliver the works to the satisfaction of all parties.

As these issues require wider changes across the historic ships sector, they are challenging to overcome. For these vessels, alternative solutions are being investigated, aiming to both negotiate these immediate challenges and provide sustainability in the long term.

Furthermore, as a relatively novel approach to historic ships, the engineered conservation method is still being developed and improved. Every vessel that it is applied to has learning opportunities for the future. An example of this is PS *Wingfield Castle,* where placing a greater emphasis on maintenance could have changed the client's approach and helped to reduce degradation and loss of historic fabric. Additionally, for the Hull Maritime project, a clearer contractual scope for changing the specification would have assisted with resolving disagreements.

4.2 CONCLUSIONS

The conservation of historic ships is a complex and dynamic process, requiring the balancing of numerous requirements and the inclusion of a range of stakeholders. Conservation projects and ongoing maintenance can be challenging for vessel owners to manage despite the strong foundation provided by NHS-UK's guidance. This can be exacerbated by a lack of the necessary technical skills, which increases the cost and complexity of work, making them more difficult to carry out in a sustainable manner.

To provide greater support for vessel owners at any stage along the Conservation Pathway, Beckett Rankine has developed the "Engineered Conservation" method. Through the practical application of engineering principles, this aims to achieve sustainability in conservation. It has been applied to several diverse vessels, with its potential demonstrated by the successful outcomes delivered. The engineered conservation method cannot eliminate the large challenges facing historic ships, but as it continues to be developed, it is vital that the experience gained on other projects is incorporated. This will enable the sustainable long-term conservation of historic vessels, preserving them for posterity.

5. ACKNOWLEDGEMENTS

We would like to thank Wynford Davies for his continued support and advice throughout our projects and in developing this methodology and paper. We would also like to thank all our Historic Ships clients for the ongoing collaboration and support, with a particular mention to the Hull City Council Project Delivery Team, the Dundee Heritage Trust, the Unicorn Preservation Society, Hartlepool Borough Council and the National Museum of the Royal Navy. Finally, we would like to acknowledge Tim Beckett and Gordon Rankine for the opportunity to work on these incredibly interesting and exciting projects.

6. REFERENCES

1. NATIONAL HISTORIC SHIPS, 'National Register of Historic Vessels', *https://www.nationalhistoricships.org.uk/the-registers/national-register-historic-vessels*, 13 September 2023.

2. NEWBERRY, D., 'National Historic Ships UK Advisory Committee', *https://www.nationalhistoricships.org.uk/register/499/hms-victory,* December 2008

3. THE NATIONAL HISTORIC SHIPS COMMITTEE, 'Memorandum submitted by the National Historic Ships Committee', 25 January 2005.

4. NATIONAL HERITAGE MEMORIAL FUND, 'Annual Report and Accounts 2021–2022', July 2022.

5. NATIONAL HISTORIC SHIPS, 'Conserving Historic Vessels', Understanding Historic Vessels Vol. 3, 2010

6. TENOLD, S., 'The Declining Role of Western Europe in Shipping and Shipbuilding, 1900–2000', *Shipping and Globalization in the Post-War Era'*, 2019.

7. UNICORM PRESERVATION SOCIETY, 'HMS Unicorn Statement of Significance', 2019

8. BECKETT RANKINE, 'PSS Wingfield Castle – Statement of Significance', 2017

9. NATIONAL LOTTERY HERITAGE FUND, '£13.6million National Lottery boost for Hull's maritime heritage', *https://www.heritagefund.org.uk/news/136million-national-lottery-boost-hulls-maritime-heritage*,2019

7. AUTHORS BIOGRAPHY

Oliver Millar is an Assistant Engineer at Beckett Rankine. He has a Master's Degree in Civil Engineering from Imperial College London. He is engaged in projects across the marine sector, including the ongoing conservation of HMS *Unicorn* and RRS *Discovery*.

Mark Thomas is a Naval Architect at Beckett Rankine. He has a Master's Degree in Maritime Engineering Science/Offshore Engineering from Southampton University and joined BR in 2019. Since then, he has been involved in many historic ship conservation projects, including HMS *Belfast* and RRS *Discovery*. Most notably, over the past three years, he has been the Owner's Representative for the restorations of the *Arctic Corsair* and *Spurn Lightship*.

John Monasta is a Principal Naval Architect at Beckett Rankine. He graduated from Strathclyde University in 2013 with a Master's Degree in Naval Architecture and Ocean Engineering. Following a career that included working in the offshore energy sector and on large infrastructure projects in London, such as the Crossrail and the Thames Tideway projects, he joined Beckett Rankine in 2018 to develop a new Naval Architecture department in the company. As part of this, he expanded the Historic Ship capability and developed this into a core business and specialism within Beckett Rankine. Has worked on the conservation of many historic ships across the UK, from HMS *Unicorn* in Dundee to HMS *Belfast* in London, to the historic trawlers *Arctic Corsair* and *Ross Tiger* on the Humber River, and others.

REVERSE ENGINEERING THE NEOLITHIC BOATS OF THE AEGEAN SEA: TYPOLOGICAL CLASSIFICATION, RECONSTRUCTION AND SEAKEEPING ASSESSMENT

P Tzovaras, University of Southampton, Centre for Maritime Archaeology, Southampton Marine and Maritime Institute, UK, **S Turnock**, Maritime Engineering, University of Southampton, **L Blue**, Maritime Archaeology, University of Southampton, **A G Vlachopoulos**, Prehistoric Archaeology, University of Ioannina, Greece, and **J Sofaer**, Archaeology, University of Southampton

SUMMARY

The hundreds of recently discovered LN-EBA rock-art images of boats from the Aegean Basin, Greece, highlighted their design and structural details. Additionally, their dating in the transition from LN to EBA offers an uninterrupted continuum of the Neolithic boat evolution. A technologically rigorous methodology was incorporated for their analysis by exploiting the concepts of reverse engineering, climatic design and the tools offered by ship science. This enabled their reconstruction and testing, providing insights into their technical traits, such as stability and resistance. These favour the view that trading, gateway communities and seagoing boats have a more dynamic history in a long Aegean tradition that goes back as far as the Neolithic.

NOMENCLATURE

[Symbol]	[Definition]	[Unit]
LN	Late Neolithic	
EBA	Early Bronze Age	
H-RTI	Highlighted-Reflectance Transformation Imaging	
EC	Early Cycladic	
FN	Final Neolithic	
EN	Early Neolithic	
MN	Middle Neolithic	
M&T	Mortice-and-tenon	
LBA	Late Bronze Age	
EH	Early Helladic	
$L/\nabla^{1/3}$	Slenderness ration	
Cu. m	Cubic metre	
t	Tonne	
WL	Waterline Length	m
L/B	Length/Breadth	m
Kg/sq. m	Kilogram square metre	
T	Draught	m
B/T	Breadth/Draught	
IMO	International Maritime Organisation	
DWL	Digital Waterline Length	m
P	Total Power	kW
kW	Kilowatt	
Fpeak	Paddle force	N
N	Newton	
m/s	metres per second	
VCG	Vertical Centre of Gravity	

1. INTRODUCTION

This research addresses *lacunae* concerning prehistoric seafaring and the boatbuilding traditions of the Aegean Basin, Greece, during LN-FN (ca 5400 / 300 – 3300 / 200 BC) and EBA periods (ca. 3200 / 3000 – 2000 BC) (Figure 1). The systematic excavations in the Aegean over the last few years bridged the LN with the EBA period. To this end, the recently discovered depictions of boat representations from Strofilas on Andros [1-2], Vathy on Astypalaia [3-5], and others significantly enhanced the database of boat imagery in the Aegean, providing important structural details, undetected in former evidence (Figures 1-2). Especially the boat evidence from Strofilas and Vathy is crucial to our endeavour since their dating from LN / FN to EC I / II (ca 4th – 3rd millennia BC) offers an uninterrupted continuum of the evolution of the boat's shape.

Due to the lack of explicit evidence in the Aegean Basin, such as actual shipwreck remains (except for the cargo remains of two EH II shipwrecks (ca. 2700 / 2650- 2200 / 2150 BC) [6-7], this endeavour focuses on the available iconographic data (Figure 2). Due to an apparent morphological similarity, all the EBA data are comparatively examined to Neolithic ones and recorded through first-hand observations and new techniques. Evidence context is also studied comparatively with the period's watercraft operating environment, raw material and sources, and technological constraints.

Overall, the scope of this research is to re-evaluate the maritime Aegean LN-EBA through a more technologically rigorous approach and examine the period's seagoing boat types and their capacities. To do so and provide an up-to-date database, the available boat evidence has been recorded via firsthand observations and through novel techniques, such as H-RTI, photogrammetry and silicone rubber for imprints' production. Eventually, shape analysis is implemented via the concept of reverse engineering. An informed morphological classification and the definition of the 'average' shape for each template is aimed. The latter are used to reproduce the boat types through ship science and based on naval architecture fundamentals, the vessel's operational environment, the period's technological constraints, and available resources.

The interpretations and analyses generated by the above-mentioned approach have produced the ability to reconstruct and test the Aegean vessels within the technological envelope in which they were developed. For their digital reconstruction, naval architecture CAD software is exploited to design the vessel types in three-dimensions, produce their lines and test their seafaring properties in simulated conditions. Outputs are further compared to the experimental reconstruction of a scaled-down, 2 m technical boat model, tested at the towing tank of Wolfson Unit (Boldrewood Innovation Campus, University of Southampton). Therefore, the produced performance envelope allows suggestions for the parameters of boats and their performance instead of subjective theories based on ambiguous iconographic data.

Figure 1: The main reference map with the administrative regions and regional units of Greece. Source: Map Generated by the author. Using: ArcGIS Pro [GIS] Version 3.0.0.

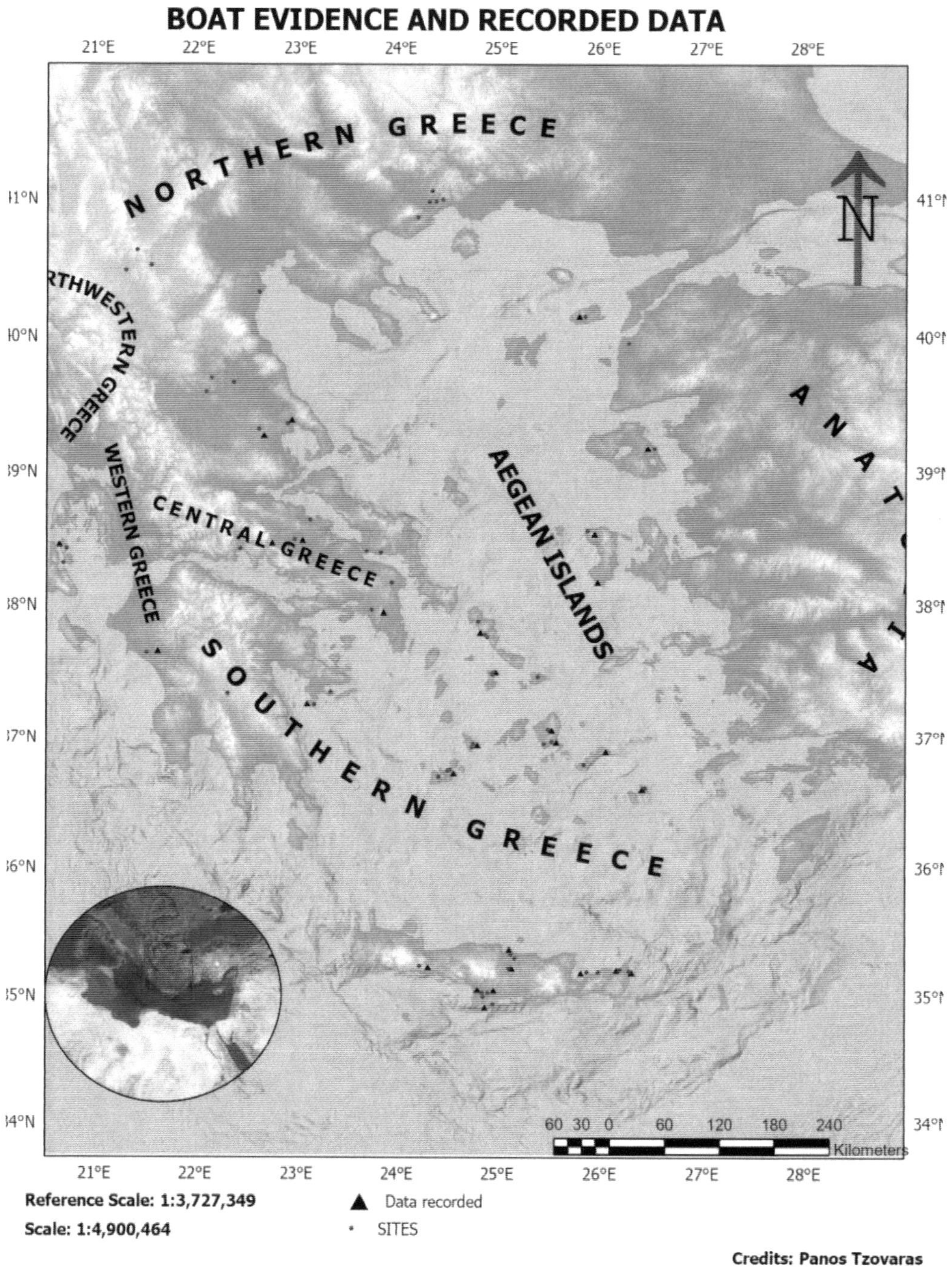

Figure 2: The location of all the extant evidence and the ones recorded by author. Source: Map Generated by the author. Using: ArcGIS Pro [GIS] Version 2.4.3.

2. CONTEXT OF RESEARCH

This research focuses on the corpus of the LN-EBA boat evidence (Figures 1-2). This period was selected due to similarities in boat morphology, artistic execution and the fact that the boats are mainly represented as mastless and possibly paddled. Nevertheless, it would be perilous to perceive the LN-EBA Aegean and mainland Greece as a uniform area (Figure 1), so we should acknowledge the chronological discrepancies for such a large extent of research as well as the ensuing cultural-geographical idiosyncrasies.

2.1 THE NEOLITHIC-EBA DATABASE OF BOATS

So far, through personal surveys, fieldwork, excavation and archival investigation, more than 270 boat and boat-shaped representations have been located spanning the whole Aegean Basin and its immediate hinterland and sites deeper in mainland Greece (Figures 1-2). Around fifty-nine sites have yielded boat evidence. Near twenty-six sites have been visited, and more than two hundred boat images and boat-shaped artefacts have been personally recorded (ca. 70% of the total number) (Figure 2). Generally, the evidence chronological horizon extends from EN to EBAIII (ca. 7th to 3rd millennium BC). Their depositional context is usually associated with settlement or burial deposits, open-air sites, and caves. However, not all the boats fit the classification criteria, selected based on the security of the context, chronology, geographical location and provenance.

The extent of the previously published and unpublished data mostly comprises of two-dimensional evidence, mainly boat representations executed on rocks (incised, pecked, engraved, relief or painted motifs) or ceramic vases (incised, relief, painted or impressed) and potters' pot-marks. As for the three-dimensional examples, these comprise of clay, wood, stone and metal boat figurines or clay/stone vases/moulds/lamps/boat-shuttles or even crucibles imitating boats and vice versa, sometimes decorated with painted motifs or incisions. To the figurine category, we may add a few boat models with anthropomorphic/zoomorphic traits and examples that constitute parts of broader assemblages. Finally, some tangible evidence exists, such as the imprints and structural remains of wooden boats or the cargo remains of actual shipwrecks.

2.2 THE CASE OF VATHY

The main case study is the boat petroglyphs from Vathy on Astypalaia (the Dodecanese, Aegean) (Figures 1, 3). On the peninsulas' rocky eastern part lies the Cape Elliniko (or Cape Pyrgos), where traces of an extensive prehistoric occupation were discovered [3-5, 8]. The salvage and systematic archaeological works on the tip of the promontory have been carried out under the auspices of the Archaeological Society at Athens and funded by the University of Ioannina [3-5, 8]. These unearthed an FN-EBAI / II citadel with a sturdy boulder-built circuit and retaining walls, as well as traces of extensive rock-art monuments [3-4]. The east coast's natural formation has been incorporated into a complex of artificial harbour installations in the form of smooth, slanted slipways, whose construction could have served the launch or retrieval of small boats; these formations develop along the coastline, and are combined and consolidated by fortification constructions [3, 8]; if correct, then it constitutes the oldest known artificial harbour constructions in the Aegean, consolidated by the extensive defensive works, offering protection from the sea.

Rock-art compositions have been identified in numerous locations of the promontory, either explicitly or implicitly related to the citadel [3, 5]. Although there are examples found at a distance from the prehistoric site, these are possibly of the same chronological extent [3, 5]. Vathy's rock-art covers a wide-ranging repertoire of motifs, either linear, schematic or figurative. Despite the ongoing detrimental effects of erosion caused by various environmental and human agencies, Vathy preserves an outstanding number of rock-art representations, enhancing the iconographic agenda of Aegean art and allowing the tracing of its roots in the Neolithic [3, 5]. These are usually carefully placed and executed on the worked surface of the rock and patently document maritime activities of the Neolithic-EBA island communities. One of the most popular rock-art motifs in the perception of Vathy's artists is the spiral and its variations, found in numerous locations [3-4, 9-10]. However, the artist Vathy chose the image of the boat to decorate a few of the citadel's most monumental and important constructions. These are mostly entailed by meticulous attention to detail and artistic proficiency. So far, eight constructions have been identified bearing at least fourteen or fifteen boat petroglyphs [11-12] (Figure 4).

Figure 3: Cape Elliniko and the FN-EBAI / II citadel. Source: Photograph taken by the author during his 2020/2021 fieldwork. Credits: Courtesy Professor A. Vlachopoulos. © Archaeological fieldwork at Vathy, Astypalaia – The Archaeological Society of Athens.

Figure 4: (Left) A detail of the representation on the Gate of Boats. Source: Photograph taken by the author during his 2017/2018 fieldwork. Credits: Courtesy Professor A. Vlachopoulos.© Archaeological fieldwork at Vathy, Astypalaia – The Archaeological Society of Athens. (Right) A drawing representation of the vessels (GoB 1-5) on the Gate of Boats. 1:50 scale. The red arrow points to boat belonging to Type I of the 'rounded category'. Drawn by the artist Nikos Sepetzoglou. Source: Vlachopoulos 2021a:128 fig. 27.

3. TECHNOLOGICAL CAPACITIES, CONSTRUCTION METHODS AND BOATBUILDING: A PATH FROM INVENTION TO INNOVATION

Though the iconographic data imply the presence of several boat types, one may wonder whether the period's technology was adequate for the design and production of seagoing vessels. By the end of the 7th millennium BC, the Neolithic craftsman was already equipped with an enhanced toolkit of polished tools, such as adzes, axes and chisels, that would have offered greater flexibility in timber felling and manipulation [13]. However, technological progress is better represented a couple of millennia later, when MN (5800-5400 / 300 BC) advancements in pyrotechnology and pottery technology gradually led to the advent of metallurgy and the mastering of metalworking by the Aegean craftsman [14-16]. Therefore, the LN copper awls, chisels, saws, scrapers and axes, attested at various Neolithic sites of the Aegean and mainland Greece, would have further facilitated the Neolithic craftsman [11-12, 17]. These technologies indicate an enhanced toolkit that would have increased the capacity of constructing more seaworthy and spacious logboat types, such as the expanded and extended ones, possibly since the beginning of the Neolithic [18-19].

Ethnographic evidence and experimental work have suggested that these tools would also enable the production and shaping of planks for the construction of more seaworthy, plank-built boats at least since LN and definitely during EBA [11, 17, 19-23]. Further answers can be sought in the Aegean carpentry and architecture, attested since the EN [24]. That general knowledge could have been applied to boatbuilding, considering that a carpenter of a Neolithic community would possibly have various tasks, from constructing tools and furniture to houses, boats and other complex architectural structures [11-12, 17, 22]. Several cases have come to light indicating early knowledge of plank-shaping and joinery, such as planks assembled with wooden pegs, shaped planks and beams, sometimes bearing evidence of mortices, timber frames with systems of joints and mortices, wooden, perforated elements joined with lashes and so on [12]. Finally, intricate wooden structures have been attested to the Neolithic lake settlement of the Amindeon Basin on Florina [25] (Figure 1).

Although speculation, these possible borrowings from housebuilding carpentry can open new horizons in Aegean boatbuilding, allowing us to assume that, if a plank-built boat was a reality, the planks could be edge-joined with the use of pegs, lashings or a combination of these [11-12, 17]. Even though there is a consensus that sewn fastening is the oldest method [18, 26-28], it is equally possible that pegged planks could have been in use. According to Chatzitoulousis [29] the technique attested to the mortices of Displio could find an application to Neolithic boatbuilding; especially the angular mortices that would have been a possibility after the advent of metallurgy. Thus, though we cannot suggest an early implementation of the M&T technique locked with pegs, we might think of the 'primitive' use of unpegged ones combined with lashings [11-12]. Besides, a similar approach has also been attested to -the admittedly later- boats of the Old Kingdom Egypt [30-33]. Interestingly, M&T joints have been used in Egyptian carpentry since at least Dynasty I [34]. As far as boatbuilding is concerned, the same technique with locked pegs first appears in the LBA Uluburun shipwreck [34-37]. Such a technique would have increased, to a certain point, the longitudinal strength of the hull against the forces of the Aegean wave regimes (see below). Interestingly, according to Agouridis [38], the forces exerted on the hulls by the waves were the main reason the Aegean boatbuilders insisted, for more than twenty-five centuries, on using the technique.

The water-tightness of the vessel could also be achieved with the well-known caulking, the insertion of any fibrous material (i.e., moss, oakum, animal hair, etc.) into any seam and its covering with an appropriate sealer [32]. This could have also been achieved by using a mixture of a clayish mud with seaweeds that does not absorb water; this water-resistant substance, known as "patelia", was used to insulate the roofs of the houses of the Aegean from prehistory until recently [12, 17]. Papoulia [39] has also pointed out the possibility of using bitumen or birch bark tar for this purpose. However, a more suitable candidate could be the mineral 'miltos' (*μίλτος*) -in the form of a red pigment- widely used in carpentry and shipbuilding [40] as a sealer and, according to Theophrastus [41], the best source was the Cycladic island of Keos [12, 17] (Map 1). For instance, Homer [42] used the epithet "*μιλτοπάριοι*", which stems from the word *μίλτος* (miltos), to describe the ships having a red-painted bow.

Regarding the construction tradition and whether these boats are logboats (simple or complex), plank-built ones, etc., this is almost impossible to be determined from iconography and our available evidence, except for some theoretical observations. It is stated that the author does not suggest that there were only plank-built boats. On the contrary, various types, such as log, bark or skin-based, would have existed according to the needs, the operational environment and available resources. Consequently, it is not indicated that any of the following examples exclusively belong to a log or plank-built tradition since it is impossible to be determined just from our dataset.

Despite what technique was used, what can be posited is that the LN inhabitants of the Aegean had a more than basic knowledge of architecture and, to a degree, engineering. This is attested to the LN-EBA double-storied houses, complex fortification systems and carefully planned layouts of settlements [24, 48]. All these technological advancements (lithic technology, pyrotechnology, metallurgy, carpentry, joinery, boatbuilding and architecture) triggered a series of processes that affected many facets of the Neolithic individual reality, such as seafaring and long-range mobility in general. The new, enhanced and advanced tools that made their appearance led to isualization , increased production and, eventually, to surplus [44-45]. Communities no longer aimed to produce enough just to survive. Hence, the need to manage and circulate their products alongside the need for access to metallurgical sources or communities that had mastered metallurgy was one of those processes that affected/drove the boatbuilding technology.

3.1 CATALOGUE OF HULL: FRAMING, SUPERSTRUCTURE AND OTHER ELEMENTS

According to Steffy [32], a good archaeological, ship report must adhere to certain aspects. One of them is the presentation of the hull catalogue, presenting a description of the timbers and hull-related artefacts. Naturally, this comes before the reconstruction and must be presented clearly, precisely and economically, containing all the relevant information your data provides [32]. Though this concerns actual boats, such as shipwrecks, it is also deemed appropriate for our endeavour. Besides, the purpose is common to offer a catalogue that could be used comparatively and as an aid to our reconstruction.

However, we should bear in mind that this catalogue cannot strictly abide by the structure proposed by Steffy [32] because our analysis does not concern a single example but hundreds, possibly of various types and construction methods; additionally, there is only a limited number of the components of the hull that can be identified with relative certainty. What follows is our modified 'catalogue of the hull', deriving from the available iconographic data, as discussed in the author's doctoral thesis [12]. Due to word limit restrictions, this is briefly listed, without being discussed in-depth. Additionally, an approximation of their earliest appearance is offered, based on the evidence's dating.

- Planking: longitudinal and/or transversal elements defining the exterior surface of the hull. Earliest iconographic appearance: MN.
- Keel/Centre plank: a longitudinal element, usually extending along the length of the bottom, protruding downwards. Earliest iconographic appearance: MN / LN
- Frames: bulkheads, transverse crossbeams, stem-stern decks, transversal framing and other transversal features (i.e., ridges, thwarts, floor timbers, etc.). Earliest iconographic appearance: MN / LN-EBA.
- Transom or transom-like formation: the vertical, athwartship element, fixed on and shaping the stern as well as joining the sides and strengthening the hull. Earliest iconographic appearance: LN.
- Fore/aft projection(s): a pointed projection that obliquely extends from one end, just above the bottom and the waterline, continuing beyond the post. It occasionally occurs on both sides. Earliest iconographic appearance: LN.
- Superstructure: Platforms (*ἰκρίον*)near the vessel's extremities, cabins/shelters, masts. Earliest iconographic appearance: LN / FN-EBA.
- Head-moulding (*ἄφλαστον*) and posts decorations: all these elements that usually appear fixed on the post or hung by it and have no structural purpose. Earliest iconographic appearance: LN.
- Vail (*θύσανος*) and banners with hanging tassels (*φᾶρος*): as such, we describe the elements typically hanging below the head-mouldings or appearing alone. They are also attached to one of the extremities and do not have any structural significance. Earliest iconographic appearance: LN / FN.
- Paraphernalia: in this category, we include the boats' equipment that is not related to its structure or superstructure, such as anchors and fishing equipment. Earliest iconographic appearance: LN / FN.
- Modes of propulsion: paddling, sailing. Earliest iconographic appearance: LN / FN.
- Steering system: single steering oar/paddle with a triangular blade, single steering oar/paddle with a leaf-shaped blade, double steering oar/paddle with a shovel-like blade (*πτυόσχημη λεπίδα*), trapezoidal steering (rudder?). Earliest iconographic appearance: LN / FN-EBA.

4. METHODOLOGICAL AND DATA ANALYSIS IMPLICATIONS: THE 'EMERGENCE' OF THE BOAT'S TYPOLOGY

Indisputably, the isualization ion of something you do not have actual remains of, but just two and three-dimensional pictorial evidence, is an uphill battle. Hence, the extant dataset has a great degree of ambiguity, considering aspects of artistic conventions, local varieties, medium, execution technique, errors, erosion and so forth. Moreover, there is always the question of the dating of the data, which can be very problematic, especially when it comes to not safely stratified findings. Hence, it was decided to find a way to assist the researcher in the process of shape segmentation, definition, and classification. To do so, reverse engineering and descriptive statistics are implemented to completely 'deconstruct' the object to try and grasp its design and architecture, considering that its function is explicitly associated with its structure and form [46]. Although reductive in its conception, this approach has the merit of giving you the ability to reproduce the product of another manufacturer through the meticulous study of its composition and by using its very same components and going back to its production cycle [46]. Hence, we should imagine the boat image as lines compiling a complex geometrical shape in a two or three-dimensional space. In fact, this method is an image analysis one and does not require any measurements, just observations of the shape under study. These geometrical shapes are analysed through isualiztive statistics, isuae the researcher describes the observed basic features to isualiz and summarise them. Then, based on the various primary and secondary morphological traits and characteristics that the vessels share as well as their differences, a dataset of variables can be generated that can find further use in an informed, typological classification and the production of templates of the 'average' boat shapes, that accumulates most of the basic morphological features defining each group/subgroup. Thereby, the produced templates represent each boat type, that can be afterwards reconstructed.

In terms of the overall description of the hull's shape, this is deduced by the morphological elements that comprise these boats. Treating, though, the two-dimensional data exactly as the three-dimensional ones could be misleading. Obviously, a two-dimensional representation primarily informs us regarding the image's dimensions, as far as the length and height are concerned; additionally, about the representation's general morphology, and most usually the profile view, but fails to inform us of the rest. Nonetheless, a two-dimensional 'drawing' can be more detailed than a three-dimensional model due to the limitations imposed by sculpting. Contrarywise, the three-dimensional boat figurines or boat-shaped vases can give us a glimpse of the missing dimensions and views and a relatively detailed understanding of the hull's architecture, superstructure, design and the elements extending into the hull's interior or exterior. Thus, the granularity degree is more significant than a two-dimensional image since it illustrates all these features more nuancedly. The next step is the implementation of the tools offered by ship science. Once the identification of general types and their parameters have been determined, several interpretations can be digitally modelled in three dimensions and tested in CAD environment. In a way, an informed computer simulation can assist in the examination of an artefact as a "dynamic object" and not passive, revealing the underlying prehistoric/ancient mechanics [46] within a certain envelope that contains aspects of the micro and macro context [11-12]. Once again, it is stated that our aim is not to reconstruct the actual boat representation but explore the specific vessel type they may represent or signify through testing various scenarios. Thereby, the vessels' lines plan and surface models can be built via MAXSURF Modeler software. The hydrostatics, stability and resistance analyses are performed through the software MAXSURF Stability and MAXSURF Resistance, respectively, in simulated conditions based on the pre-defined environmental factors of the period. By isualiza the same software, various proposed hypotheses for the structure of the vessels' representation can be tested and validated accordingly. Complementary, the results are compared to the ones produced during the towing tank tests of the 2 m technical boat model.

4.1 DATA ANALYSIS TECHNIQUE – REVERSE ENGINEERING ITINERARY

As mentioned, to perform a shape analysis and extract the missing information it was decided to implement the concept of reverse engineering. This concept has been recently applied in Archaeology, where an itinerary has been proposed that will assist in going back to its development cycle; the following data analysis technique partly adheres to the proposed steps [46] (Figure 6).

Step 1: Data digitization

The first step is to isualiz the raw data (archaeological artefact), when possible, through personal recordings.

Step 2: Data segmentation/extraction

Each boat shape should be segmented and differentiated from the others via descriptive statistics to be assessed as geometrical shapes. For the segmentation to be relevant, a few variables have been determined based on the aspects that most commonly appear in each representation. Inevitably, the selection of variables and their description represent two major parts of this methodology with a substantial level of potential arbitrariness.

Step 3: Data typology

The 'disassembled' components should be 'reassembled' through an informed frequency analysis that assesses the frequency with which a trait appears in a dataset. This not only helps in the identification of patterns in our dataset but also the possibility of errors in our data by providing us with a cumulative percentage for each variable. Hence, the next step contains the typological classification; to provide a hierarchical structure and assist in the isualization of the base types and their variations, a decision tree has been produced for each group. Each is assigned to a representative master type.

Step 4: Data integration and computer simulation

The next step is to produce a template for each boat type based on the determined variables, demonstrating the 'average' shape, as well as reproduce the product (3D surface and models and a physical one) (Figure 6a-b). Additionally, to integrate further data into these in order to produce the 'climatic design' that assesses aspects of the operational environment [47].

Step 5: Specifications

This step concerns the defined boat types' stability, resistance and performance tests that highlight their specifications.

Step 6: Recontextualisation

The final step contains the discussion and re-contextualisation of the data based on the produced results.

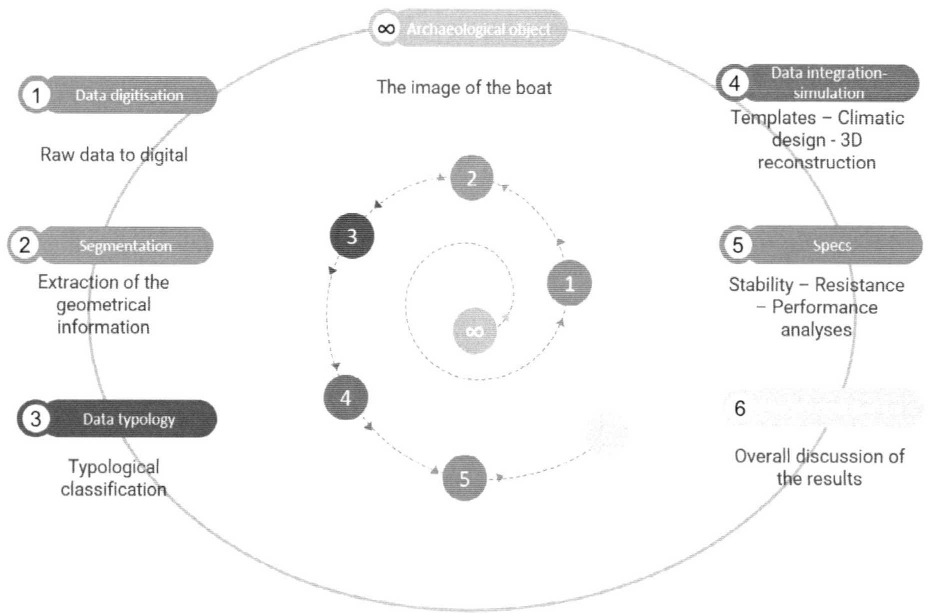

Figure 5: The itinerary exemplifying the application of reverse engineering in order to move backwards to the development cycle of the archaeological object of concern. Source: the author.

4.2 RESULTS: DATA CLASSIFICATION AND AVERAGE SHAPES

Based on the discussed methodology, the boat evidence has been 'deconstructed' and classified based on certain morphological features. There are three major groups based on the hull's general shape and symmetry/asymmetry of the extremities. Each group was further divided into subgroups based on the transverse plane and plan view shapes. Each subgroup contains several variations based on the formation of the posts and the bottom as well as the presence/absence of the oblique, pointed projection and a keel/centre plank.

4.2 (a) The templates of shapes of the LN-EBA boat types

Based on the preceding classification analysis, there is a significant number of variations. The author attempted to illustrate this variability by accounting for any minute detail that differentiates a boat from another in order to point out the richness of the dataset. Even though some of the examples bear some resemblance to others, they always have a small attribute that differentiates them. Hence, a more generalised approach is followed for the production of the shape templates due to this variability and the ambiguity of the dataset. Thereby, it was decided only to attempt to define the ones that are better represented regarding the population number and the state of preservation. Interestingly, most of the templates of the boats are represented both in the LN and EBA period, suggesting a continuity of the local boatbuilding tradition.

Two main categories of templates are discerned from the previous analysis. The first contains the rounded designs, and the second the angled ones. The variables picked for each one comply with their general design and the classified data. Hence, each template contains the design and a simplified, rationalised version of the plan, profile views, and body plans (Figure 6).

Rounded shapes (Types I-V of the rounded category)

- Template 1: convex bottom, symmetrical posts, one convex and another obtuse. U-shaped transverse plane and elliptical/lanceolate plan view.
- Template 2: convex bottom and symmetrical convex posts. U-shaped transverse plane and elliptical plan view (meniscus-like/double-ended).
- Template 3: convex bottom and asymmetrical, convex posts. U-shaped transverse plane and lanceolate plan view.
- Template 4: convex bottom, asymmetrical posts, one concave and another vertical. U-shaped transverse section and lanceolate/ogive plan view.
- Template 5: flattened bottom and symmetrical, convex posts. U-shaped transverse plane and elliptical plan view.

Angled shapes (Types I-VI of the angular category)

- Template 6: flat/flattened bottom, asymmetrical posts, one concave and another oblique. Rectangular/trapezoidal transverse section and secant-ogive/lanceolate plan view.
- Template 7: flat/flattened bottom and oblique, asymmetrical posts. Rectangular/trapezoidal transverse section and secant-ogive/lanceolate plan view.
- Template 8: V-shaped bottom, an oblique post and another vertical. V-shaped transverse section and pentagonal plan view.
- Template 9: flat/flattened bottom, asymmetrical posts, one convex and another oblique. Rectangular/trapezoidal transverse section and lanceolate plan view.
- Template 10: flat/flattened bottom, symmetrical, obtuse posts. Rectangular/trapezoidal transverse section and rectangular/oval plan view.
- Template 11: flat/flattened bottom, symmetrical, obtuse posts. Rectangular/trapezoidal transverse section and elliptical plan view (double-ended).

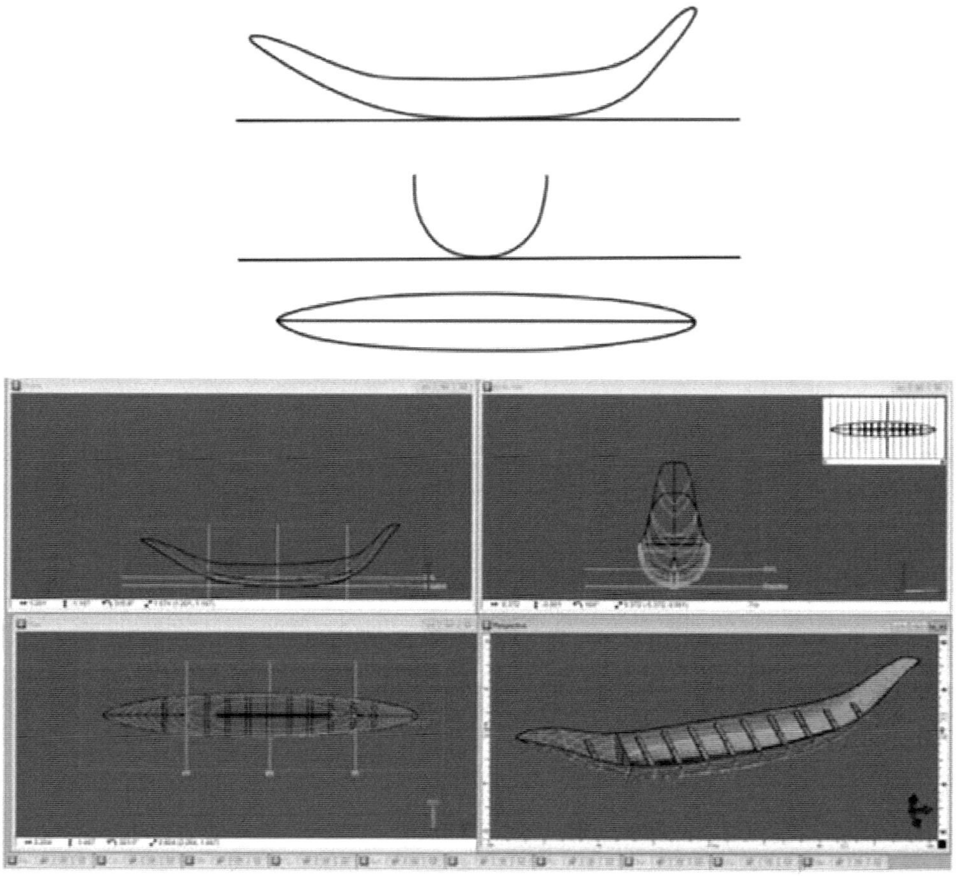

Figure 6: (Top) Template 1 of the rounded boat shapes. The simplified, basic profile and plan views as well as the body plan view that was digitally reconstructed in MAXSURF. Source: digitally drawn the author, using CorelDRAW 2020. (Bottom) A 3D surface model and the lines of the Type I boats of the 'rounded' shapes (Template 1). With framing and keel/centre plank. Displacement: 6 t, plus 1.5 t extra cargo capacity. Source: produced by the author in MAXSURF Modeler.

5. SHIP DESIGN AND RECONSTRUCTION

This section uses the templates produced to reconstruct and assess the viability of the defined boat types through computational means as well as ship science and naval architecture. It mainly focuses on seagoing, presumably paddled vessels, not on sailing ones. However, before reconstruction, some environmental parameters that could have affected the boats' characteristics (i.e., dimensions) and performance must be addressed. This is based on what has been termed 'climatic design' and concerns the operational environment and how a boat's structure would be affected [47]. Thereby, their validity would be evaluated through their inclusion within a performance, an environmental and technological envelope that will facilitate suggestions for the parameters of boats and their seaworthiness.

5.1 DESIGN PARAMETERS, RECONSTRUCTION AND TESTING

The theoretical estimation and selection of the main dimensions of a ship before the reconstruction process is termed "Concept Design" [48]. Thus, the following analysis for the approximation of the boats' dimensions is also based on acclaimed ship science and design endeavours that offer some typical ratios to which any seacraft should adhere, based on physics, experimentation and empirical observations [47, 49-51]

Our design scheme is primarily influenced by the forces of specific operational environments that affect a boat in seaways, as related to its geometric shape and size [47, 49, 52]. Subsequently, the design parameters laid out by Wilson [51] are considered. Next, the parameters concerning us are the technological capacities of the LN-EBA period, timber sources and construction methods.

Certainly, some major phenomena would have concerned the seafarers of the period, especially during heavy weather circumstances. According to the International Code of Intact Stability [53], a ship in these conditions can be 'trapped' in critical stability situations and extreme pitch/roll angles. Therefore, these phenomena would have a bearing on the vessel's performance and, most importantly, stability. It has been argued that the most fearsome aspect of the Aegean Sea would have been the force of the breaking wave; these would have been a usual phenomenon in the Aegean due to the short period of the wave trains [12]. Moreover, we can infer that most boats would have launched from a beach, especially considering the apparent absence of harbour installations (except for Vathy). Therefore, the issue of breaking waves should have been of concern since it could have easily damaged or capsized an inadequate boat.

It has been argued that a boat' length should measure more than 30-40% of the height of the wave [50]; thus, this could be a rough guide to estimate their length. It has been observed that the climatic and meteorological conditions of the LN-FN would be similar to today's [12]. If we consider that the mean annual wave height is between ca. 0.8-2 m alongside the 30-40% limit, a boat of more than 5 m long would be safe on an ordinary day in the Aegean; a 10 m one would also survive most of the rough seas as well as a 15 m and a 20-25 m long boat.

Another aspect we should account for is the slenderness ratio ($L/\nabla^{1/3}$), providing a good "hydrodynamic length", that should be in the range of 5.5-8 [47]. To determine this, we need to approximately calculate the displacement of the abovementioned sizes and their WL. This must correspond to the estimated weight of the hull and the crew (1 cu. M = 1 t). We first need to calculate the vessel's breadth for accurate results. As suggested by Basch [54] and other scholars [12, 56] and indicated by the Aegean iconography, the most probable position of paddlers would be this of working abreast or in a single row. So, if operating side by side or in a staggered formation, their transversal separation must be about 0.85 m to 1 m [12]. Of course, this will not be enough since it would result in extreme slenderness and lack of transverse stability. A good L/B ratio usually ranges between 5-7 [47] which can be estimated by using the previously determined lengths (5, 10, 15, 20 and 25 m) based on the height of a wave versus the length of a boat. A 5 m boat should have a maximum beam of no more than 1 m to have an L/B ratio of 5. A 10 m boat must have no more than a 2 m maximum beam to score an L/B ratio of 5. A 15 m boat should have a 3 m maximum beam for an L/B of 5; a beam of 2 m gives an L/B of 7.5, above that threshold. A 20 m boat needs to be 3 or 4 m wide to have an L/B of 6.5 or 5, respectively. Finally, a 25 m boat needs to have a breadth of 4 or 5 m to score an L/B of 6.2 and 5 m, respectively. However, too much beam can also adversely affect the boat's stability and overall performance. Having approximately identified various sizes, as far as the length and breadth of the boats are concerned, we may return to our initial problem, the calculation of the $L/\nabla^{1/3}$ ratio. This should be within the range of 5.5-8 [47]. If we consider the paddler's interscalmium (ca. 0.75-1 m) [11-12, 54-57] a 5 m boat can accommodate ca. 2-3 paddlers on each side, a 10 m boat ca. 7-8 paddlers, a 15 m boat around 10-12 paddlers, a 20 m boat around 12-15 paddlers and a 25 one ca. 15-20 paddlers. Thereby, the weight of the hull for boats of 5 m, 10 m, 15 m and 25 m can be estimated at upwards of 25 kg/sq. m, including other woodwork at 15 kg/sq. m (Weight of hull = Surface area of hull x Weight per sq. m). Thereby, at least approximately, a 5 m boat (WL ca. 3 m) would displace around 0.5 t, a 10 m boat (WL ca. 5.5 m) about 1.5-17 t, a 15 m boat (WL ca. 8 m) around 4 t, a 20 m boat (WL ca. 10.7 m) around 5 t and a 25 m one (WL ca. 13.5 m.) around 7 t. Hence, the $L/\nabla^{1/3}$ ratio for a 5 m boat is 3.7; for a 10 m boat, it is around 4.8; for a 15 m boat is around 5; for a 20 m boat, is around 6.2; for a 25 m boat is around 7 (the basic hydrostatics have been approximated in MAXSURF). Regarding the freeboard and T, these can be estimated by accepting paddling as the mode of propulsion and the above-estimated displacement values. If paddling was employed by paddlers being sat or kneeled for increased stability, then, based on ethnographical and archaeological data, the lowest height of the sides would be between 0.8-1 m above the waterline (freeboard), at least in normal see states [12]. These freeboard numbers have been successfully confirmed in the Aegean Sea context with the paddling of the Bracciano replica during the Monoxylon III expedition [58]. Hence, this freeboard roughly yields a T of ca. 0.2-05 m. Accordingly, the overall height of the sides (the vertical distance from the gunwale to the bottom of the boat) should be around 1.2-1.5 m or more. Another aspect included is the B/T ratio. It has been proposed that a moderate beam should be no more than three to four times the body depth of the hull or have a B/T ratio of 2.5-5 [47, 59].

Based on these estimations, we can exclude from our reconstruction process the boats of ≤ 5 m since their ratios would be lower than the proposed ones [47, 59], at least for open sea boats able to safely transport people and cargo. The boats that measure 10 x 2 m check some of the abovementioned criteria. The 15 x 3 m long ones, indicate some of the best examples, scoring well in all ratio categories. The 20 x 3 m would also be a fine example. As far as the 25 m long, this requires a 4 or 5 m breadth. However, a 5 m beam seems extreme and would make the boat considerably stiff. Consequently, a 25 x 4 boat appears a more plausible option, as far as the abovementioned ratios are concerned.

Next, it remains to be seen whether these four sets of longitudinal dimensions (10 m, 15 m, 20 m and 25 m) were technologically viable or supported by the timber availability and the boat evidence of the period. In terms of technology, we proposed that the necessary knowhow and toolset to fell and process a log in order to construct an extended/expanded logboat or shape it into planks is evident at least since the LN period; likewise, the necessary joinery techniques [11-12]. In terms of timber availability, it has been recommended that the tree species required to construct a ≥ 10 m logboat were present, such as *Pinus nigra* and *brutia* or oaks; similarly, adequate timber for shaping the planking of a plank-built hull [11-12]. Hence, at least theoretically, one could produce quite long, solid planks from a single log. However, since there were adequate joinery techniques in the LN-EBA Aegean, considerable lengths could have been attained by joining shorter planks. Regardless, there should have been a limitation in a boat's length, either built from a log or planks exclusively, especially if we accept that the most expensive aspect of boatbuilding is to increase the boat's length [47]. Possibly, in our spatiotemporal context, one should also consider the increase in breadth and depth as expensive. However, in our case, the expensiveness might not be that much associated with economic incentives but with demographic ones. Thereby, we should also consider Broodbank's [21-22] estimates about the considerable labour intensiveness of big-sized boats, as well as Tzovaras' [17] discussion on the logistics and risk-to-benefit ratio for the construction of long vessels (more than 25 m), whose sinking can lead to a demographic and economic disaster. Furthermore, we should also consider that the longer the boat, the more power is required, hence more paddlers. Did the settlements of the period have the available 'manpower' to operate a more than 20-25 m long boat? In terms of the ones that could measure 25 x 4/5, they seem too excessive. By decreasing the beam of a 25 m long boat to below 4 m, it would be too narrow, tender and unstable in heavy seas; if we maintain its beam to 4-5 m, it would be considerably beamy and stiff.

Thus, based on the rationale of moderation, we can infer the following. First, the boats whose dimensions vary from 10x2 m to 15 x 2 m seem more than feasible, considering that 10-15 m would be the estimated length of the EH II Dokos shipwreck [60] (Figures 1-2). Though the 20x3/4 m boats would also be massive for the period (see below), we decided to digitally reconstruct and test them for the sake of experimentation.

Based on these criteria, the eleven defined templates were modified and digitally reconstructed. There are two versions of them, one without framing or a keel/centre plank and another with frames. The line plans and digital models were produced in MAXSURF Modeler (Figure 6). Each has been reconstructed x 3 times based on the previously defined set of dimensions. Thereby, the author digitally reconstructed and tested sixty-six variations of the eleven boat types. The 15 m long digital design 'Type I' of the rounded shapes (represented by the rock-art boat images from Vathy, Figure 6) has been scaled down in order to physically reconstruct a 2 m long technical boat model (scale: 1:7:5), to be tested in the towing tank of the University of Southampton (Boldrewood Campus, Wolfson Unit) (the exemplary reconstruction has been carried out by RFH PIERCE) (Figure 7).

Figure 7: The testing of the 2 m technical boat model in the towing tank of the University of Southampton (Boldrewood Campus, Wolfson Unit). Lightship condition.

5.1 (a) Stability

The resulting values from the stability analysis are evaluated based on the Stability Criteria Evaluation data loaded in the MAXSURF software, which adheres to international codes and regulations on stability. Hence, we decided to compare the overall stability with the criteria set by the IMO and Code of Intact Stability. More specifically, IMO A.749 (18), Chapter 3, design criteria applicable to all ships, which provides a pass or fail to certain stability categories. Additionally, via the STIX Stability Index (ISO 12217-2:2002 €), intended for boats measuring between 6 to 24 m in length and having a beam greater than 1 m. This provided a rating on the overall stability of the boats and enabled its assignment to certain design categories (A-D), specifying the wind and wave conditions that the boat can sustain (see below). The higher the rating, the better, indicating safe and seaworthy crafts with increased dynamic stability. The STIX Stability Index (via MAXSURF) was also implemented with success in the reconstruction project of the Min of the Dessert, a prehistoric seagoing boat from the period of the reign of Hatshepsut (early New Kingdom) [61]; thereby, it has been suggested that, though these criteria concern modern boats and ships, they can still be beneficial comparators of stability of ancient crafts [62].

It was decided to test our digital models in fully laden conditions to assess how well or poorly these boats can react if we stretch them to their capacity limits (see below). The extent of the loadcase condition is determined in MAXSURF based on the stability tests and by making sure that the DWL is always below the margin line.

5.1 (b) Resistance

The resistance analysis is carried out in MAXSURF Resistance through analytical methods. We build on the premise that we can determine an average and maximum speed by observing the resistance curve of each boat type. So, we examine the curve's 'humps' and 'hollows' produced by the wave interference triggered after the vessel's speed increases. When the boat reaches a certain speed and the hump starts getting too steep, the drag and, consequently, the P (kW) required to overcome it would surpass the power output of the crew of a paddled boat. Therefore, this could be used as an initial threshold. To further validate how realistic the produced speed figures are, we examined whether a certain number of paddlers can achieve these based on the energy expenditure required to move a boat. Thus, the number of paddlers necessary to move a boat at a specified speed requires an x amount of power (kW).

Thereby, the number of paddlers can be estimated by dividing *P* by the paddler's peak power, which can be estimated as *Fpeak* multiplied by a relative paddle speed -applied to the paddle- per paddler. Typically, this paddle speed will be related to the stroke frequency, e.g., how many 'paddles' occur in a given time. The relative speed is found from the stroke length divided by the active stroke phase minus the boat speed. As boat speed rises, the paddler needs to raise frequency and stroke length.

For example, suppose a 10 m boat requires thirty paddlers in order to reach the specified speed by expending an x amount of power. In that case, one can logically infer that this number is unrealistic since they cannot fit in a 10 m-long boat, considering the limitations of the paddler's interscalmium. The MAXSURF Stability software can determine the *P*. For the *Fpeak*, we need to resort to experimental and empirical data concerning paddling performance and the paddler's physiology and to studies on kinematics and biomechanics. Besides, the speed of a boat is explicitly associated with the power output delivered by the paddler, transferred to the water via the paddle's blade in order to overcome the hull's aerodynamic and hydrodynamic drag and instigate forward movement [63-66]. Several studies indicate that *Fpeak* averages around 300N and is also adopted by the present research [63-65,67-69].

Therefore, reaching and maintaining average or high speeds requires significant amounts of energy, and a wrong 'strategy' can lead the crew to overexert itself unnecessarily and, consequently, to fatigue [66]. Several stages of loss of energy have been defined during this process [12, 66, 70] and are detailed as follows:

Human fuels themselves with food → Loss of energy in food in the conversion process → Muscles use energy to pull the paddle → Oxygen uptake and heart rating increases → Loss of energy in muscles (heat) into force through arms/legs/back → Lever force on the paddle to force in water → Minor loss of energy in mechanical system → Paddle blade → Significant loss of energy in converting force into paddle into axial thrust as well as stirring up the water. Thus, it is also critical to determine the mechanical efficiency of paddling, "the ratio between the energy expended by muscle contraction and the mechanical work done" [67]. We can also understand it as an *eta* value describing the difference between mechanical power loss and power input. Obviously, this value cannot be the same as the efficiency of conventional ship powering. Instead, it should be determined -and limited- based on the mechanical efficiency of muscular work when paddling. Several studies have been undertaken, indicating an efficiency value ranging from 16-30% [67, 70-73]. Thus, this range is opted for by this endeavour.

These data were loaded in MAXSURF Resistance. The analytical slender body method was chosen as described by Couser [74] and Couser et al. [75]. The resistance is calculated by the form factor derived from Holtrop's formulation and ITTC'57 friction coefficient, and the correlation allowance is 0.0004 [76] The efficiency chosen is the one defined above (ca. 30%). Regarding the range of speeds, we tested the boats in a range of 0-10 Knotts. In our case, this maximum speed value might be extreme, and we know that it is almost impossible to achieve by our boat types. Nevertheless, we required an upper-speed threshold based on empirical and experimental observations on modern paddling [64]. The author provides two sets of maximum and average speed estimates, one based on the possibility of two rows of paddlers working abreast and another of a single/staggered row of paddlers. The maximum speeds are validated via the resistance curve and the method described above. The average ones are estimated based on the flatter parts of the curve and before the major humps, where the drag is minimal and, consequently, the effort and the power output. Thus, the following data are summarised in Figure 8a-d.

In terms of the technical boat model, a series of tests were performed. Overall, we completed 26 runs. Each run examined the model's drag (N), heave (mm) and trim (°) at a predefined speed (carriage speed) and loading condition (Runs 1-10 (lightship condition): 0.3-2.3 m/s, Runs 10-15 (laden-ballasts): 0.7-1.9 m/s, Runs 16-20 (trim by the bow/stern): 0.9-1.6 m/s, Runs 21-26 (regular-irregular sea states): 1.3 m/s) (Figure 9).

SPEED - RESISTANCE - POWER OF THE ROUNDED TYPES (I-V) - SINGLE ROW OF PADDLERS					
Length (m)	**Average Range of Speed (Knots)**	**Drag (kN) -Per Average Speed**	**Maximum Speed (Knots)**	**Drag (kN) -Per Maximum Speed**	**Maximum Power (kW)**
Type I					
10	1.7 - 3 (μ=2.35)	0.0 - 0.2	3.8	0.5	3.000
15	2 - 3.2 (μ=2.6)	0.1 - 0.3	4.2	0.6	4.549
20	3 - 4.2 (μ=3.6)	0.2 - 0.4	5	0.7	5.800
Type II					
10	1.7 - 3 (μ=2.35)	0.0 - 0.2	3.7	0.4	2.630
15	2 -2.7 (μ=2.35)	0.1 - 0.3	3.9	0.6	4.500
20	3 - 3.9 (μ-3.45)	0.2 - 0.4	5	0.7	5.781
Type III					
10	2 - 3 (μ=2.5)	0.0 - 0.2	4.1	0.4	2.900
15	3 - 3.7 (μ=3.35)	0.1 - 0.3	4.7	0.6	4.726
20	3.2 - 4.2 (μ=3.7)	0.2 - 0.4	5.4	0.6	6.000
Type IV					
10	1.7 - 3.5 (μ=2.6)	0.0 - 0.2	4.2	0.4	2.996
15	3 - 4 (μ=3.5)	0.2 - 0.3	4.5	0.6	4.433
20	3.2 - 4.7 (μ=3.95)	0.2 - 0.4	5.5	0.6	6.012
Type V					
10	1.7 - 3 (μ=2.35)	0.0- 0.2	3.6	0.4	3.000
15	2.5 - 3.2 (μ=2.85)	0.1 - 0.3	4	0.7	4.858
20	3 - 3.7 (μ=3.35)	0.2 - 0.4	4.5	0.8	6.023
Average Values	2.99	0.2	4.4	0.6	4480.5

	AVERAGE VALUES OF ALL THE ROUNDED TYPES (I-V)				
Length	**Average Speed (Knots)**	**Drag (kN) -Per Average Speed**	**Maximum Speed (Knots)**	**Drag (kN) -Per Maximum Speed**	**Maximum Power (kW)**
10	2.43	0.1	3.88	0.42	2.905
15	2.93	0.2	4.26	0.62	4.613
20	3.61	0.3	5.08	0.68	5.923

SPEED - RESISTANCE - POWER OF THE ROUNDED TYPES (I-V) - TWO ROWS OF PADDLERS					
Length (m)	**Average Range Speed of Speed(Knots)**	**Drag (kN) -Per Average Speed**	**Maximum Speed (Knots)**	**Drag (kN) -Per Maximum Speed**	**Maximum Power (kW)**
			Type I		
10	2 - 3.1 (μ=2.55)	0.1 - 0.2	5	0.7	5.796
15	3 - 3.7 (μ=3.35)	0.2 - 0.4	4.8	1	9.068
20	3.5 - 4.5 (μ=4)	0.3 - 0.6	5.7	1.1	12.000
			Type II		
10	2 - 3.1 (μ=2.55)	0.2 - 0.3	4.25	0.8	5.500
	2.6 -3.5 (μ-3.05)	0.2 - 0.4	4.7	1.1	8.991
20	3.5 - 4.5 (μ=4)	0.3 - 0.6	5.6	1.1	12.000
			Type III		
10	3.2 - 3.9 (μ=3.55)	0.2 - 0.3	5.2	0.6	5.850
15	3.2 - 4 (μ=3.6)	0.2 - 0.4	5.1	1	8.900
20	3.7 - 4.7 (μ=4.2)	0.3 - 0.5	6	1.2	12.242
			Type IV		
10	2 - 4 (μ=3)	0.1 - 0.3	5.5	0.7	6.184
15	3 - 4.3 (μ=3.65	0.2 - 0.4	5.3	1	9.000
20	4 - 4.8 (μ=4.4)	0.3 - 0.5	6.1	1	12.000
			Type V		
10	2 - 3.1 (μ=2.55)	0.1 - 0.3	4.3	0.8	6.000
15	2.7 - 3.4 (μ=3.04)	0.2 - 0.4	4.4	1.1	9.000
20	3.2 - 4 (μ=3.6)	0.3 - 0.5	5.2	1.3	11.913
Average Values	3.4	0.3	5.1	1.0	8.963

	AVERAGE VALUES OF ALL THE ROUNDED TYPES (I-V)				
Length (m)	**Average Speed (Knots)**	**Drag (kN) -Per Average Speed**	**Maximum Speed (Knots)**	**Drag (kN) -Per Maximum Speed**	**Maximum Power (kW)**
10	2.84	0.2	4.85	0.72	5.866
15	3.338	0.3	4.86	1.04	8.992
20	4.04	0.4	5.47	1.14	12.031

Length (m)	Average Range of Speed	Drag (kN) -Per Average Speed	Maximum Speed (Knots)	Drag (kN) -Per Maximum Speed	Maximum Power (kW)
Type Ia (lower post as the bow)					
10	1.7 - 2.7 (μ=2.2)	0.0 - 0.2	3.5	0.5	3.085
15	2.5 - 3 (μ=2.75)	0.2 - 0.3	3.9	0.6	4.500
20	3 - 3.6 (μ=3.3)	0.2 - 0.4	4.5	0.7	6.000
Type Ib (lhigher post as the bow)					
10	1.7 - 2.7 (μ=2.2)	0.0 - 0.2	3.9	0.4	3.016
15	2.5 - 3 (μ=2.75)	0.2 - 0.3	3.9	0.6	4.500
20	3 - 3.6 (μ=3.3)	0.2 - 0.4	4.6	0.7	6.062
Type II					
10	1.7 - 2.7 (μ=2.2)	0.1 - 0.2	3.3	0.5	2.995
15	2.5 - 3 (μ=2.75)	0.2 - 0.3	3.5	0.7	4.500
20	3 - 3.5 (μ=3.25)	0.2 - 0.4	4.1	0.9	5.900
Type III					
10	3.2 - 4.7 (μ=3.95)	0.1 - 0.2	6.5	0.3	3.014
15	4.2 - 5.5 (μ=4.85)	0.2 - 0.3	6.1	0.4	4.500
20	4.5 - 6 (μ=5.25)	0.2 - 0.4	6.7	0.5	5.920
Type IV					
10	2 - 3.2 (μ=2.6)	0.0 - 0.2	3.9	0.4	3.093
15	3 - 3.3 (μ=3.15)	0.2 - 0.3	4	0.5	3.437
20	3 - 4 (μ=3.5)	0.3 - 0.4	6.8	1.7	6.026
Type V					
10	3 - 4.7 (μ=3.85)	0.1 - 0.2	5.6	0.3	3.000
15	3.5 - 4.7 (μ=4.1)	0.2 - 0.3	5.6	0.4	4.500
20	3.2 - 4.5 (μ=3.85)	0.3 - 0.5	5.6	0.6	6.135
Type VI					
10	2 - 3.5 (μ=2.75)	0.0 - 0.2	4.5	0.4	3.080
15	3 - 3.7 (μ=3.35)	0.2 - 0.3	4.6	0.5	4.500
20	3.5 - 4.5 (μ=4)	0.3 - 0.5	5.2	0.7	6.116
Average Values	3.3	0.2	4.8	0.6	4470.4

	AVERAGE VALUES OF ALL THE ANGULAR TYPES (I-VI)				
Length	Average Speed (Knots)	Drag (kN) -Per Average Speed	Maximum Speed (Knots)	Drag (kN) -Per Maximum Speed	Maximum Power (kW)
10	2.82	0.1	4.46	0.4	1.820
15	3.38	0.2	4.51	0.53	4.348
20	3.77	0.3	5.36	0.83	5.370

SPEED - RESISTANCE - POWER OF THE ANGULAR TYPES (I-VI) - TWO ROWS OF PADDLERS					
Length (m)	Average Range of Speed (Knots)	Drag (kN) -Per Average Speed	Maximum Speed (Knots)	Drag (kN) -Per Maximum Speed	Maximum Power (kW)
Type Ia (lower post as the bow)					
10	2 - 3 (μ=2.5)	0.1 - 0.3	4.2	0.8	6.065
15	2.5 - 3.2 (μ=2.85)	0.2 - 0.4	4.7	1	8.394
20	3 - 4 (μ=3.5)	0.2 - 0.5	5.6	1.2	12.020
Type Ib (lhigher post as the bow)					
10	2 - 3 (μ=2.5)	0.1 - 0.3	4.3	0.8	6.000
15	2.5 - 3.2 (μ=2.85)	0.2 - 0.4	4.8	1	8.857
20	3 - 4 (μ=3.5)	0.2 - 0.5	5.65	1.2	12.002
Type II					
10	2 - 2.8 (μ=2.4)	0.2 - 0.3	4	0.8	5.400
15	2.5 - 3.1 (μ=2.8)	0.2 - 0.4	4	1.3	7.031
20	3 - 3.6 (μ=3.3)	0.2 - 0.5	4.7	1.4	11.772
Type III					
10	3.7- 5.7 (μ=4.7)	0.1 - 0.2	7.7	0.4	6.078
15	5 - 6.2 (μ=5.6)	0.2 - 0.5	8	0.6	8.806
20	5 - 6.7 (μ=5.85)	0.3 - 0.5	8.7	0.8	11.784
Type IV					
10	2.7 - 3.4 (μ=3.05)	0.1 - 0.3	4.8	0.7	6.086
15	3 - 4 (μ=3.5)	0.2 - 0.5	4.8	1.1	9.097
20	3 - 4.2 (μ=3.6)	0.3 - 0.5	9	2.5	11.753
Type V					
10	3 - 4.7 (μ=3.85)	0.1 - 0.2	7.5	0.5	5.892
15	4 - 5.5 (μ=4.75)	0.3 - 0.5	7	0.7	8.706
20	3.5 - 5.1 (μ=4.3)	0.3 - 0.5	7.5	0.9	12.180
Type VI					
10	2.2 - 3.2 (μ=2.7)	0.1 - 0.2	5.9	0.6	6.001
15	3 - 4.2 (μ=3.6)	0.2 - 0.4	5.7	0.9	9.134
20	3.5 - 4.7 (μ=4.1)	0.3 - 0.5	6.5	1.1	12.174
Average Values		0.3	6.0	1.0	8820.6

	AVERAGE VALUES OF ALL THE ANGULAR TYPES (I-VI)				
Length	Average Speed (Knots)	Drag (kN) -Per Average Speed	Maximum Speed (Knots)	Drag (kN) -Per Maximum Speed	Maximum Power (kW)
10	3.24	0.2	5.49	0.66	5.932
15	3.70	0.3	5.57	0.94	8.575
20	4.02	0.4	6.81	1.3	11.955

Figure 8: a) The speed, resistance and power values of the boat types of the rounded shapes, generated through the MAXSURF Resistance software. This table demonstrates the values of boats propelled by a single row of paddlers. b) the values of the rounded boats propelled by a two rows of paddlers working abreast. c) the values of the angular boats propelled by a single row of paddlers. d) the values of the angular boats propelled by two rows of paddlers working abreast. The average values have been calculated in Microsoft Excel. Source: produced by the author, by using MAXSURF Resistance and Microsoft Excel.

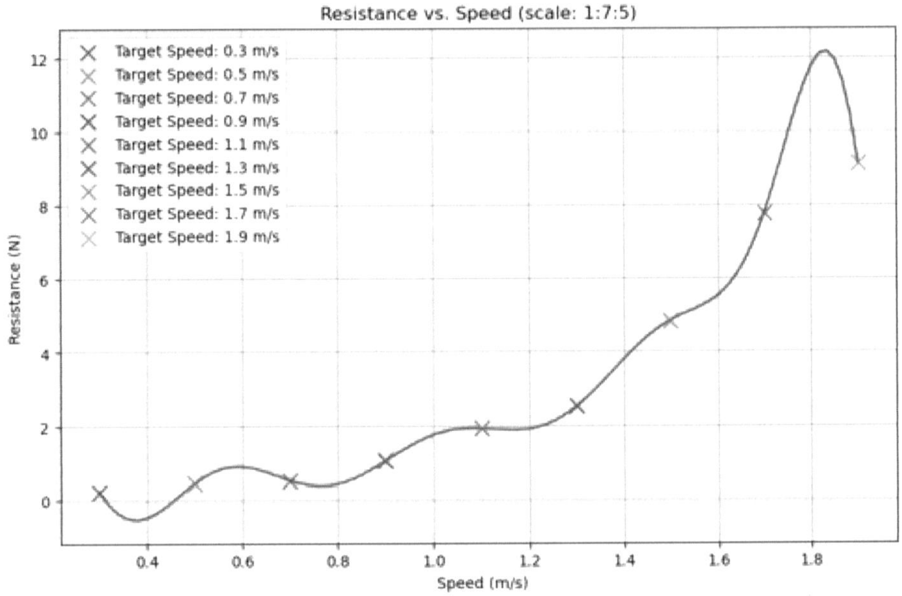

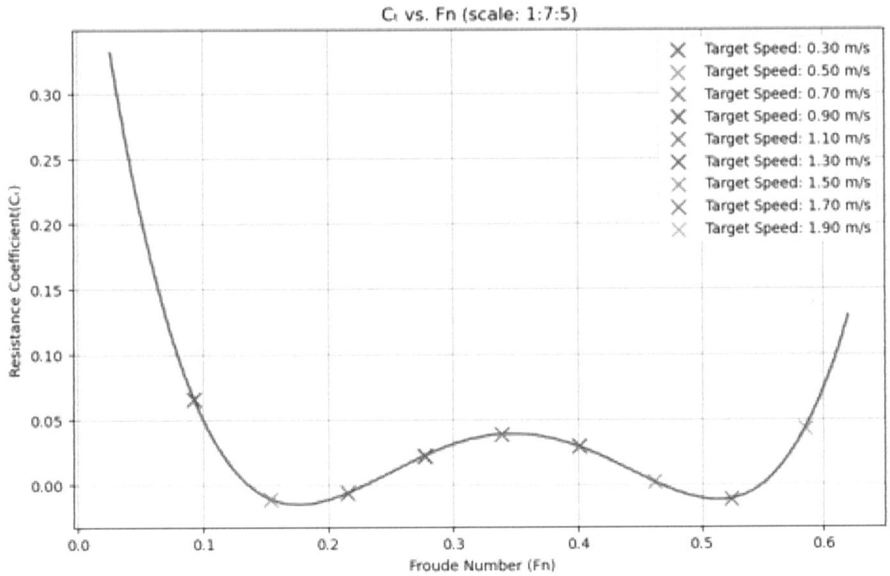

Figure 9: (Top) Resistance (N) vs Speed (m/s). (Bottom) Resistance Coefficient vs Froude Number. Source: the author.

5 CONCLUSION: AN OVERAL DISCUSSION OF THE RESULTS

All the boat types passed the IMO and STIX criteria, mostly achieving 'Category A' and have an exemplary range of positive stability and an adequate maximum righting moment (Figure 10). In terms of performance, the resistance analysis demonstrated no significant differences among the types of rounded and angular shapes. However, the latter yielded slightly better values (Figure 8a-d). It is interesting to point out that the difference between the results of boats paddled by a single row of paddlers and those by two rows of paddlers working abreast is insignificant (Figure 8a-d). This is probably due to the great drag imposed on the vessels, which would require much power to overcome (Figure 8a-d). More specifically, the average speed of both categories, with a single row of paddlers, varies from ca. 2-4 knots (maximum speed ca. 4-5.5 knots) while for those with two rows from ca. 3-4 knots (maximum speed ca. 4.5-6.8 knots) (Figure 8a-d). Hence, it would be sensible to assume that these boats would most likely be paddled by a single row of paddlers, which would have minimised the demographic risks for a community in case of death at sea or the risks imposed by their absence for an extensive period. In any case, by looking at all these values, we can suggest that they align with the speeds suggested by other scholars based on their observations of archaeological, empirical, experimental and ethnographical data [22, 58, 77]. Interestingly, the results produced by the technical boat model are not that far to MAXSURF-generated ones (Figures 8-9).

Returning to stability, we can also point out that the ones with framing and a keel/centre plank usually have better results due to the changes in VCG and the addition of a keel/centre plank (Figure 10). Although the differences in the stability results among all the types are minute, a few stand out. Regarding the increased displacement, we notice that the most balanced results are usually produced by the 15 m boats, requiring minimal ballast in the case of a trimmed bow/stern. At the same time, the bigger vessels have performance issues due to increased drag. For instance, there are cases where the smaller boats (i.e., of 10 m) yield equal or even better speed values with less drag in comparison to the bigger ones (Figure 8a-d). Thus, we can conclude that it is unnecessary for all the boat types to be more than 10 m long to achieve good speeds. Besides, there is little to no evidence of boats more than 15 m in length in the prehistoric Aegean; even the shipwrecks of the LBA are hardly over 15 m (restored length), such as the Uluburun one [37]. However, the bigger vessels usually have better stability results since they are heavier, lengthier and beamier. Additionally, we can observe a more extensive range of average speeds (Figure 8a-d). Still, though, that does not mean that they existed.

Having in mind that all the types fall within Category A of the STIX criteria we can suggest that all the variations of the rounded and angular shapes are seaworthy and able to complete extended, long-range voyages and survive in extreme conditions, even in fully laden states, as far as their stability is concerned. This is underpinned by their performance that advocates the possibility that some vessel types would not have been required to travel along the coasts or depend on island hopping. This observation is further underlined by the Monoxylon III expedition, during which the Bracciano replica covered the distance between the islands of Santorini-Dia (ca. 120 km) without any other intermediary stops [77] (Figure 1). Therefore, at least theoretically, some of the boats with the best performance could cover the same distance within a day, even if they just travel during the daylight. Based on these scenarios, one can ponder how conceptionally far were other distant destinations, when these could be reached within a few days or so. Thereby, a vessel with all these elements identified in the catalogue of the hull, such as a keel/centre plank, frames, a sound steering system, anchors, and other auxiliary aspects (i.e., cabins, shelters, platforms, etc.), would have even more increased capabilities for long-range seafaring. Moreover, the necessary capacities to transport with safety all these commodities that were of prime importance and in demand by the Aegean communities. The development of these practices and the level of their distribution is embedded in and in accordance with other emerging LN / FN phenomena, such as metallurgy and maritime specialization. Consequently, our results open a new window into Aegean prehistory, not only about the regional maritime networks of the LN-EBA periods but also the 'international' ones.

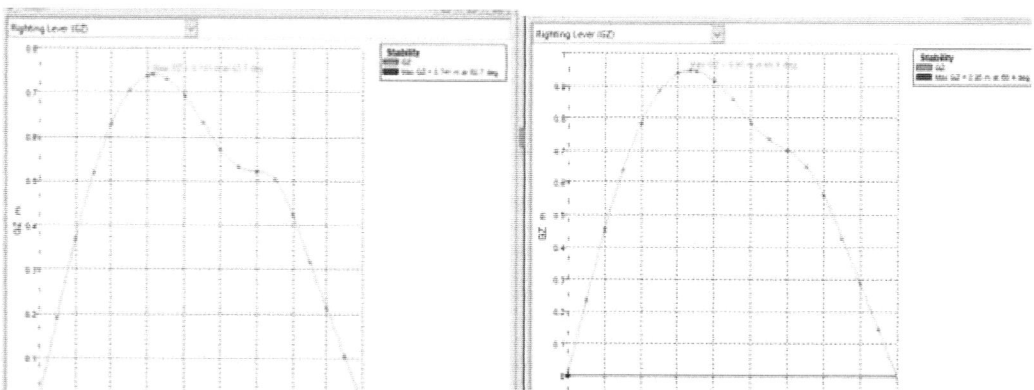

Figure 10: (Left) Type I of the rounded shapes. The 'unframed' 15 m long boat variation in fully laden condition. STIX score: Category A, 58.1 (margin +81.56%). Maximum righting moment: 0.741 m at 62.7°. (Right) Type I of the rounded shapes. The 15 m long boat variation with frames – keel/centre plank. Fully laden condition. STIX score: Category A, 63.9 (margin +99.53%). Maximum righting moment: 0.95 m at 66.4°. Source: Generated by the author by using MAXSURF Stability.

6 ACKNOWLEDGEMENTS

The author would like to thank Professor Stephen Turnock for his valuable support and guidance. Many thanks are also due to Professor Lucy Blue, Professor Andreas Vlachopoulos and Professor Joanna Sofaer.

7 LIABILITY

The opinions expressed are solely those of the author.

8 REFERENCES

1. TELEVANTOU, CHR A., 'Strofilas: a Neolithic settlement on Andros', *Horizon. A Colloquium on the Prehistory of the Cyclades,* 2008.

2. TELEVATNOU, CHR A, 'The Neolithic Settlement at Strofilas, Andros', *Papyri - Scientific Journal 8*, 2019.

3. VLACHOPOULOS, A G., 'Βαθύ Αστυπάλαιας: Έρευνα και μελέτη σε ένα διαχρονικό παλίμψηστο της νησιωτικής Αρχαιολογίας/Vathy, Astypalaia. Research in a diachronic palimpsest of the Aegean', *Δωδεκανησιακά Χρονικά ΚΘ,* 2021.

4. VLACHOPOULOS, A G (ed.)., *Δέκα χρόνια έρευνας σε ένα διαχρονικό παλίμψηστο του Αιγαίου. Τόμος II. Η Διεπιστημονική Έρευνα στο Βαθύ / Vathy, Astypalaia 2011-2021. Ten years of research (2011-2020) on a diachronic palimpsest of the Aegean. VOLUME II. Archaeological Fieldwork at Vathy*, forthcoming.

5. VLACHOPOULOS, A G., 'Vathy Astypalaia. An Early Cycladic site in the Dodecanese', *AM* 2021.

6. PAPATHANASOPOULOS, G., 'Το Πρωτοελλαδικό ναυάγιο της Δοκού', *Αρχαιολογικά Ανάλεκτα εξ Αθηνών IX*, 1976.

7. EVAGGELISTIS, 'Γιαγάνα' *Αρχαιολογικόν Δελτίον 55*, 2000.

8. VLACHOPOULOS A G., 'Αρχαιολογικές έρευνες πεδίου στο Βαθύ Αστυπάλαιας' *Δωδώνη (Ιστορία-Αρχαιολογία) V. ΜΓ'-ΜΔ'*, 2017.

9. VLACHOPOULOS, A G., 'The rock-art fleet of Astypalaia', *Conference presentation. Nostoi II. Traveling in the Eastern Mediterranean. Sea and Inland Routes from the Early Bronze to the End of the Early Iron Ages*, 2022

10. VLACHOPOULOS, A G. and ANGELOPOULOU, A., 'Early Cycladic figurines from Vathy, Astypalaia', *Beyond the Cyclades. Early Cycladic sculpture in context from mainland Greece, the North and East Aegean*, 2019.

11. TZOVARAS, P., 'Technological and typological analysis of the ships in the rock-art representations at Vathy, Astypalaia', *Δέκα χρόνια έρευνας σε ένα διαχρονικό παλίμψηστο του Αιγαίου. Τόμος II. Η Διεπιστημονική Έρευνα στο Βαθύ / Vathy, Astypalaia 2011-2021. Ten years of research (2011-2020) on a diachronic palimpsest of the Aegean. VOLUME II. Archaeological Fieldwork at Vathy,* 2023.

12. TZOVARAS, P., *The boatbuilding tradition of the Aegean during the Late Neolithic – Early Bronze Age periods. Typological classification, digital reconstruction and seakeeping assessment*, 2023

13. MOUNDREA-AGRAFIOTI, A., 'Η λάξευση του λίθου στην Παλαιολιθική Εποχή', *Αρχαιολογία και τέχνες 61*, 1996.

14. ZACHOS, K., 'Metallurgy', *Neolithic Culture in Greece*, 1996.

15. ZACHOS, K., 'The Neolithic background: A reassessment', *Metallurgy in the Early Bronze Age Aegean. Sheffield archaeological monographs Volume 7, Sheffield studies in Aegean archaeology*, 2007.

16. PAPANGELOPOULOU, T., 2016. 'Metallurgy-Metalwork', *Cycladic Society. 5000 years ago*, 2016.

17. TZOVARAS, P., 'Before 'Thalassocracies': Reconstructing the 'longboat' and rethinking its use and social implications in the 4th and 3rd millennium South Aegean', *Ships, Boats, Ports, Trade, and War in the Mediterranean and Beyond. Proceedings of the Maritime Archaeology Graduate Symposium 2018*, 2020.

18. MCGRAIL, S., *Ancient boats in North-West Europe: The archaeology of water transport to AD 1500*, 1987.

19. MCGRAIL, S., *Boats of the world: from the Stone Age to Medieval times*, 2001.

20. BROODBANK, C., *The making of the Middle Sea. A history of the Mediterranean from the beginning to the emergence of the Classical world*, 2013.

21. BROODBANK, C., 'The longboat and society in the Cyclades in the Keros-Syros Culture' *American Journal of Archaeology 93(3)*, 1989.

22. BROODBANK, C., *An Island Archaeology of the Early Cyclades*, 2000

23. TZOVARAS P., 'Nejstarsi Zobrazena plavidla', *Epos o předchůdcích Gilgameše: archeologický kontext expedice Monoxylon III*, 2022.

24. PALIVOU, C., 'Οικοδομική Τεχνολογία των προϊστορικών χρόνων', *Αρχαιολογία και Τέχνες 94*, 2005.

25. GIAGKOULIS, T., *The pile-field and the wooden structures from the Neolithic lakeside settlement Anarghiri IXb, Western Macedonia, Greece. Vol. I.*, 2019.

26. LANDELS, J G., *Engineering in the Ancient World*, 1978.

27. GREENHILL, B., *The archaeology of the boat. A new introductory study*, 1976.

28. ERKURT, G., 'Experimental Archaeology: Project for the re-animation of Cycladic boat', *Across: The Cyclades and Western Anatolia during the 3rd Millennium BC*, 2011.

29. CHATZITOULOUSIS, S. 'Woodworking technology at the Neolithic lakeside settlement of Dispilio, Kastoria', *Anaskamma 1*, 2008.

30. WACHSMANN, S., *Seagoing ships and seamanship in the Bronze Age Levant*, 1998.

31. JOHNSTONE, P., '*The sea-craft of Prehistory*, 1980.

32. STEFFY, R J., *Wooden ship building and the interpretation of shipwrecks*, 1994

33. MORRISON, J., 'Ships and boats of the Mediterranean from 3000-500 BC', *The archaeology of boats & ships. An introduction*, 1995.

34. PULAK, C., 'Mortise-and-tenon joints of Bronze Age seagoing ships', *Boats, Ships and Shipyards/ Proceedings of the Ninth International Symposium on Boat and Ship Archaeology, Venice 2000*, 2003.

35. PULAK, C., 'The Uluburun shipwreck: An overview,' *The International Journal of Nautical Archaeology 27*, 1998.

36. PULAK, C., 'The Late Bronze Age shipwreck at Uluburun: aspects of hull construction', *The Point Iria Wreck. Interconnections in the Mediterranean ca. 1200 BC*, 1999.

37. PULAK, C., 'The Uluburun shipwreck and Late Bronze Age trade', *Beyond Babylon: Art, Trade,*

and Diplomacy in the Second Millennium B.C, 2006.

38. AGOURIDIS, C., 'Sea routes and navigation in the third millennium Aegean' *Oxford Journal of Archaeology 16(1)*, 1997.

39. PAPOULIA, CHR., 'Seaward dispersals to the N.E. Mediterranean islands in the Pleistocene. The lithic evidence in retrospect,' *Quaternary International 431*, 2006.

40. PHOTOS-JONES, E., COTTIER, A., HALL, A J. and MENDONI L G., 'Kean Miltos: The well-known iron oxides of antiquity', *The Annual of the British School at Athens, Volume 92*, 1997.

41. THEOPHRASTUS, *De Lapidibus, trans. Apostolodis, N., Chr.*, 1975.

42. HOMER, 'Iliad, trans. Murray, A. T. Revised by Wyatt, W. F., 1924', *Loeb Classical Library 170*, 1924.

43. PALIVOU, C., 'Η πολυωροφία στο Αιγαίο κατά τους προϊστορικούς χρόνους', *Ήρως κτίστης, Μνήμη Χαράλαμπου Μπούρα,* 2018.

44. RENFREW, C., *The emergence of civilization. The Cyclades and the Aegean in the third millennium B.C.*, 1972.

45. PERLÈS, C. and VITELI, K D., 'Craft specialization in the Neolithic of Greece', *Neolithic Society in Greece. Sheffield Studies in Aegean Archaeology, Volume 2,* 1999.

46. DE ALMEIDA, V M., *Towards functional analysis of archaeological objects through reverse engineering processes*, 2013.

47. MOLLAND, A., TURNOCK, S., and HUDSON, D., *Ship Resistance and Propulsion: Practical Estimation of Propulsive Power*, 2011

48. PAPANIKOLAOU, A., S*hip design. Methodologies of preliminary design*, 2014.

49. RAWSON, K J., and TUPPER, E C., *Basic ship theory. Vol. 1. 4th ed.*, 1994.

50. CLAUGHTON, A., 'The stability of yachts in large breaking waves', *Adlard Coles' heavy weather sailing. 6th edition,* 2008.

51. WILSON, P A., *Basic Naval Architecture. Ship stability*, 2018.

52. WILLERTON, P F., *Basic shiphandling for masters, mates and pilots*, 1980.

53. INTERNATIONAL CODE OF INTACT STABILITY, 'Resolution msc.267(85) (adopted on December 4 2008). Adoption of the international code on intact stability (2008 IS CODE) [online]'. *Available at:* https://assets.publishing.service.gov.uk/government/uploads/system/uploads/attachment_data/file/915778/MSIS43_Intact_Stability_R05.20.pdf [Access March 10 2022], 2008.

54. BASCH, L., *Le musée imaginaire de la marine antique*, 1987.

55. WEDDE, M., *Towards a hermeneutics of Aegean Bronze Age ship imagery*, 2000.

56. MARINATOS, S., 'La marine Créto-Mycénienne', *Bulletin de correspondance hellénique, Vol. 57*, 1933.

57. MARINATOS, S., 'Das Schiffsfresko von Akrotiri, Thera', *Seewesen. Archaeologia Homerica, Band I Kapital G*, 1974.

58. TICHY, R., 2020., 'Monoxylon III: Return of the logboat to the Eastern Mediterranean', *Revue de l'Archéologie du Bateau*, 1, 2020.

59. STEPHENS, O., 'Yacht design and construction for heavy weather', *Adlard Coles' heavy weather sailing. 6th edition*, 2008.

60. PAPATHANASOPOULOS, G., VICHOS G., CHATZIDAKI, E., and LOLOS, Y., 'Δοκός: Ανασκαφική περίοδος 1990', *Ενάλια II* (3/4), 1992.

61. WARD, C., 2012., 'Building pharaoh's ship: Cedar, incense and sailing the Great Green', *British Museum Studies in Ancient Egypt and Sudan 18*, 2012.

62. COUSER, P., WARD, C., and VOSMER, T., 'Hypothetical reconstruction of an ancient Egyptian sea-going vessel from the Reign of Hatshepsut, 1500BCE', *International Conference on Historic Ships. The Royal Institution of Naval Architects, London, United Kingdom, 19-20 November 2009*, 2011.

63. JACKSON, P S., 'Performance prediction for Olympic kayaks', *Journal of Sports Sciences 13*, 1995.

64. MICHAEL, J S., SMITH, R., and ROONEY, K B., 'Determinants of kayak paddling performance', *Sports Biomechanics 8(2),* 2009.

65. ONG, K., ELLIOTT, B., ACKLAND, T., and LYTTLE, A., ' Performance tolerance and boat set-up in elite sprint Kayaking', *Sports Biomechanics 5(1),* 2006.

66. HIGGENS, A., CONWAY, L., BANKS, J., TAUNTON, D., HUDSON, D., and TURNOCK, S., 'Development of a kayak race prediction including environmental and athlete effects', *11th Conference of the International Sports Engineering Association (ISEA 2016). 1-14 July 2016, Delft, Netherlands. The Engineering of SPORT 11. Procedia Engineering Volume 147*, 2016.

67. FUKUNAGA, T., MATSUO, A., YAMAMOTO, K., and ASAMI, T., 'Mechanical efficiency in rowing', *European Journal of Applied Physiology 55*, 1986.

68. GOMES, B B., RAMOS, N V., CONCEIÇÃO, F A V., SANDERS, R H., VAZ, M A P., and VILAS-BOAS, J P., 'Paddling Force Profiles at Different Stroke Rates in Elite Sprint Kayaking', *Journal of Applied Biomechanics 31*, 2015.

69. TORNBERG, Å B., HÅKANSSON, P., SVENSSON, I., and WOLLMER, P., 'Forces applied at the footrest during ergometer kayaking among female athletes at different competing levels - a pilot study,' *BMC Sports Science, Medicine and Rehabilitation* 11(1), 2009.

70. HENDERSON, Y., and HAGGARD, H W., 'The maximum of human power and its fuel', *American Journal of Physiology 72,* 1925.

71. CUNNINGHAM, D A., GOODE, P B., and CRITZ, J B., 'Cardiorespiratory response of exercise on a rowing and bicycle ergometer', *Medicine and Science in Sports and Exercise* 7, 1975,

72. HAGERMAN, F C., CONNERS, M C., GAULT, J A., HAGERMAN, G R., and POLINSKI, W J., 'Energy expenditure during simulated rowing', *Journal of Applied Physiology 45*, 1978.

73. ASAMI, T., ADACHI, N., and YAMAMOTO, K., 'Biomechanical analysis of rowing performances,' *Biomechanics VII-B*, 1981.

74. COUSER, P., *An investigation into the performance of high-speed catamarans in calm water and waves*, 1996.

75. COUSER, P., WELLICOME, J F., and MOLLAND, A F., 'An improved method for the theoretical prediction of the wave resistance of transom-stern hulls using a slender body approach', *International Shipbuilding Progress*, 1998.

76. MAXSURF, *User manual. MAXSURF Resistance*, 2018.

77. TICHY, R., 'Archaeology of the wind natural conditions and early sea navigation in the Eastern Mediterranean,' *ŽIVÁ ARCHEOLOGIE – REA*, 2021.

9 AUTHOR'S BIOGRAPHY

Panos Tzovaras is a post-doctoral researcher at the University of Ioannina. The author studies the impact of early boats and seafaring on the colonisation and neolithisation of the eastern Mediterranean, through the lenses of archaeology, paleogeology, ship science and engineering. The author's previous experience includes a doctoral thesis (University of Southampton) on the Neolithic boatbuilding tradition of the Aegean and the classification, digital reconstruction and testing of the period's boat types. Additionally, the design, reconstruction and towing tank testing of the technical boat model. Finally, he has participated in various excavations works (i.e., at Strofilas on Andros, Vathy on Astypalaia, etc.) and recorded numerous, two and three-dimensional boat representations.

A CASE STUDY OF HISTORIC SHIPS IN CANADA: THE STEAM TUG *MASTER*

R G Allan, P. Eng.(ret), FRINA, HMSNAME, Robert Allan Ltd., Canada

SUMMARY

The paper describes the lack of coherent or consistent policies within Canada which would enable the recognition and ultimately the preservation of worthy vessels of historic importance. Current designations of vessels as "National Historic Sites" are very few, and the responsible departments within government at all levels lack a clear set of criteria for such designation. The impact of that state of affairs on the attempts at preservation and restoration of the century-old, historic steam tug *Master* in British Columbia is described.

1. INTRODUCTION

Canada has a less than stellar record for recognizing and supporting the preservation of its maritime history. Subject to the vagaries of political influence, there is no clear pathway, nationally or regionally, to identifying, cataloguing, or preserving vessels of historic importance to the country.

In British Columbia, the century-old tugboat *Master* is believed to be the only surviving wooden-hulled, steam-powered tugboat afloat in the world. She is the only remaining steam tug from British Columbia of the last century still in her original working configuration. However, time and weather are taking their toll, and extensive restoration is needed to ensure *Master*'s survival.

The history of this iconic BC tugboat has been well documented and the boat itself has been as well-cared for as the limited resources of a small, committed group of volunteers permit, but without any Government support whatsoever. The vessel's present condition is precarious and urgent restoration work is required. If the monies for proper restoration are not found the only recourse may be a comprehensive digital recreation, an option under serious consideration.

The quest for a full restoration of this important little ship continues in earnest, coupled to the creation of a formal registry of historic vessels at either a provincial or national level.

2. HISTORIC VESSELS IN CANADA

Canada is not generally identified as a maritime nation, despite having by far the longest oceanic coastline of any country on the globe at 243,042 km, facing three oceans and four of the Great Lakes. However, we do have a long and sometimes even slightly distinguished history of ship and boatbuilding, and despite the railways being typically considered as the ties that bound this vast place into a unified country, boats and small ships of diverse types have always played a significant role in the exploration and development of the nation.

Regionally, some parts of Canada are considered more "maritime" than others. The Atlantic provinces are collectively referred to as "the maritimes", based largely on its historical significance as a major fishing area, whereas the west and arctic coasts have no such popular designation. The Great Lakes are a vital trade route for grain and ore from central North America to tidewater, but despite their significant size are generally not noted as being of maritime interest, other than for the large number of shipwrecks that exist there.

Successive Canadian governments have failed to acknowledge in any meaningful way the role played by vessels of any type in the exploration of and the economic development of the country. Recognition of individual vessels is done in an ad hoc manner in the absence of any formal national policy, and typically is granted only after the persistent pressure of individuals or small dedicated groups who advocate for recognition of particularly important ships. Accordingly, it is possible to only identify a small number of vessels which currently merit designation as a "museum ship" (i.e., a ship or boat in the custody and care of a registered museum) in Canada. Figure 1 overleaf identifies 37 such vessels.

Ship	Launched	Type	Location/Custodian	Place	Province/ Territory	Notes
Tilikum	1850	Dugout Canoe	Maritime Museum of BC	Victoria	B.C.	
Dorothy	1897	Sailng Yacht	Maritime Museum of BC	Victoria	B.C.	2023: restored & afloat
SS Moyie	1898	Sternwheel Passenger Ship	Kootenay Lake Historical Society	Kaslo	B.C.	world's oldest intact passenger sternwheeler
SS Naramata	1914	Tugboat	S.S. Sicamous Marine Heritage Society	Penticton	B.C.	
SS Sicamous	1914	Sternwheel Pass'r/Cargo Ship	S.S. Sicamous Marine Heritage Society	Penticton	B.C.	largest surviving paddle wheeler in Canada
B.C.P. 45	1927	Fishing Vessel	Campbell River Maritime Heritage Centre	Campbell River	B.C.	Classic wooden seiner, portrayed on Canadian $5 bill from 1972 to 1986
RCMP St. Roch	1928	Arctic Patrol Vessel	Vancouver Maritime Museum	Vancouver	B.C.	First vessel to transit the NW passage from west to east and first to circumnavigate North America
Persephone	1965	Log Salvage Boat	Gibson's Landing	Gibson's	B.C.	
R.V. Ben Franklin	1966	Submersible	Vancouver Maritime Museum	Vancouver	B.C.	
SS Keenora	1897	Steam Passenger Ship	Marine Museum of Manitoba	Selkirk	Manitoba	
CCGS Bradbury	1915	Fisheries Patrol Vessel	Marine Museum of Manitoba	Selkirk	Manitoba	
M.S. Chickama II	1942	Passenger Ship	Marine Museum of Manitoba	Selkirk	Manitoba	
MS Lady Canadian	1944	Fishboat	Marine Museum of Manitoba	Selkirk	Manitoba	
Peguis II	1955	Tugboat	Marine Museum of Manitoba	Selkirk	Manitoba	
Joe Simpson	1963	Lake Freighter/Ferry	Marine Museum of Manitoba	Selkirk	Manitoba	
CSS Acadia	1913	Hydrographic Research Vessel	Maritime Museum of the Atlantic	Halifax	Nova Scotia	first vessel specifically designed and built to survey Canada's northern waters. Still maintained afloat.
HMCS Sackville	1941	Flower Class Corvette	Maritime Museum of the Atlantic	Halifax	Nova Scotia	Canada's oldest warship. A Naval Memorial and combat veteran of WWII. The last of Canada's 123 corvettes. Still maintained afloat
Radium King	1937	River Tugboat	Northern Life Museum	Fort Smith	NWT	
MS Normac	1902	Ferry		Port Dalhousie	Ontario	
James Whelan	1906	Tugboat	Kaministiquia Riverfront Heritage Park	Thunder Bay	Ontario	
SS Keewatin	1907	Passenger Ship		Kingston	Ontario	
HMCS Haida	1942	Destroyer	*HMCS Haida* National Historic Site	Hamilton	Ontario	Last of the Tribal class destroyers that served in WWII, Flagship of the Royal Canadian Navy . Still maintained afloat
Ned Hanlan	1943	Tugboat		Hanlan's Point,Toronto	Ontario	
MS Norgoma	1950	Ferry		Sault Ste. Marie	Ontario	
Ancaster	1951	Tugboat	Owen Sound Marine & Rail Museum	Owen Sound	Ontario	
CCGS Alexander Henry	1958	Icebreaking Buoy Tender	Lakehead Transportation Museum Society	Thunder Bay	Ontario	
Flecia	1923	Tugboat	Musee Maritime de Charlevoix	Saint Joseph de la Rive	Quebec	
Marie Clarisse II	1923	Fishing Schooner	Musee Maritime de Charlevoix	Saint Joseph de la Rive	Quebec	
R.F. Grant	1934	Tugboat	Parc Maritime de Saint-Laurent	Saint-Laurent-Ile-d'Orleans	Quebec	
CCGS Ernest Lapointe	1940	Icebreaker/Nav-Aids Vessel	Musee Maritime du Quebec	L'islet	Quebec	
St. Andre	1956	Motor Schooner	Musee Maritime de Charlevoix	Saint Joseph de la Rive	Quebec	
Jean-Yvan	1958	Fishboat	Musee Maritime de Charlevoix	Saint Joseph de la Rive	Quebec	
HMCS Bras d'Or	1968	Hydrofoil Patrol Vessel	Musee Maritime du Quebec	L'islet	Quebec	
J.E. Bernier II	196_	Sailboat	Musee Maritime du Quebec	L'islet	Quebec	
SS Keno	1922	Sternwheel Cargo Ship	S.S. *Keno* National Historic Site	Dawson City	Yukon	built to move ore from Mayo Landing on the Stewart River to Stewart Island on the Yukon River. Important for its completeness as a fast water shallow draft sternwheel steamer
SS Klondike	1937	Sternwheel Cargo Ship	S.S. *Klondike* National Historic Site	Whitehorse	Yukon	specifically designed and constructed to carry ore sacks from the Mayo silver mining district upriver to Whitehorse.
Dredge #4	1912	Placer mining Dredge	*Dredge No. 4* National Historic Site	Bonanza Creek	Yukon	Was the largest wooden hulled bucket line dredge in North America. Worked in the Klondike Valley until 1940 then relocated to Bonanza Creek until 1959

- refers to a vessel designated as a National Historic Site in Canada

Figure 1: "Museum Ships" identified in Canada

Six of those vessels (approximately one-sixth of the total and more than any other single institution) are on display at the Marine Museum of Manitoba (MMM) in Selkirk, Manitoba, the province in Canada farthest away from both the Pacific and Atlantic oceans. The MMM's mandate is "to tell the story of the development and the operation of transportation on Lake Winnipeg and the Red River."

Kudos to those in Manitoba who have made that happen and who have thus succeeded in drawing attention to the local marine history of this prairie province. It is however a pity that similar stories cannot be told in the other maritime areas of Canada.

The largest and most noteworthy of maritime museums in Canada is the Maritime Museum of the Atlantic in Halifax, Nova Scotia, which, as the star feature of the city's waterfront, hosts the 1913 hydrographic survey ship CSS *Acadia*, three working local sailing vessels and about 70 diverse small craft under shelter. Nearby, the replica Schooner *Bluenose II,* the original of which for many decades adorned our 10-cent coin, is operated by the Lunenburg Marine Museum Society on behalf of the Province of Nova Scotia. The replica of *Bluenose*, built in 1963, was constructed by private business funds, but later "gifted" to the Province of Nova Scotia. After some years that replica needed substantial repairs, carried out in 2011 to 2013. Those repairs evolved into an almost complete restoration of a ship barely 50 years old, receiving more government support ($24M) than any other maritime artefact in the country, and leaving a bad taste in the mouths of the politicians who originally supported it. [1]

The mid-19th century gold rush era in western and northern Canada gave rise to the construction of a significant number of shallow-draft paddle steamers, and three of these have been saved in land-locked facilities, each receiving designation as a "National Historic Site" in Canada. The *SS Klondike* (Figure 2) is a fine example of these unique craft.

Figure 2: SS *Klondike* in Whitehorse, Yukon Territory (Fritz Mueller Photo)

The collection of museum ships contains a number retired from Government service, a very few warships, and the majority of others represent a diverse mixture of vessels typically of very local interest. Among the most noteworthy vessels to receive the National Historic Site designation in western Canada are (a) the RCMP patrol vessel *St. Roch*, (Figure 3) built in 1928, and the first vessel to transit the Northwest Passage from west to east (1940 to 1942), the first to complete the passage in one season (1944) and the first to circumnavigate North America, and (b) the wooden fishing seiner *B.C.P. 45*, (Figure 4) built in 1927, and most noted as the vessel which adorned the Canadian $5 dollar bill from 1972 to 1986.

Figure 3: Arctic patrol vessel *RCMP St. Roch* (courtesy of Vancouver Maritime Museum)

Figure 4: Historic Salmon Seiner *B.C.P. 45* (courtesy Maritime Heritage Centre, Campbell River, BC)

In addition to those vessels listed above which are associated with formal maritime museums or dedicated societies, there are numerous smaller local associations or societies who work to preserve specific vessels or vessel types, usually on a regional basis. A good example of that type of organization is the Ladysmith Maritime Society in Ladysmith, BC. Based solely on the work of dedicated volunteers, the society has done superb work in restoring and showcasing the lovely little 1938-built tugboat *Saravan* and a small collection of similar local workboats.

Bureaucrats consistently advise that vessels afloat are impossible to designate as heritage as they are "movable" assets. However, Figure 1 shows clearly that three former government ships are maintained afloat in eastern Canada and also enjoy status as National Historic Sites. Absent from the list are those commercial or privately-owned vessels which have, like the Saravan, through the commitment of their individual owners or caretaker societies, managed to be maintained intact and afloat for many decades, and which are very likely to have historical, technical, and/or artistic merit.

3. JURISDICTION

The designation of historic vessels in Canada is not part of the mandate of any specific government department, neither provincially or federally, and thus falls into a "no man's land" where neither responsibility, recognition nor funding for support exist.

In British Columbia we have been blessed by the more than 50 years of significant effort of Mr. John Macfarlane, FRGS, who has rigorously and virtually single-handedly created *Nauticapedia*, (https://www.nauticapedia.ca), a compilation of information of virtually every vessel ever built in British Columbia. That database now holds 88,238 vessel histories (with 15,295 images and 11,485 records of shipwrecks and marine disasters). His work has also included efforts to establish a registry of historic vessels in BC. In 2017 Macfarlane wrote [2]

> "*The British Columbia Vintage Vessel Registry (1990) was one of the first formal [vessel] heritage designation programs in existence in the world. Designation was voluntary in the sense that it could be conferred only after application by the owner. A number of vessels that met rigorous criteria were chosen to be documented by the Maritime Museum of British Columbia.*
>
> *An annual catalogue of vessels in the year's uptake was published to showcase the new designations and to share their pedigrees. Unfortunately, after about six years, the volunteer drive to keep the program going was insufficient to sustain it. The program is currently dormant – but it could be revitalized when the Museum considers it appropriate to do so.*"

This writer has recently taken up this same cause [3] and continues to attempt to create a comparable registry but the pathway to achieving that is complex and convoluted.

The responsibility for Canadian Heritage rests within the Department of Canadian Heritage, also referred to as Heritage Canada. However, for reasons that escape logic, the recognition of historic places and events is the responsibility of Parks Canada, a department within the Ministry of Environment and Climate Change. Parks Canada is in fact responsible for all National historic sites, National Parks, National marine conservation areas, National urban parks, Nature and Science, and "Culture and History". National historic designations within Canada are presently classified as follows:

- National Historic Sites
- National Historic persons
- National Historic Events
- Heritage Lighthouses
- Heritage railway Stations
- Grave Sites of Canadian Prime Ministers
- Residential schools system in Canada

> *A national historic site designation is a place that has a direct association with a nationally significant aspect of the history of Canada. For example, it can be an archaeological site, a structure, a building or group of buildings, a district, or a cultural landscape.*
>
> *A national historic event may be designated if it occurred at least 40 years ago and if it represents a defining action, episode, movement, or experience in the history of Canada.*

Funding assistance is available to support the conservation and presentation of federally designated heritage "places". Movable assets such as ships still afloat (which could reasonably be considered as "structures") do not appear to have a clearly defined home in the mandates of either Heritage Canada or Parks Canada, despite the aforementioned three government vessel exceptions. The remaining vessels in the list which have status as a National Historic Site (see Figure 1) are all dry-berthed. All require significant maintenance but do qualify for federal funding and grants.

In contrast, buildings which "may" have historic significance ("Built Heritage") have a clear mandate for protection and conservation under Parks Canada, and are supported by a team of heritage architects, advisors, and historic craftspeople "*who provide conservation services for heritage buildings and cultural resources across Canada.*" The marine community continues its quest for a level of recognition and support even remotely comparable to that for buildings.

Parks Canada states that *"National historic designations illustrate many defining moments in Canada's history and the diverse and complex stories that are shared. They weave together narratives of who we are and connect us to the past, enriching an understanding of ourselves, each other, and this country. Under the National Program of Historical Commemoration, Parks Canada supports the Historic Sites and Monuments Board of Canada in the commemoration of these places, persons, and events of national historic significance."* One would be forgiven for thinking that some of those narratives would include the vessels that were critical to the development of coastal cities and industries across the nation.

Beyond the National Historic Sites are "National Historic Events" which are recognized by the display of a commemorative plaque, but as they relate to a "moment in time" and not to "built heritage" are not eligible for financial support. Most germane to the current discussion among these events is the 2014 recognition of the importance of the tug and barge industry to the growth and commerce of the Canadian west coast. That recognition is documented as follows:

Recognition Statute:	Historic Sites and Monuments Act (R.S.C., 1985, c. H-4)
Designation Name:	**Tugboating along the West Coast**
Importance:	It has been of fundamental importance to the early and ongoing economic development and settlement of British Columbia and its main industries, forestry, mining, and fishing.
Existing Plaque:	Granville Island Public Market Dock, Vancouver, British Columbia (Figure 5)
Text:	Since the mid-19th century, tugboats have been crucial to the development of the main industries of British Columbia, particularly forestry, mining, and fishing. They tow loads to and from remote businesses and communities, assist large ships in port and through difficult coastal waters, and supply many companies with goods and materials. West Coast tugboats, such as *SS Master*, have evolved into a distinct vessel type, well-suited to the topography, water, and weather conditions. Called "the railway of the sea", they link local, regional, national, and transnational shipping and transportation networks.

Figure 5: Commemorative Plaque for "Tugboating" on the BC coast
(background mounting designed by the author)

That most awkward term, "tugboating", was devised by Parks Canada presumably to avoid being vessel specific in the recognition of this "event", an activity which has been present on the BC coast since the earliest days of steam power and which continues to this day. However, as this designation was achieved through the significant efforts of the SS Master Society (about whom more will follow), the steam tug *Master* is specifically cited as an example of the tugs that deserve this recognition.

Beyond the federal jurisdiction, at present there is no formal registry or recognition for historically significant vessels in any Province or Territory of Canada.

In British Columbia the Heritage Conservation Act ([RSBC 1996] CHAPTER 187) defines the land or objects which may qualify for heritage designation and the processes to achieve such designation. A "heritage object" means, "*whether designated or not, personal property that has heritage value to British Columbia, a community or an aboriginal people.*" That definition certainly could encompass floating vessels. More credibility is given to shipwrecks than to ships under this act. A "Heritage wreck" is specifically defined and can be designated if two or more years have passed from the date of sinking and the wreck has been abandoned by its owner. It makes no mention about a limiting age for designation of the vessel at time of sinking.

Heritage Branch, part of the BC Ministry of Tourism, Arts, Culture and Sport (TACS), "supports and facilitates the conservation of heritage values on the land base in B.C…." (emphasis added). The fact that the definition specifically addresses the land base, by default appears to exclude any floating object. Similar limiting descriptions are found in the mandates of local heritage at the municipal level.

Navigating these channels of bureaucracy in an attempt to gain formal recognition for our maritime heritage, and ideally some supportive funding has proven difficult for those who have a passion for the maritime "artifacts" which grace our waters. This author has, for more than 15 years, been associated with the SS Master Society, a non-profit society dedicated to the preservation of the steam tug *Master*, the last remaining steam tug in BC waters. Her story follows.

4. THE STEAM TUG *MASTER*

4.1 WORKING HISTORY

Since the earliest days of European settlement in B.C., but especially in the period between the great wars, the coast of British Columbia on Canada's west coast was alive with new and growing industries. But this coastline, with its high, wild mountains and rugged deep inlets is no host for the likes of roads or rail lines. For almost its entire length, from the 49th parallel to the Alaskan border, the only means of moving goods along the BC coast is by sea. By great fortune almost the entire length of that coast is protected from the open Pacific by a barrier of islands, creating, amongst the world's most favoured cruising grounds, a marine highway of well-protected navigable channels, known as "The Inside Passage".

This passage is the home for myriad tugboats and barges moving the products of the forest, fishery, and mining companies from the north and mid-coasts to the mills and factories of the big cities of Vancouver and Victoria close to the US border. It is also the route for one of the world's busiest cruise ship routes.

In the 1920's and 1930's literally hundreds of wooden tugs and barges were built to serve these industries. This was not glamorous work; towing rafts of logs at 1 to 2 knots or cumbersome wooden scows at maybe 4 or 5 knots, but it was important and valuable work. It was not always easy either and many lives were lost at sea in those pioneering days.

The tugs of this era were typically from 75 to 100 feet and sometimes as much as 150 feet in length. They were universally single-screw and fitted with steam engines of modest power, between 300 and 500 HP.

After WWII many of these wooden tugs were retro-fitted with diesel engines, but only a few of them survived through the 1960's when a new fleet of higher-powered, steel-hulled diesel tugs were built to replace them. Some were converted to yachts or charter vessels or even to fishboats. Only one from this golden age of BC steam tugs survives today as an authentic coastal tugboat in her original guise. That notable exception is SS *Master*, (Figure 6) which celebrated her centennial in 2022 and according to all our research is the sole surviving wooden-hulled, steam-powered tug afloat IN THE WORLD.

Figure 6: SS *Master* under steam in Vancouver's False Creek

Master was designed and built by noted shipwright Arthur Moscrop on the shores of False Creek in Vancouver. Between 1899 and 1926 Moscrop built 48 tugs, 46 scows and 2 small freighters at his modest Beach Avenue Shipyard. He was retained in 1928 by the Burrard Drydock Company to supervise the construction of the famed RCMP patrol ship *St. Roch*, (see Figure 3) the first vessel to navigate the Northwest Passage from west to east and the first to circumnavigate North America.

The original Builder's Certificate shows that *Master* was built for Capt. Herman Thorsen of Vancouver. Thorsen retained full ownership until 1927 when the Master Towing Company was incorporated and took title of the ship along with a mortgage back to Thorsen. This mortgage was transferred to the Home Oil Company in 1933. First working for Fraser Mills and later chartered to the Lamb Logging Company, she was put in general log and barge towing service from up coast to the mills in False Creek (Vancouver) and elsewhere. In 1940, she was purchased by the Marpole Towing Company. The stack was painted with the Marpole colours, black diamonds on a white band on an orange stack. *Master* wears these colours to this day. The black diamonds, which had been the insignia of Marpole Towing since shortly after the turn of the century, signified the towing of coal barges from Vancouver Island to the company's plant in Coal Harbour, Vancouver. In 1947, control of the Marpole Towing Company was assumed by Evans, Coleman, and Evans (Evco) – although actual title to the ship was not transferred until 1959.

Around 1951, she had become part of the operations of the Gilley Bros. fleet, another subsidiary of Evans, Coleman, and Evans, but her Marpole colours remained unchanged. Evco ultimately became part of the Ocean Cement Group, which was later absorbed by the Lehigh Cement company, now Lehigh-Hansen Inc., and itself now part of the international Heidelberg Cement Group. All those companies have changed; *Master* remains as her original self!

Her duties consisted mostly of coastwise towing of logs and barges laden with sand, gravel, limestone, and coal. That essential but unsung work of moving bulk raw materials, save coal transportation, remains much the same today.

4.2 CONSTRUCTION

Master is a stoutly built wooden tugboat, constructed almost entirely of Douglas Fir, the wood that was largely responsible for the early exploration and development of BC. The strength and resilience of this abundant wood attracted the interest of the Royal Navy for use as masts in its sailing ships, and one of the world's dominant forest industries grew as a result.

Douglas Fir was the timber of choice in heavy duty wooden boatbuilding in BC, being plentiful, durable and with excellent strength to weight characteristics, and easily worked. No specific drawings of *Master's* construction have survived, if in fact there ever were any, as boatbuilding in the 1920's was still very much based on the skill of the shipwright. The cutaway illustration in Figure 8 gives some sense of the robust nature of her hull. The hull is constructed with 3″ outer planks, 8″ deep ribs, and 2″ inner planks. The gap between the inner and outer planks was reportedly originally filled with salt.

During her restoration in the early 1960's some framing, and planks were replaced, however it is estimated that today Master is still about 80% original. As will be discussed later however much of that wooden structure is now in urgent need of repair/replacement.

Figure 8: Cutaway view of SS *Master*

4.3 MACHINERY

Master runs on steam. That system comprises a Scotch boiler and triple expansion steam engine rated at approximately 332 BHP. The *Master*'s engine, produced in 1916 in the UK by Speedwell Iron Works Co. in Coatbridge, Scotland, was originally built for a British mine sweeper. The engine was never used and after the war the engine was sold as surplus and fitted to *Master*. The engine is now 107 years old and after towing over a million miles, is still in original condition and with considerable TLC is running reasonably well.

The steam engine details are as follows:

- High Pressure Cylinder Bore - 9.625"
- Mid Pressure Cylinder Bore - 15.5"
- Low Pressure Cylinder Bore - 26"
- All cylinders - 18" stroke
- Nominal Horsepower - 332 BHP at 100 rpm

Many of the *Master*'s systems are also run by steam, some requiring continuous steam pressure and others on an on-demand basis. The boiler produces the steam, and each system is fed via a series of high-pressure steam tubes, valves and return tubes. The engine is powered by the main high-pressure steam tube and all other shipboard steam systems run from the auxiliary steam tubes.

The steam-driven systems aboard *Master* include the steering engine, electrical generator, boiler fuel system, main condenser, towing winch, anchor windlass, condenser/boiler water transfer pumps, and sea water cooling pumps for the condenser.

Fuel for the boiler, bunker oil or diesel, is carried in a 6300-gallon tank forward under the main deck. Raising steam from cold takes two days, as the boiler has to be "warmed through" slowly to prevent stressing the fire tubes, etc.

5. SAVING THE *MASTER*

At the end of her commercial working life in 1959, like her counterparts, supplanted by steel, diesel-powered tugs, *Master* was left to deteriorate. However, in 1962 the local branch of the World Ship Society saw the promise in this fine little ship, purchased her for a nominal sum and over the ensuing approximately five years restored her to a semblance of her past glory. That work, and full ownership of the vessel, has subsequently been taken over by the SS Master Society, a small group of admittedly aging volunteers, who for more than 50 years have cared for her, kept her fully operational, and shown her off at classic boat festivals and similar events up and down the coast, educating residents and visitors alike about the important role of tugboats in the development of the B.C. coastal economy, and showing off the magic of steam power. *Master* is the iconic working vessel of the BC west coast today, representing the legacy of towboats, shipbuilding and design that has played such an important role in the economic development and character of the BC coast. However, at the current age of 101 years, *Master* can no longer be preserved by a handful of committed volunteers; far greater support and resources are required. Sun, rain, and salt water have taken their toll on the boat over the years and our small group are unable to fund major repairs. Major work is required, especially on the decks and topsides. Below waterline she is actually not in bad shape (Figure 9). Essential repairs required include replacement of the aft bulwarks, rim timber, deck beams, possibly the horn timber, replacement of upper deck, deck beam repairs, and other repairs to deckhouse and wheelhouse. Also required are overhauls/refurbishing of the boiler, the steam engine, and the steam-driven towing winch and anchor windlass.

Figure 9: *Master* below the waterline – 2022 docking

Estimates for the full scope of repairs required are in the range of Canadian $5,000,000 in 2023 dollars, and it is anticipated this work could easily take 2 to 3 years to complete, after the time required to procure, dry and cure the necessary volume of timber. This is likely at least a five-year project.

Recognizing the need for these major works, the Society outlined a restoration plan in about 2017, dubbed the SS *Master* Centenary Project, with the ambitious goal of raising the money needed and having the work completed in time for *Master*'s centennial in 2022. Initial efforts at fundraising stalled at about $200,000, and ultimately the goal was re-stated to have raised enough money to start the work by the centennial year. That goal too has slipped by, and we are presently looking carefully at our options. One major event early in 2023 did raise enough funds to order a full weatherproof winter cover for the boat, something we wish we had a decade or more ago, as the BC climate is tough on wooden boats.

Current plans include a staged process for preservation and restoration:

Step 1: Would be to remove the boat from the water and get it under a shelter to avoid further deterioration, and to remove the risk of sinking should a major leak develop. This would also enable the society to double-down on fundraising efforts without the worry of the tug not surviving another winter.

Step 2: Would then be to create a complete 3D digital model of the tug, emulating the work done by Cooper et al for the Qatar Museum's watercraft collection [4]. This critical step would document *Master* inside and out for historical purposes and for any future research work. If our ultimate objective of a complete restoration were to fail, we would at least have a photo-realistic 3D digital model of the complete vessel and all her equipment and systems.

Step 3: Subject to full funding, would be the full restoration of the vessel into a full working ship, to carry on the important education and goodwill work she has done for so many years.

Throughout this process the Master Society has had some success in obtaining the support of the marine community for specific mini projects. Seaspan Vancouver Shipyards has been very generous over the years in offering their facilities and services for quadrennial dockings and inspections. (Figure 10).

Figure 10: Seaspan Vancouver Shipyard workers showing their support for the *Master*

International Paint has been equally generous in regularly providing all required coatings. A local and iconic machine shop, Burrard Iron Works, has just recently undertaken the complete restoration of the winch and windlass on a no-charge basis. The maritime press, both locally and internationally, have been very generous in publishing the story of *Master* and our efforts to save her. But finding the much-needed cash to fund the hard work of sourcing timber and doing the necessary repairs has proven a most challenging task.

6. WHY BOTHER?

So why should we keep an old tug afloat? It has to do with the recognition and appreciation of our Province's maritime heritage, celebrating those who came before us, building the foundations upon which the abundant life we enjoy today is based. It has to do with recognizing publicly and prominently the vital role that the coastal towing industry has played and continues to play in the commerce of the province.

Master is fully engaged with the public for much of the year at Granville Island in Vancouver or at the Historic Britannia Shipyard in Steveston, BC. Visitors from all over the world are thrilled to experience this vital part of BC history and get a sense of the wonder of a steam-powered vessel. *Master* has historically journeyed to many other places in BC to attend maritime events and festivals as an ambassador of the BC shipbuilding and towing industries. Currently unable to raise steam due to the need for boiler repairs, she has been immobile for the past 2-3 years.

Most great cities in the world feature an important historical vessel on their waterfront, and Vancouver certainly likes to think of itself now as a "world city". Think *Cutty Sark, HMS Victory, USS Constitution, Queen Mary, Kon Tiki, Fram* and now *Maud* and many more. These vessels all have community and civic (and frequently national) support. San Francisco has *Hercules*, a steel steam-powered tug and *C.A. Thayer*, a west coast lumber schooner rebuilt at a cost of $14 million. San Diego has *Star of India* and others. Washington State has the steamship *Virginia V* (a $6.5 million restoration), *WT Preston* in Anacortes and the *Sand Man* tugboat in Olympia. The inner harbour of Rotterdam is alive with an entire fleet of historic tugs. In each case communities and societies have raised money, put in a huge effort, and preserved a critical part of their local maritime history and identity. These vessels are too important to our heritage and identity to let them become derelict. Vancouver has no historic working vessel other than the *Mast*er that is alive and on the water; original, functional, meaningful to local maritime history, and accessible to the public.

7. CONCLUSIONS

Enabling the official recognition of historic vessels at any level of government is an elusive goal in Canada. The designation of "National Historic Site" to marine vessels is assigned on an extremely limited and very ad-hoc basis but does include vessels afloat as well as the more typical "dry-berthed" display.

The century old coastal steam tug *Master*, a beautiful example of her type, is unique in the world today as a wooden-hulled, steam powered tug afloat in original configuration, and is certainly most-deserving of this type of heritage designation and the access to monies for restorative work associated with it. Finding the required support within various levels of bureaucracy and from an elected official who might champion the cause is a serious challenge. Attempts to fund this restoration entirely from private and corporate donations have fallen far short of the target, and it appears that only government intervention will save Master from a sorry fate.

Critical next steps for the SS Master Society involve ensuring *Master* stays watertight, above and below waterline, conducting a complete 3D digital survey of the vessel, and then a hard press to find the much-needed funding to complete her restoration. While those "hardware" efforts continue, creating a National (or even a Provincial) Register of Historic Vessels and obtaining formal recognition of *Master* as a most important example of her type and classification as a "National Historic Site" remain as key objectives.

8. REFERENCES

1. (https://www.cbc.ca/news/canada/nova-scotia/bluenose-ii-bluenose-2-heritage-ship-schooner-rebuild-boondoggle-rudder-1.4219098)

2. MACFARLANE, J M., (2017) 'Recognizing and Preserving Floating Heritage in British Columbia', *Nauticapedia.ca* 2017. http://nauticapedia.ca/Articles/Floating_Heritage.php

3. ALLAN, R G., and MACFARLANE, J M., (2019) 'The Case for an Historic Vessel Registry for British Columbia', *Nauticapedia.ca* 2019. http://nauticapedia.ca/Gallery/FILE.php

4. COOPER, J P., WETHERELT, A., ZAZZARO, C., and EYRE, M., University of Exeter, 2018. 'From Boatyard to Museum: 3D laser scanning and digital modelling of the Qatar Museums watercraft collection, Doha, Qatar', *The International Journal of Nautical Archaeology* (2018) 47.2: 419-442 doi: 10.1111/1095-9270.12298

9. AUTHOR'S BIOGRAPHY

Robert G. Allan is a third-generation naval architect, currently retired from active duty, but still serving as the Executive Chairman of the Board of the consulting firm Robert Allan Ltd., established in 1928 by his grandfather. He is internationally recognized as a leading authority on the subject of tug and specialized workboat design and has received major peer awards for this work, including the 2005 Royal Institution of Naval Architects Small Craft Group Medal for Significant Lifetime Achievements in Naval Architecture. He is a Fellow of RINA and an Honorary Member/Fellow of SNAME.

He was a long-term Trustee of the Vancouver Maritime Museum and is a member of the SS Master Society.

BUILDING THE REPLICA THAMES BARGE BLUE MERMAID

R D Titchener and **P Spooner**, Sea-Change Sailing Trust and Paul Spooner Design, UK

SUMMARY

The building of a replica steel engineless Thames Sailing Barge, the *Blue Mermaid*, based on an original design from 1930; the design process; doing this within the modern regulatory framework; observations on engineless operation and the developing sail cargo initiative.

1. INTRODUCTION

In 1951, shortly before he secured the *Cutty Sark* for the nation, the first Curator of the National Maritime Museum, Frank Carr, published the second edition of *Sailing Barges*. In the postscript he said the remaining fleet of barges in trade would be a useful training ground for the merchant service now that it had become expensive to maintain square-rigged vessels for the job. He had already been a founding member of the Thames Barge Sailing Club in 1947 to keep at least one sailing barge operational. Now they are a trust, and he would be pleased to learn they have two. In 1957 the loss of the *Pamir* with many of her crew and trainees ended deep sea cargo as a means of training young people, and the subsequent resurgence of sail training has been without it. When the trustees of Sea-Change set about raising funds to replace vessels they hired this history informed their thinking. And so it was that they launched an appeal to build a replica of the last full-sized barge built for trade in 1930. They did so just as the world reeled from the 2008 banking crisis and it took some years to realise the project. This is the story.

2. THE PROJECT

The Sea-Change Sailing Trust started in 2007 and works afloat each year with around 200 young or vulnerable people and seamanship trainees, using a Thames sailing barge. One of the barges used before *Blue Mermaid* was the engineless *Cambria*, which had carried the last cargo under sail in 1970. She traded between the Trent and Cornwall and to France in earlier days. Restored with a grant from the Heritage Lottery Fund, she was hired to trial successfully engineless sail training, recreating her last cargo between Tilbury and Ipswich in 2012. She visited Norwich in 2014 and was photographed leaving Yarmouth pierheads mirroring a similar view in the 1960s. She may be the last significant vessel to visit Norwich as Network Rail wish to have a fixed bridge.

At the start of the appeal MCA was involved, and it is very pleasing to be able to report their representative, Rob Shaw, was still there in 2019 to oversee the inclination test. He had just started in 2007 and it is just possible he was sent rather than a more senior surveyor as a measure of the likelihood of success. At this stage many asked why not build under 24 metres and use the Workboat Code. The original was 87 feet and 10 inches on deck being 27 metres and that code limits cargo. It was also suggested building abroad and indeed prices from ten Polish yards revealed the median being 10% cheaper, but we are proud to have used a UK yard.

Rear Admiral David Snelson kindly agreed to chair the appeal, not the subject of this paper, but suffice to say a contract was awarded to C Toms and Sons at Polruan in Cornwall and the hull was delivered in June 2016, for £293,000. It was then fitted out in Maldon by Jim Dines at TS Rigging who rerigged the *Cutty Sark* for the Diamond Jubilee. As a builder of fishing boats, for Toms it was a novel venture. They had to obtain weld shop approval from Germanischer-Lloyd who were our class society, and they did a fantastic job. The GL surveyor said he had seen better welding but only once before and that was in Korea. The eventual total appeal target was £750,000.

Why build new and why GL? We had tried to buy two existing sailing barges and recognised it would cost a great deal of money to make them fit for purpose; and that was because we wished to be able to load cargo as well as trainees. It did not seem right to subject a 100-year-old hull to the rigours of loading, nor did it seem likely the regulator would be willing for this to happen. It seemed much simpler to build to class and take a completion certificate to the MCA and seek a loadline. There was also the desire to save on annual maintenance costs and avoid chipping rust.

We had spent considerable time sailing on one or other of the seven barges built by Frederick Horlock in rivetted steel between 1923 and 1930 at Mistley in Essex, and found them speedy, handy and fun to sail, combined with the advantages of a steel structure for strength so we set about finding drawings for the last two built, *Resourceful* and *Blue Mermaid*. The former remains as a houseboat and the latter was lost in 1941 to a magnetic mine. Her crew of two are commemorated on the Merchant Navy Memorial on Tower Hill and their descendants have given their blessing to the project.

Six Class Societies were invited to participate and two responded. One would not look at what we were then considering, an existing replica, whereas GL took the view that anything was possible, especially Peter Conneely of the Orpington office. Peter has stuck with us right through as an advisor.

3. TECHNICAL ASPECTS

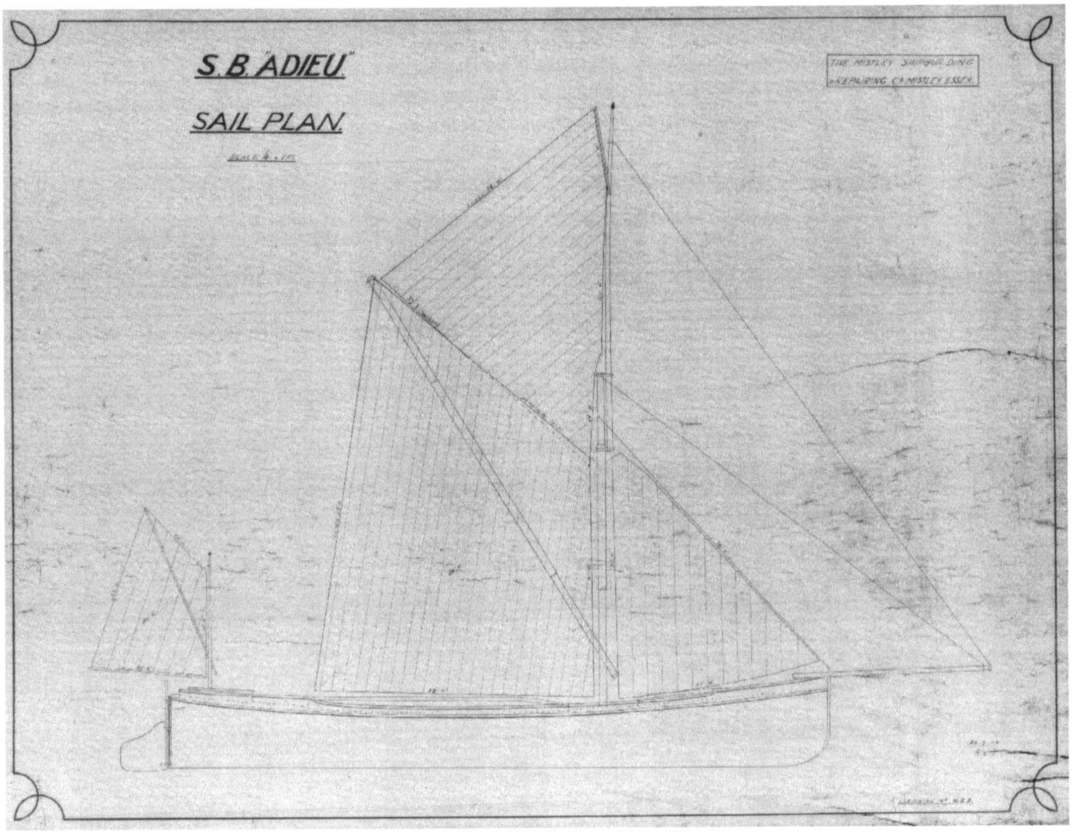

Figure 1: Original Sail and Rig Plan

Many historic craft have been replicated over the years, particularly as yachts, and many of them are commercially registered, even if solely to create the opportunity of running charters and raising revenue to assist with the running costs.

However, when building a replica vessel for use in the modern commercial world it is necessary to meet modern rules and standards and jump through certain hoops that may mean the finished vessel cannot be the entirely authentic replica that was originally dreamed of.

When thinking about replica vessels it is important to consider, not only the aesthetics, structures and historical detail but also the way it is operated and sailed. Does it feel like you are aboard a historic ship when underway or has it been inundated with mod cons and other engineering systems such that it is really no different to operate than a modern vessel.

These were all things that were hugely important to the Sea-Change Sailing Trust when the *Blue Mermaid* concept was developed. And although a few modern adjustments were required to meet current regulations these were kept to a minimum and do not affect the operation of the vessel in any way.

At a particular meeting with the MCA early in the project it was suggested that the fitting of a small engine would make it easier to gain a cargo license, but the idea was thrown out very quickly as the overall traditional operation of the barge would have been compromised simply by having the option of running an engine at difficult times. So how did we go about the technical aspects of the project?

4. STRUCTURAL DESIGN AND ENGINEERING.

The starting point for any project is a historical search for original drawings. Where possible this gives us the precise shape of the vessel and as much detail about the structures and rig as is available. In the case of the *Blue Mermaid* this was a huge amount of information. A full lines plan was found as well as a sail plan, detailed structural drawings and even an arrangement for the aft accommodation. And for a cargo carrying barge this is pretty much everything that would have been detailed originally and is thus enough information for a yard to build.

However, there are quite a few aspects of the original boat that do not simply slot into any of the current rules and regulations from any of the Classification Societies or the MCA.

Put simply, the Classification Societies, such as Lloyds, ABS or in our case DNVGL deal with the structural and engineering aspects of the vessel whilst the MCA deal with the stability and other safety aspects, and to tie this together the MCA rules always state that the ship must comply to Classification Society rules for structure and engineering.

There are several Classification Societies however in our case Germanischer Lloyd (now DNVGL) seemed the most appropriate due to their work on traditional sailing rigs. So, we set about trying to adapt the structures of the Blue Mermaid to match their rules.

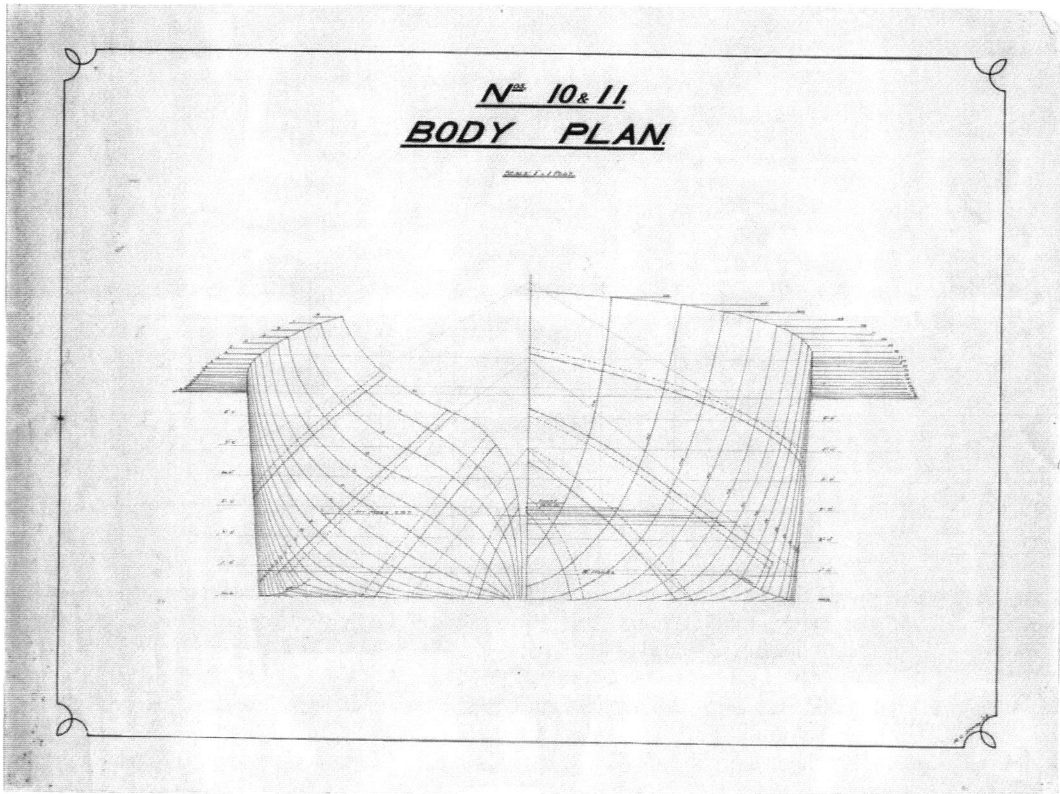

Figure 2: Original Body Plan

The first thing to go was the riveting as all current rules state that structures must be welded. This simplified the structure greatly as in order to rivet two pieces of steel together there must be two 'back-to-back' flanges to rivet through, whereas with welding one flange is removed. Keeping the frame spacing as original we found a single angle section that was the closest metric equivalent to the original Imperial sections and which was large enough to fit the modern rule.

For the same reason the deck beam sections are reversed with the toe of the angle welded to the underside of the deck plate, rather than the flange being uppermost to rivet to the plate. This arrangement also forms a stronger overall section.

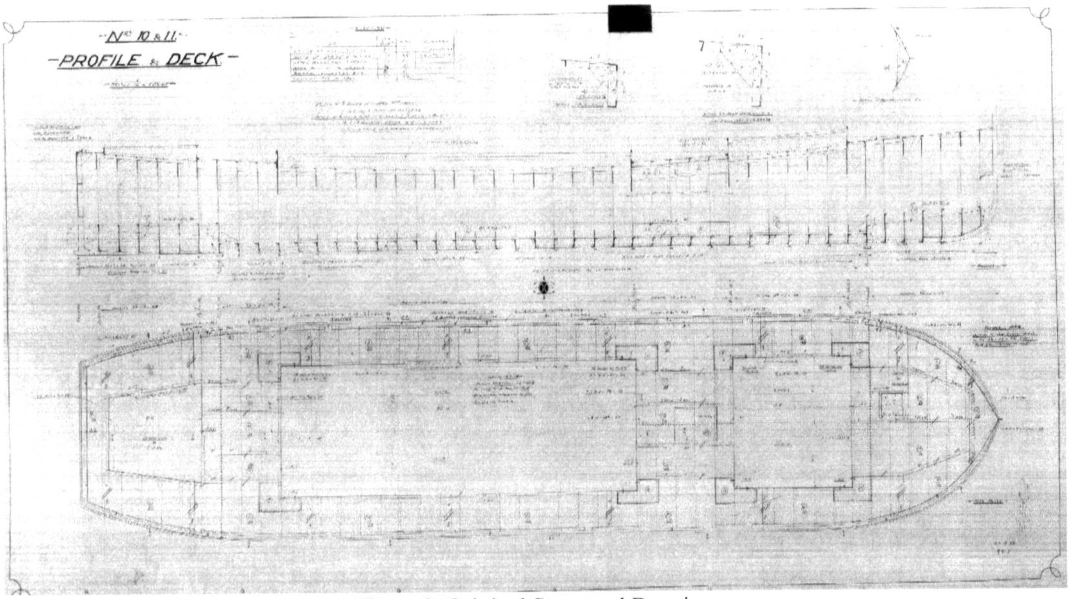

Figure 3: Original Structural Drawings

Other issues to overcome were square hatch corners on the main deck; a cause of stress concentration in the steelwork and a recognised failure point, this was overcome by locally thickening the deck around the hatch corners.

Another point of discussion was the requirement for pillars under the four hatch corners down to the hull, which would cause an obstruction in the hold and which simply were not present in the original barge. Unfortunately, we were not able to persuade them that these were not necessary and so they have been fitted to the new barge.

In the same way the low bulwarks now have vertical stiffening webs along their length which were not present in the original, unfortunately creating mud and dirt traps along the deck edge. The new barge also has a slightly neater installation for the chainplates which were originally riveted to the shell but are now welded into the transverse webs, which has proved to be an improvement.

Overall, these modifications were extremely small and the current structure is largely very similar to that of the original, with the exception of welding, and the changes required have not made any difference to the operation or overall displacement of the vessel. The hull structures were signed off by GL prior to build and were surveyed during construction to ensure that everything was on track and the quality of workmanship satisfactory.

Once the overall steel arrangement was drawn in 2D, an entire steelwork model was created virtually in 3D which was subsequently developed into cutting and shaping files for every separate piece of steel. The steel kit was then formed by CNC directly from the computer files at a yard in the Netherlands and delivered to Tom's shipyard at Polruan in Cornwall for assembly.

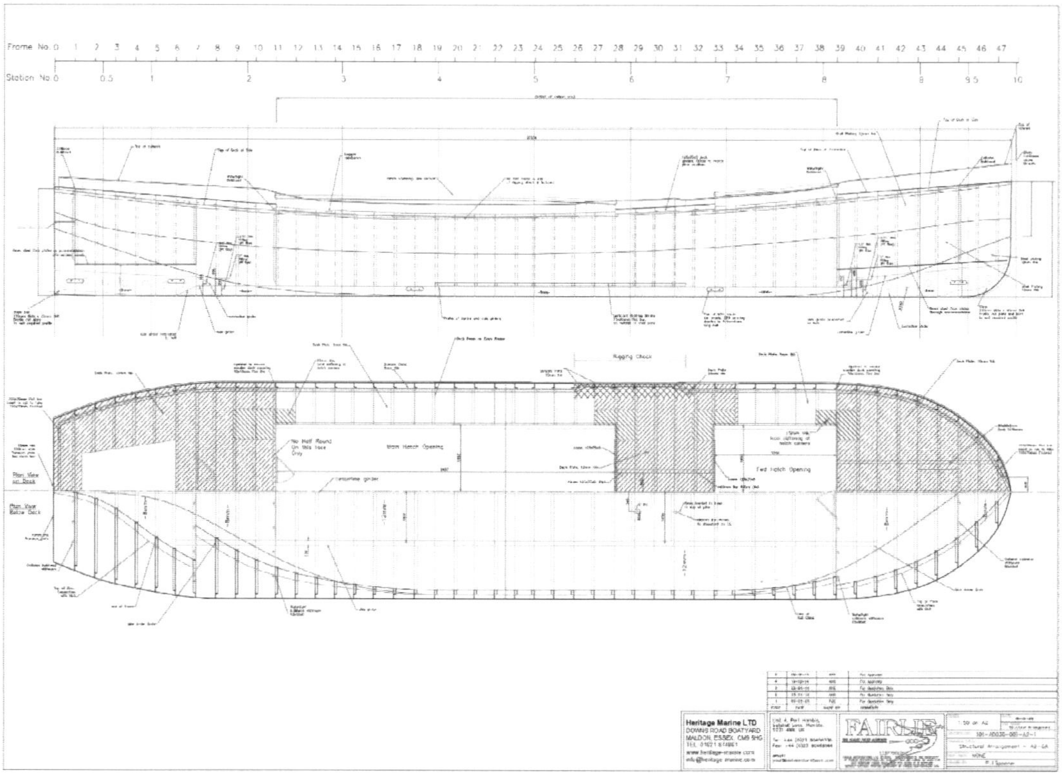

Figure 4: 2D Steelwork Plans

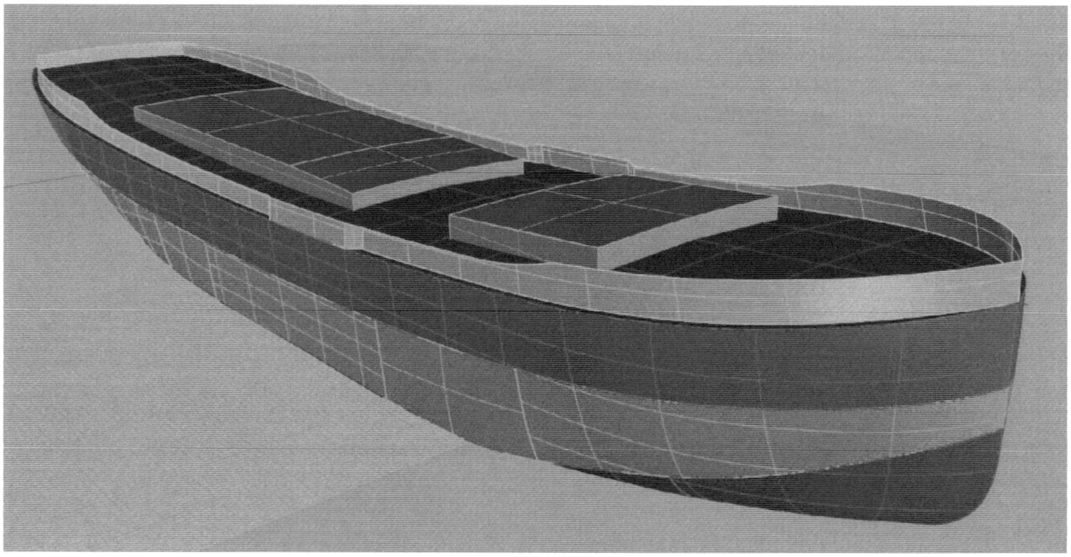

Figure 5: 3D Steelwork Model

The hull was built and launched before being towed to Downs Road Boatyard in Maldon for rigging and fit-out.

5. OPERATION AND SAFETY

As with the structures the safety of a vessel must be governed by a set of modern rules and is overseen by the MCA. There are a number of rules governing sailing craft of different types however all rules currently designed for sailing vessels state very clearly on the front page that they are for vessels that 'DO NOT CARRY CARGO' and no original regulations for the operation of Thames Sailing Barges were ever found. We did find a reference to Sailing Barges in the 1954 Loadline Regulations which fixed a minimum freeboard of 11" when loaded. Our final agreement with the MCA now sees our minimum freeboard of 50cm which is only very slightly more.

Stability rules for Sailing Vessels are all quite similar and were developed from a traditional deep keeled hull with a relatively narrow beam and a high ballast ratio. The barge hull is obviously the complete opposite to this having a wide beam, shallow draft and no ballast, so it was by no means certain that she would comply.

Added to this is the fact that most sailing vessels operate between Laden and Unladen displacements which may vary by up to 5% of their overall displacement, mainly through Fuel and Water capacity. However, the Thames Sailing Barges operated between 60 and 200t with freeboards ranging between 1.4m and 0.3m.

There are original Thames Sailing Barges in operation now, however as they do not load cargo, they are all operating with relatively light displacement and with plenty of freeboard and relatively un-changing static stability data.

In the end we decided to work with the current Large Yacht Code as the stability regulations are well developed, and having created a virtual stability model for the barge we set about detailing a range of load cases that were based on potential cargos and capacities in the hope that we could prove that our barge hull could pass regulations in all operating conditions.

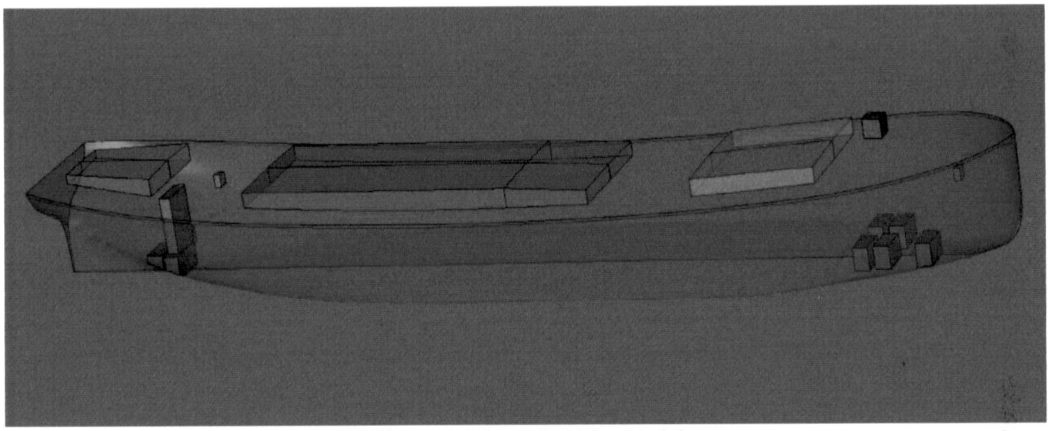

Figure 6: Virtual Stability Model

The various Loadcases tested were as follows;

LC	Condition	Crew/Pax	Tanks	Cargo
1	Departure	14	100%	0
2	Arrival	14	10%	0
3	Lightship	3	10%	0
4	Cargo 1	14	100%	56t Palletised
5	Cargo 2	14	100%	83t Palletised
6	Cargo 3	14	100%	64t Liquid Containers
7	Cargo 4	14	100%	97t Liquid Containers
8	Cargo 5	14	100%	102t Crushed Stone
9	Cargo 6	14	100%	100t Soya Bean Meal
10	Cargo 7	14	100%	128t Soya Bean Meal
11	Cargo 8	14	100%	20t General Cargo
12	Cargo 9	14	100%	125t Cement Bags

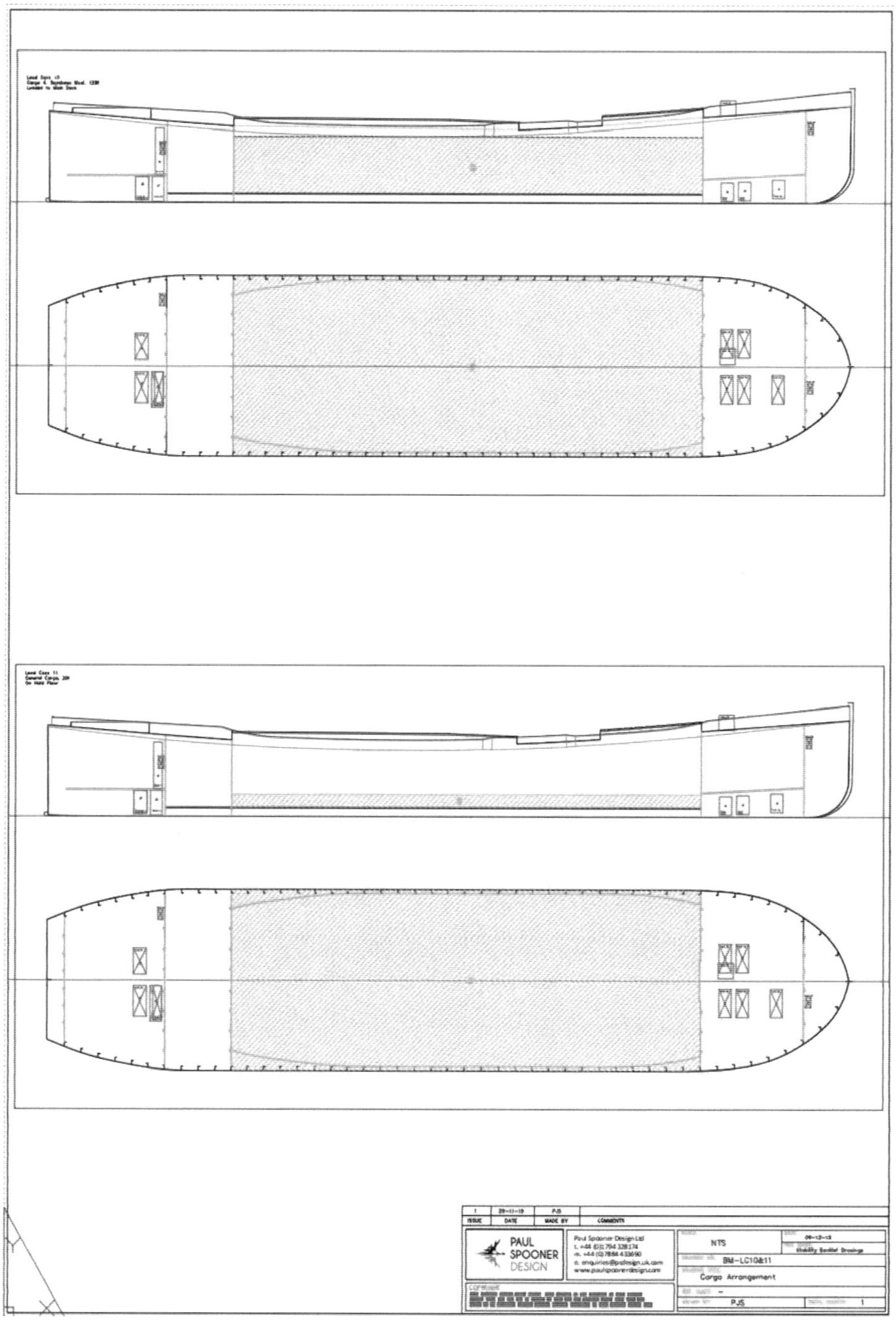

Figure 7: Loading Configurations Varying

Different cargos not only affect the weight, but also the VCG (vertical centre of gravity) due to their density and how high it is stacked. We decided early on that all cargos would be bagged or palletised for ease of handling and to avoid any issues with shifting when under sail. This seemed to put the MCA at ease as well.

The stability curve for a wide shallow hull form like a barge is entirely different from that of a deep keeled, high ballasted hull. It has a large amount of initial stability at small angles shown by the steepness of the start of the curve, but this stability is lost much earlier with a relatively short range of positive stability (in our case around

70 degrees for the unladen hull, whereas the deep hull will stay positive until well past 90 deg and sometimes the full 180 deg.

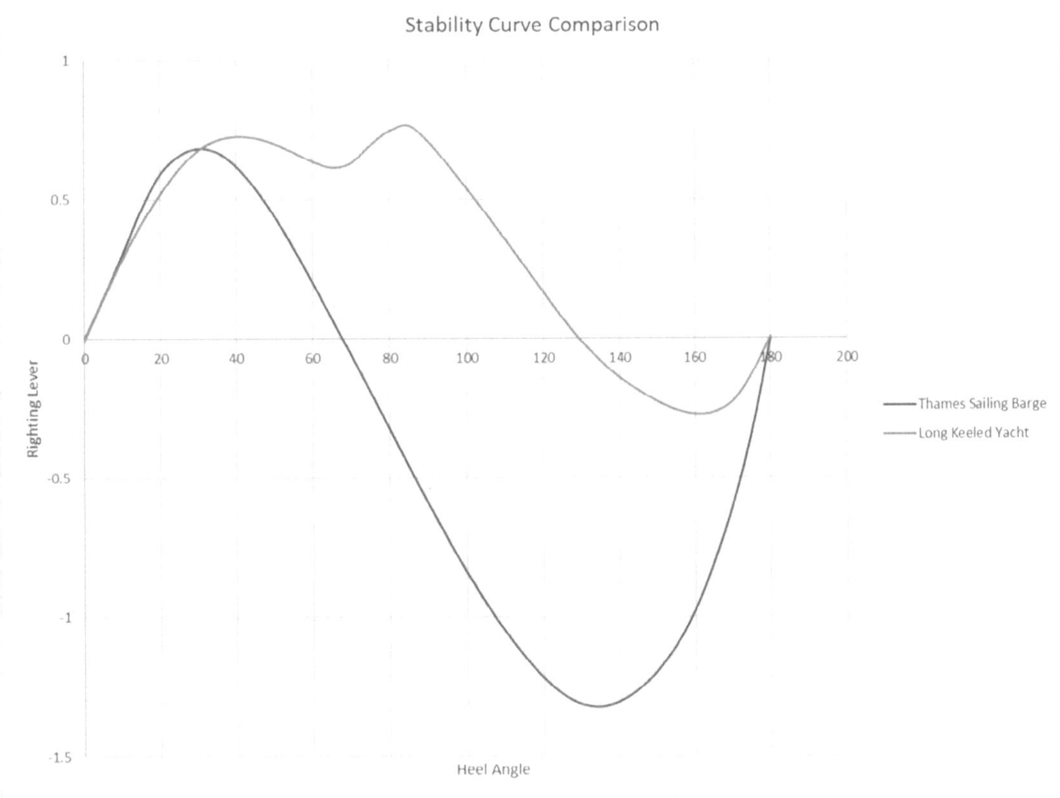

Figure 9: Stability Curve Comparison

As the barge is loaded however, the displacement increases and the overall VCG moves down relatively, so the curve widens increasing the range of stability but reducing the initial stability making her more tender. The curves below show the differing stability curves for the load cases tested. The rules require the following criteria to be met which are all derived from these Stability Curves.

- Range of Positive Stability – Greater than 70 deg
- The angle at which righting moment is greatest – Greater than 25 deg
- The area under the stability curve up to 30 deg heel and 40 deg heel – Greater than 0.055m radians and 0.09m radians respectively.
- The initial GM (Metacentric Height)– Greater than 0.35m
- The angle of heel created by a steady breeze that would lead to downflooding in a gust of 2x the steady wind strength. – Greater than 15 deg

Fortunately, when all these criteria were tested for our load cases they were found to be acceptable in all but one (though it was very close in others); the overall range of stability was slightly under 70 degrees when the barge was operating in Lightship, unladen, condition. However, after discussions with the MCA we were granted an exemption on this point as the barge would be operating in the restricted waters of the Thames Estuary.

When the barge was completed and rigged, an Inclining Test was carried out and the precise stability data used to re-check the rules. Happily, our predictions were correct and the MCA agreed to issue a Loadline Exemption for the *Blue Mermaid* allowing her to load. In the event that a particular cargo is booked that does not match any of our tested load cases, it will be relatively straight forward to set up a new load case and check compliance.

The *Blue Mermaid* is therefore now available for cargo carrying within the Thames Estuary under her Loadline Exemption. She operates entirely engineless and in precisely the same way as they did in the Heyday of the Thames Sailing Barges when 1000's plied their trade up and down the Thames.

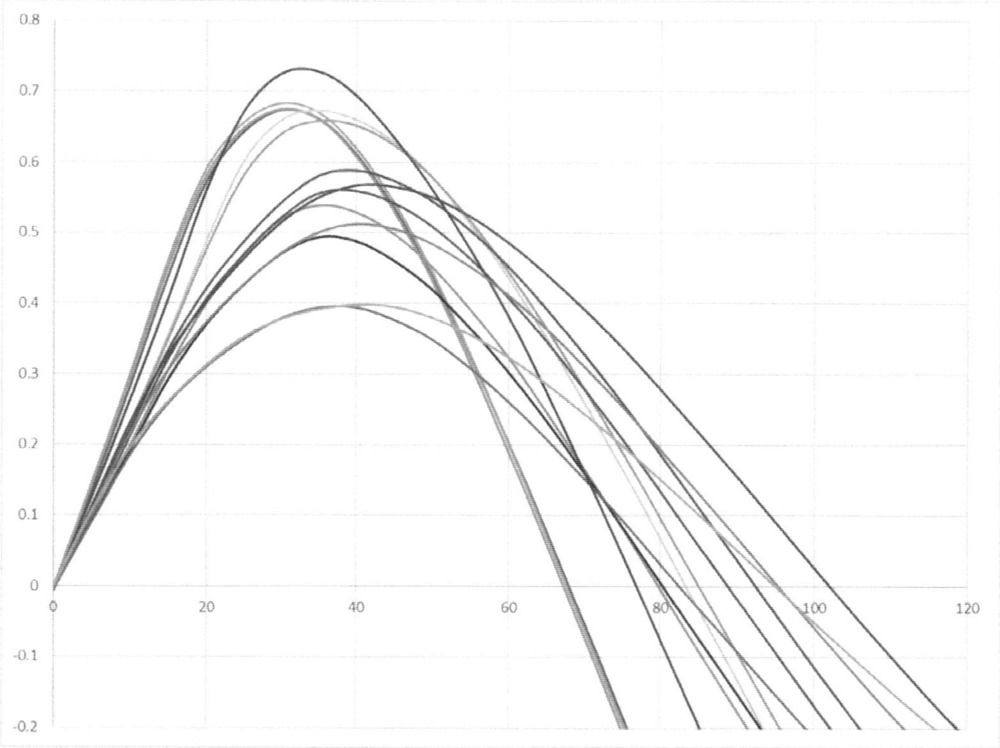

Figure 10: Loadcase Stability Curves

6. COMPROMISES

The first compromise was to extend the accommodation aft enough to fit a two-berth cabin and a galley in what would have been a cupboard to the hold in the original. On deck the appearance was unchanged as the final hatch board lifts to give storage space for the manual pumps, made by Jim Dines in stainless but to the original nodding design. Forward the accommodation was extended one frame for a four-berth cabin for trainees with accommodation for the mate and a reasonable-sized toilet. Each end has separate solar panels and batteries, with a second battery aft for navigational equipment, and separate water supplies manually operated both in and out. Consequently, we use far less water than expected and have over-catered. Provision was made for black tanks as the rules are changing. At this stage we felt we had to abide by what regulations we could and seek exemption from what we could not, so we fitted one shower although two were theoretically needed. As it turned out the MCA exempted the accommodation from the regs, although to do so we needed to obtain the support of the seamens' union that we did not dilute standards generally. Everything in the hold, including six pipe cots, is demountable and there is permanent accommodation elsewhere for six trainees and three crew. On deck forward, there is a header tank served by a manual pump in the heads. After three attempts we managed to get the hot water to thermally circulate through a Reflex diesel heater obviating the need for a circulating pump. This took so long and cost so much, I fear the true cost of each shower is very high. The rig was made new, although the anchor windlass was from *Ena*, and some other winches were restored from other original vessels. The steering gear was from the *Felix*. As much as possible is galvanised including the mastcase (or tabernacle), the leeboard winches, newly fabricated, and the bitts. The leeboards are a galvanised steel structure with wood inserted to provide strength and shape without the trouble of leaks from hollow steel. They have been moved forward two feet since commissioning to balance the ship. There are no chain plates to avoid the damage caused to them by the leeboards and the rig is attached to the deck edge by means of shackles to the webs. I should say, we looked at riveting but decided against it due to time and cost, as the charity is not a preservation organisation and is more about operational skills. The brown sails are in man-made Clipper Cloth, and the jibs of cotton with polyester strands. The Worshipful Company of Weavers put a great effort into weaving flax in the UK for sail making for the first time in twenty-five years, but in the end the process was too slow for the owner of the loom who also used it for silk.

7. CONCLUSION

From the inclination test in July 2019 to the award of a load line exemption in September 2022, *Blue Mermaid* became a successful sail training vessel for up to twelve voyage crew on the southeast coast under the General Exemption applicable to Thames Barges. She is a Gold member of the PLA's Green Scheme. Apart from mainstream work with young people and vulnerable adults, she has sailed with courses each year for seamanship trainees supported by Trinity House.

It soon became clear that there are no regulations to cover a sailing cargo vessel today. The last time they are mentioned was in the 1954 Load line regulations and although we obtained a copy and tried hard the MCA would not accept them. Had they done so, we would have had a load line 11 inches below the weather deck as can be seen etched into several original craft. With the growth in sustainable transport, there is a case for a set of regulations for sail cargo vessels.

This year the search for cargo has proved more challenging than expected. So far we have delivered some small, high value or token cargoes like beer and Caribbean produce, but the load line allows us to carry 110 tons which is four lorries. Even offering the 1931 freight rate has not netted any significant loads as yet, although we keep looking and I am confident we will succeed.

Those vessels engaged in sail cargo, and there are a few now, are mainly carrying small amounts of high value goods, where people will pay extra for delivery by sail. There is a sail cargo network and perhaps the most successful driver of this is Alex Guildenhuys whose New Dawn Traders at Falmouth is making a name for itself. *Tres Hombres* has been sailing between Europe and the Caribbean for twelve years and has the right to be known as the true trail blazer here. She takes paying guests and up to 40 Tons of cargo and is engineless, although she has a rib with an outboard much as we have an outboard on our bargeboat. Mind you we also have two 28 feet long sweeps for when it is calm.

One of the Tres Hombres, Jorne Langelaan, has a new project called Ecoclipper to build an engineless three masted ship to carry 500 tons of cargo. Like us, he is finding a slow buy in to his ideas and thinks there is a need for people at government level to be persuaded of the true potential of sail cargo.

Sea-Change supports the International Windship Association which is active in many areas from kites and rotor developments to influencing at government and similar level.

8. AUTHORS BIOGRAPHY

From Maldon, **Richard Titchener** is Executive Officer and Skipper for the Sea-Change Sailing Trust. He had a first career in HR in UK manufacturing before working in sailing barges from 1994, turning a hobby in traditional vessels into a profession. He established Sea-Change with others in 2007, using hired vessels to work afloat with young or vulnerable people and managed the *Blue Mermaid* project to commissioning in 2019 and into operation.

Living in the New Forest, **Paul Spooner** (CEng, MRINA) has over 30 years' experience as a naval architect. Since being a director of Fairlie Restorations where he helped restore many of the classic yachts gracing our waters today, he has run his own firm at Paul Spooner Design. Paul acted as designer for the *Blue Mermaid* project and produced stability and other data required to bring the vessel to operation.

The Royal Institution of Naval Architects
8-9 Northumberland Street
London, WC2N 5DA
United Kingdom

ISBN 978-1-7138-9562-6
9 781713 895626

CURRAN ASSOCIATES INC.
proceedings.com